T0192091

Lecture Notes in Computer Science   13116

David Mohaisen · Ruoming Jin (Eds.)

Computational Data and Social Networks

10th International Conference, CSoNet 2021
Virtual Event, November 15–17, 2021
Proceedings

 Springer

Editors
David Mohaisen 🄳
University of Central Florida
Orlando, FL, USA

Ruoming Jin
Kent State University
Kent, OH, USA

ISSN 0302-9743   ISSN 1611-3349 (electronic)
Lecture Notes in Computer Science
ISBN 978-3-030-91433-2   ISBN 978-3-030-91434-9 (eBook)
https://doi.org/10.1007/978-3-030-91434-9

LNCS Sublibrary: SL1 – Theoretical Computer Science and General Issues

This Springer imprint is published by the registered company Springer Nature Switzerland AG
The registered company address is: Gewerbestrasse 11, 6330 Cham, Switzerland

Preface

The 10th International Conference on Computational Data and Social Networks (CSoNet 2021), held online, is a premier interdisciplinary forum that brings together researchers and practitioners from all fields of big data and social networks, such as billion-scale network computing, social network/media analysis, mining, security and privacy, and deep learning. CSoNet 2021 aims to address emerging, yet important computational problems with a focus on the fundamental background, theoretical technology development, and real-world applications associated with big data network analysis, modelling, and deep learning. CSoNet 2021 welcomed both the presentation of original research results, the exchange and dissemination of truly innovative theoretical advancements, as well as outcomes of practical deployments and real-world applications in the broad area of information networks.

The core research topics include: theories of network organization; influence modeling, propagation, and maximization; adversarial attacks of network; NLP and affective computing; computational methods for social good; and security, trust, and privacy, among others. We have selected 18 regular full papers, 8 short papers, along with 4 two-page extended abstracts for presentation and publication. An additional 4 invited papers from active researchers in the related fields are also included. Similar to previous years, a few special tracks with specific themes were organized. A special track on Information Spread in Social, Data and Economic Networks accepted 2 papers and 2 abstracts, and the 2nd International Symposium on Fact-Checking, Fake News and Malware Detection in Online Social Networks (OSNs) accepted 3 papers along with 2 abstracts. A number of selected best papers were invited for publication in the Journal of Combinatorial Optimization, IEEE Transactions on Network Science and Engineering, and Computational Social Networks.

This conference would not have been possible without the support of a large number of individuals. First, we sincerely thank all authors for submitting their high quality work to the conference, especially as the Covid-19 pandemic continues to ravage communities and countries around the world. We fully understand the unique challenges facing authors during the pandemic. Our thanks also go to all Technical Program Committee members and sub-reviewers for their willingness to provide timely and detailed reviews of all submissions. Their hard work during the pandemic made the success of the conference possible. We also offer our special thanks to the Publicity and Publication Chairs for their dedication in disseminating the call and encouraging participation in such challenging times, in addition to the preparation of the proceedings. Special thanks are also due to the Special Tracks Chair, Finance Chair and the Web Chair. Lastly, we acknowledge the support and patience of Springer staff members throughout the process.

November 2021

Ruoming Jin
David Mohaisen

Organization

General Chairs

Long Le	University of Quebec, Canada
Jun Pei	Hefei University of Technology, China

Technical Program Committee Chairs

Ruoming Jin	Kent State University, USA
David Mohaisen	University of Central Florida, USA

Publicity Chairs

Duy Nguyen	San Diego State University, USA
Khoa Phan	La Trobe University, Australia
Yang Zhou	Auburn University, USA
Alfred Zimmermann	Reutlingen University, Germany
Gwanggil Jeon	Incheon National University, South Korea

Web Chair

Abdellah Chehri	University of Quebec, Canada

Steering Committee

My T. Thai (Chair)	University of Florida, USA
Kim-Kwang Raymond Choo	University of Texas at San Antonio, USA
Zhi-Li Zhang	University of Minnesota, USA
Weili Wu	University of Texas, Dallas, USA

Technical Program Committee Members

Thang Dinh	Virginia Commonwealth University, USA
Zhi Liu	University of North Texas, USA
Akrati Saxena	Eindhoven University of Technology, The Netherlands
Weizhi Meng	Technical University of Denmark, Denmark
Jinyoung Han	Sungkyunkwan University, South Korea
Zhenming Liu	College of William and Mary, USA
Ting Hua	Virginia Tech, USA
Tho Quan	Ho Chi Minh City University of Technology, Vietnam
Marwan Omar	Saint Leo University, USA

Rajdeep Bhowmik	Binghamton University, USA
Mohammed Abuhamad	Loyola University, USA
Donghyun Kim	Georgia State University, USA
Vamsi Paruchuri	University of Central Arkansas, USA
Ralucca Gera	Naval Postgraduate School, USA
Pavel Kromer	VSB – Technical University of Ostrava, Czech Republic
Yuan Yao	Nanjing University, China
Hien Nguyen	Banking University of Ho Chi Minh City, Vietnam
Rhongho Jang	Wayne State University, USA

List of Subreviewers

Vitaly Belik
Vladimir Boginski
Ngoc-Tu Huynh
Antonina Podusova
Kirill Yanchuk
Qipeng Zheng

Contents

Measurements of Insight from Data

Complex Networks Analytics

**Special Track: Fact-Checking, Fake News and Malware Detection
in Online Social Networks**

Combinatorial Optimization and Learning

Streaming Algorithms for Maximizing Non-submodular Functions on the Integer Lattice

Bin Liu[1][(✉)], Zihan Chen[1], Huijuan Wang[2], and Weili Wu[3]

[1] School of Mathematical Sciences, Ocean University of China, Qingdao, China
binliu@ouc.edu.cn
[2] School of Mathematics and Statistics, Qingdao University, Qingdao, China
[3] Department of Computer Science, The University of Texas at Dallas,
Richardson, TX 75080, USA

Abstract. Submodular functions play a key role in combinatorial optimization field. The problem of maximizing submodular and non-submodular functions on the integer lattice has received a lot of recent attention. In this paper, we study streaming algorithms for the problem of maximizing a monotone non-submodular functions with cardinality constraint on the integer lattice. For a monotone non-submodular function $f : \mathbf{Z}_+^n \to \mathbf{R}_+$ defined on the integer lattice with diminishing-return (DR) ratio γ, we present a one pass streaming algorithm that gives a $(1 - \frac{1}{2^\gamma} - \epsilon)$-approximation, requires at most $O(k\epsilon^{-1} \log k/\gamma)$ space and $O(\epsilon^{-1} \log k/\gamma \cdot \log \|\mathbf{B}\|_\infty)$ update time per element. To the best of our knowledge, this is the first streaming algorithm on the integer lattice for this constrained maximization problem.

Keywords: Streaming algorithm · Cardinality constraint ·
Non-submodular maximization · Integer lattice

1 Introduction

A set function $f : 2^E \to \mathbf{R}$ with a ground set E is called *submodular* if for any $A, B \subseteq E$, it holds that $f(A) + f(B) \geq f(A \cup B) + f(A \cap B)$. There is an equivalent definition of submodularity which called diminishing marginal return property, i.e. for any $S \subseteq T \subseteq E$ and $e \in E \setminus T$, we have $f(S \cup \{e\}) - f(S) \geq f(T \cup \{e\}) - f(T)$. We say a set function f is *monotone* if for any $S \subseteq T \subseteq E$, it holds $f(S) \leq f(T)$. Submodular functions play a key role in combinatorial optimization, as they capture many instances such as rank functions of matroids, cuts functions of graphs and covering functions [1,8]. The few decades have seen a proliferation of works on submodular maximization. In particular, there are many algorithms for maximizing a submodular function subject to various

This work was supported in part by the National Natural Science Foundation of China (11971447, 11871442), and the Fundamental Research Funds for the Central Universities.

D. Mohaisen and R. Jin (Eds.): CSoNet 2021, LNCS 13116, pp. 3–14, 2021.
https://doi.org/10.1007/978-3-030-91434-9_1

constraints such as greedy algorithms, random sampling algorithms and local search algorithms which achieve constant factor approximation guarantees.

Due to the phenomenon of big data, constrained submodular maximization has found many new applications, including data summarization [2,3], generalized assignment [4] and influence maximization in social networks [5–7]. In some of the above applications, the amount of input data is much larger than the data that the individual computers can store. This issue motivates us to use streaming computation approach to process the data which uses only a small amount of memory and only a single pass over the data ideally. That is, when each item in the ground set $E = \{e_1, ..., e_n\}$ arrives, the streaming algorithm must decide whether to keep the current item before the arrival of the next item. Generally, there are four indicators to measure the streaming algorithm, which are approximation ratio, query complexity, memory complexity and the number of passes to scan all data. Up to now, much work has been done on submodular maximization in the streaming model [9–12].

Set functions are powerful tools to describe the problem of elements selection. However, in practice, we sometimes face situations that cannot be solved with set functions such as problems that allow multiple choices of an element in the ground set. Thus it is nature for us to generalize a submodular function from set E to integer lattice $\mathbf{Z}^E$. Recently, much work has studied the generalization of submodular functions on bounded integer lattice [13–15]. But in instances, functions with many problems are non-submodular [16,17]. To solve this problem, some parameters were proposed to describe the closeness of non-submodular function and submodular function, such as diminishing-return ratio, submodularity ratio and generic submodularity ratio [16,21,22]. In this paper, we describe a streaming algorithm for non-submodular maximization with a cardinality constraint on the integer lattice. Let $\mathbf{B} \in \mathbf{Z}_+^n$ be an integer vector and $[\mathbf{B}] = \{\boldsymbol{x} \in \mathbf{Z}_+^n | 0 \leq x(i) \leq B(i), \forall 1 \leq i \leq n\}$ be an integer lattice domain, where $x(i)$ denotes the i-th component of vector $\boldsymbol{x}$. Our problem is described as follows

$$\max f(\boldsymbol{x})$$
$$\text{s. t. } \|\boldsymbol{x}\|_1 \leq k, \tag{1}$$
$$\boldsymbol{x} \leq \mathbf{B}.$$

where $f : \mathbf{Z}_+^E \to \mathbf{R}$ is a non-negative monotone non-submodular function with $f(\mathbf{0}) = 0$, and k is a positive integer.

Our Contribution. In this paper, we focus on maximizing non-submodular functions subject to a cardinality constraint on the integer lattice. Inspired by the Sieve-Streaming algorithm introduced by Badanidiyuru et al. [9], we propose a one pass streaming algorithm. For each arriving item with its copies, we employ a modified binary search algorithm (cf. Sect. 2) to determine the amount of the current item that should be kept. Finally, we give a $(1 - \frac{1}{2\gamma} - \epsilon)$-approximation algorithm with memory $O(k\epsilon^{-1} \log k/\gamma)$ and update time $\tilde{O}(\epsilon^{-1} \log k/\gamma \log l_{max})$ per element, where γ is the diminishing return (DR) ratio (cf. Definition 1) for non-submodular functions on the integer lattice. To the best of our knowledge, this is the first streaming algorithm on the integer lattice for this constrained maximization problem.

1.1 Additional Related Work

The research on submodular and non-submodular optimization is too extensive to give a comprehensive description. In the following, we only describe the related work of this paper.

Integer Lattice. As a generalization of submodular set functions, a function is called *lattice submodular function* if for any $x, y \in \mathbf{Z}^E$, we have $f(x) + f(y) \geq f(x \vee y) + f(x \wedge y)$, where $\vee$ and $\wedge$ are coordinate-wise max and min. However, unlike set functions, submodularity on integer lattice is not equivalent to diminishing returns property. A function $f : \mathbf{Z}^E \to \mathbf{R}$ satisfying $f(x + \mathcal{X}_e) - f(x) \geq f(y + \mathcal{X}_e) - f(y)$ is called *diminishing-return(DR) submodular function* for any $x, y \in \mathbf{Z}^E$ with $x \leq y$ and $e \in E$, where $\mathcal{X}_e$ denotes the unit vector with coordinate e being 1 and other components are 0. Note that lattice submodularity is weaker than DR-submodularity in general [14]. For maximizing a monotone submodular function subject to a knapsack constraint on the integer lattice, Soma et al. [14] proposed a pseudo-polynomial-time algorithm with approximation ratio $1 - 1/e$. Later, the running time is significantly improved. Soma et al. [13] proposed polynomial-time algorithms which use threshold greedy technique to achieve an arbitrarily close to $1 - 1/e$ approximation for both lattice and DR-submodular maximization under a cardinality constraint, DR-submodular maximization under a polymatroid constraint and a knapsack constraint.

Streaming Model. For monotone submodular maximization subject to a cardinality constraint, Badanidiyuru et al. [9] presented the first one pass $1/2$-approximation streaming algorithm named Sieve-Streaming with memory $O(k \log k/\epsilon)$ and update time $O(\log k/\epsilon)$ per element. In contrast, Buchbinder et al. [10] designed a streaming algorithm with a lower ratio of $1/4$ but an improved memory complexity $O(k)$. For this problem, Norouzi-Fard et al. [11] proved that with memory $O(n/k)$, the best approximation ratio of the one pass streaming algorithm for this problem is $1/2$. Kazemi et al. [12] described a streaming algorithm called Sieve Streaming++, which obtained a approximation of $1/2$ with memory $O(k/\epsilon)$. Following this vain, [17] studied the non-submodular function with cardinality constraint and give a $(1 - \frac{1}{2\gamma} - \epsilon)$-approximation streaming algorithm with memory $O(k \log(k/\gamma)/\epsilon)$ and update time $O(\log(k/\gamma)/\epsilon)$ per element.

In the streaming setting, besides the set functions we mentioned above, there are also works considering submodular maximization on the integer lattice. Zhang et al. [19, 20] gave $(1/2 - \epsilon)$ algorithms for DR-submodular and lattice submodular maximization with cardinality constraint, respectively. For DR-submodular maximization subject to knapsack constraint, Tan et al. [18] proposed a $(1/3 - \epsilon)$-approximation algorithm with a single pass.

Non-submodular Functions. Das and Kempe [21] gave the concept of *submodularity ratio* γ_s to describe how close a function is from being submodular. Kuhnle et al. [16] proposed the *diminishing-return (DR) ratio* γ_d on the integer lattice, and generalized the definition of submodularity ratio γ_s in [21] from set functions to the integer lattice. Later, Nong et al. [22] defined the *generic submodularity ratio* γ to measure the diminishing return property of a set function. DR ratio defined in [16] is the extension of generic submodularity ratio defined in [22] from set to the integer lattice. For maximizing a monotone non-submodular function subject to cardinality constraint, Nong et al. [22] proved that standard greedy algorithm achieves a $(1 - e^{-\gamma})$-approximation with query complexity $O(nk)$. Wang et al. [17] introduced a streaming algorithm with approximation $1 - \frac{1}{2^\gamma} - \epsilon$. In addition, Kuhnle et al. [16] utilized threshold greedy technique to obtain a algorithm for maximizing monotone non-submodular functions on the integer lattice whose approximation ratio arbitrarily approaching $(1 - e^{-\gamma_s \gamma_d})$.

The rest of this paper is organized as follows. The necessary notations and subproblems that our algorithms need to solve are introduced in Sect. 2. In Sect. 3 we first propose a streaming algorithm with known optimal value. Then in Sect. 4 we introduce a streaming algorithm with known value of the unit standard vector. Finally, we present the one pass streaming algorithm for the non-submodular functions in Sect. 5.

2   Preliminaries

In this section, we will define the DR ratio and introduce some notations and properties about non-submodular function on a bounded integer lattice.

Notations. Denote $\mathbf{Z}_+$ and $\mathbf{R}_+$ be the non-negative integers and non-negative reals, respectively. Let E represent a finite ground set of size n. For each $e \in E$, we use $x(e)$ to denote the component of a vector $\boldsymbol{x} \in Z_+^E$ corresponding to element e. Let $\mathbf{B} \in \mathbf{Z}_+^n$ be an integer vector and $[\mathbf{B}] = \{\boldsymbol{x} \in \mathbf{Z}_+^n | 0 \leq x(i) \leq B(i), \forall 1 \leq i \leq n\}$ be an integer lattice domain. Specially, $[k] = \{1, 2, ..., k\}$ for any integer $k \in \mathbf{Z}_+$. For any $e_i \in E$, let $\mathcal{X}_i$ denote the unit vector in which the i-th component is 1 and the other components are 0. The zero vector is represented by $\mathbf{0}$. For vectors $\boldsymbol{x}, \boldsymbol{y} \in \mathbf{Z}_+^E$, we define $f_{\boldsymbol{y}}(\boldsymbol{x}) = f(\boldsymbol{y} + \boldsymbol{x}) - f(\boldsymbol{y})$. For a vector $\boldsymbol{x} \in \mathbf{Z}_+^E$, let $supp^+(\boldsymbol{x})$ be the set $\{e \in E | x(e) > 0\}$, and $\{\boldsymbol{x}\}$ be the multiset corresponding to vector $\boldsymbol{x}$. Finally, we define $\boldsymbol{x} \vee \boldsymbol{y}$ to be the vector whose i-th coordinate is $\max\{\boldsymbol{x}(i), \boldsymbol{y}(i)\}$, and $\boldsymbol{x} \wedge \boldsymbol{y}$ to be the vector whose i-th coordinate is $\min\{\boldsymbol{x}(i), \boldsymbol{y}(i)\}$.

Using this notations we can now define DR ratio as follows.

Definition 1 (DR Ratio [16]**).** *Let function* $f : \boldsymbol{Z}_+^E \to \boldsymbol{R}$, *the diminishing-return (DR) ratio of* f, γ, *is the maximum value in* $[0, 1]$ *such that for any* $e \in E$, *and for all* $\boldsymbol{x} \leq \boldsymbol{y}$, *such that* $\boldsymbol{y} + \mathcal{X}_e \leq \boldsymbol{B}$,

$$\gamma f_{\boldsymbol{y}}(\mathcal{X}_e) \leq f_{\boldsymbol{x}}(\mathcal{X}_e).$$

Next, we describe a subproblem of our algorithm need to solve, namely, binary search pivot subproblem.

BinarySearchPivot. Integer lattice can be represented as a multiset, where elements can be contained repeatedly. In a streaming algorithm on the integer lattice, when element e and its copies arrive, any l satisfying $(1) f_{\boldsymbol{y}}(l\mathcal{X}_e) \geq$

$l\tau$; $(2) f_{\boldsymbol{y}}((l+1)\mathcal{X}_e) < (l+1)\tau$ is named as a *pivot* with respect to $\boldsymbol{y}, e, \tau$. The subproblem is described as follows: given a threshold τ, determine a valid pivot which satisfies both (1) and (2). In the literature, Soma et al. [13] studied the maximization of DR-submodular functions f on the integer lattice, and they used the standard Binary Search algorithm to obtain a valid pivot. Meanwhile, when f is non-submodular, BinarySearchPivot algorithm of Kuhnle et al. [16] ensures the average marginal contribution of elements added exceeds τ. In this paper, we focus on non-submodular functions, thus we analyze our algorithms using the BinarySearchPivot as a subroutine. The full algorithm for BinarySearchPivot is given in Appendix A. Next, we use the following lemma of Kuhnle et al. [16].

Lemma 1 ([16]). *BinarySearchPivot finds a valid pivot* $l \in \{0,..,l_{max}\}$ *in* $O(\log l_{max})$ *queries of* f, *where* $l_{max} = \min\{B(e) - y(e), k - \|\boldsymbol{y}\|_1\}$.

3   Streaming Algorithm with Known $\boldsymbol{OPT}$

In this section, we assume that the optimal value of the problem (1) is known. Then we present an algorithm that obtains a constant ratio with polynomial query complexity. Procedure for BinarySearchPivot is desired; see section 2 for an analysis of this subproblem.

Overview of Algorithm. Inspired by [9], we present a threshold greedy algorithm in the streaming model. Suppose that a parameter v with $\lambda OPT \leq v \leq OPT$ is known, where $\lambda \in [0,1]$. When element e_i and its $B(e_i)$ copies arrive, we utilize the BinarySearchPivot algorithm with threshold $\tau = \frac{\gamma v}{2\gamma k}$ to obtain a valid pivot l_i. That is, when algorithm runs to element e_i, we employ Algorithm BinarySearchPivot to obtain a appropriate l_i which satisfies

$$\frac{f_{\boldsymbol{y}}(l_i\mathcal{X}_i)}{l_i} \geq \frac{\gamma v}{2\gamma k}.$$

and

$$\frac{f_{\boldsymbol{y}}((l_i+1)\mathcal{X}_i)}{l_i+1} < \frac{\gamma v}{2\gamma k}.$$

Finally, we add $l_i\mathcal{X}_i$ to the current solution $\boldsymbol{y}$.

Algorithm 1. Streaming-Know-OPT

Require: function f, cardinality k, ground set E, **B** and v such that $\lambda OPT \leq v \leq OPT$.

1: $\boldsymbol{y} \leftarrow \boldsymbol{0}$
2: **for** $i = 1, 2, ..., n$ **do**
3:     **if** $\|\boldsymbol{y}\|_1 \leq k$ **then**
4:         $l_i \leftarrow$ BinarySearchPivot$(f, \boldsymbol{y}, B(e_i), e_i, k, \frac{\gamma v}{2\gamma k})$
5:         $\boldsymbol{y} \leftarrow \boldsymbol{y} + l_i\mathcal{X}_i$
6:     **end if**
7: **end for**
8: **return** $\boldsymbol{y}$

Lemma 2. *For the i-th iteration, Algorithm BinarySearchPivot considers the element e_i and its copies $B(e_i)$. Let $\boldsymbol{y}_i$ be the current solution $\boldsymbol{y}$ at the end of iteration i of Algorithm 1. Then for each vector $\boldsymbol{y}_i \leq \boldsymbol{v} \leq \boldsymbol{w} \leq \boldsymbol{B}$, where $i = 1, 2, ..., n$, we have*

$$f(\boldsymbol{w}) - f(\boldsymbol{v}) \leq \frac{1}{\gamma} \sum\nolimits_{e_i \in supp^+\{\boldsymbol{w}-\boldsymbol{v}\}} (w(e_i) - v(e_i)) f_{\boldsymbol{y}_i}(\mathcal{X}_i).$$

The above lemma is proven in Appendix B. From Lemma 2, it is clearly to see that for any vector $\boldsymbol{y} \leq \boldsymbol{v} \leq \boldsymbol{w} \leq \boldsymbol{B}$, we also have

$$f(\boldsymbol{w}) - f(\boldsymbol{v}) \leq \frac{1}{\gamma} \sum_{e_i \in \{\boldsymbol{w}-\boldsymbol{v}\}} f_{\boldsymbol{y}}(\mathcal{X}_i). \tag{2}$$

Lemma 3. *Let $\boldsymbol{y}_i$ be the vector $\boldsymbol{y}$ following the i-th update of Algorithm 1, then it satisfies*

$$f(\boldsymbol{y}_i) \geq \frac{\gamma v \|\boldsymbol{y}_i\|_1}{2^\gamma k}. \tag{3}$$

Proof. The proof is by induction. First, since f is normalized, we have $f(\boldsymbol{0}) = 0$. Next, we suppose that for vector $\boldsymbol{y}_{i-1}$ the result holds, that is, $f(\boldsymbol{y}_{i-1}) \geq \frac{\gamma v \|\boldsymbol{y}_{i-1}\|_1}{2^\gamma k}$. Let l_i be the valid pivot returned by Algorithm BinarySearchPivot during the i-th iteration of Algorithm 1, then we have $\boldsymbol{y}_i = \boldsymbol{y}_{i-1} + l_i \mathcal{X}_i$. We now show inequation 3 holds.

By Algorithm BinarySearchPivot we know

$$f(\boldsymbol{y}_{i-1} + l_i \mathcal{X}_i) - f(\boldsymbol{y}_{i-1}) \geq l_i \frac{\gamma v}{2^\gamma k}. \tag{4}$$

Using the above assumption and the equality 4, we have

$$\begin{aligned}
f(\boldsymbol{y}_{i-1} + l_i \mathcal{X}_i) &\geq f(\boldsymbol{y}_{i-1}) + l_i \frac{\gamma v}{2^\gamma k} \\
&\geq \frac{\gamma v \|\boldsymbol{y}_{i-1}\|_1}{2^\gamma k} + l_i \frac{\gamma v}{2^\gamma k} \\
&= \frac{\gamma v (\|\boldsymbol{y}_{i-1}\|_1 + l_i)}{2^\gamma k} \\
&= \frac{\gamma v \|\boldsymbol{y}_i\|_1}{2^\gamma k}.
\end{aligned}$$

Thus the lemma follows.

Theorem 1. *Let f be a non-submodular function with DR ratio $\gamma \in [0, 1]$. For any given $\lambda \in [0, 1]$, denote $\boldsymbol{y}$ be the output of Algorithm 1. Then we have*

- *$f(\boldsymbol{y}) \geq \min\{\lambda\gamma/2^\gamma, 1 - 1/2^\gamma\} OPT$, where OPT is the optimal value.*
- *Algorithm 1 requires one pass, at most k space and $O(\log \|\boldsymbol{B}\|_\infty)$ update time per element.*

Proof. The proof splits into two cases depending on the cardinality of solution y returned by Algorithm 1.

1. At the end of the algorithm we have $\|y\|_1 = k$, then by Lemma 3 we get

$$f(y) \geq \frac{\gamma v \|y\|_1}{2\gamma k} = \frac{\gamma v}{2\gamma} \geq \frac{\lambda \gamma}{2\gamma} \cdot OPT.$$

2. We consider the case $\|y\|_1 < k$. Suppose that y^* is the optimal solution of Algorithm 1, then it is easy to see that $\|y^*\|_1 = k$. Let $x = (y^* - y) \vee 0$. Denote y_i be the vector at the end of i-th iteration of Algorithm 1, $y_i = l_1 \mathcal{X}_1 + ... + l_i \mathcal{X}_i$. Then by Lemma 1 and Algorithm BinarySearchPivot, we have

$$f_{y_i}(\mathcal{X}_i) < \frac{\gamma v}{2\gamma k}. \tag{5}$$

Next, from 5, the monotonicity of f, Lemma 2 and the choice of v, we have that

$$\begin{aligned}
OPT - f(y) &\leq f(y^* \vee y) - f(y) \\
&\leq f(y + x) - f(y) \\
&\leq \frac{1}{\gamma} \sum_{e_i \in supp^+\{x\}} x(e_i) f_{y_i}(\mathcal{X}_i) \\
&\leq \frac{1}{\gamma} \sum_{e_i \in supp^+\{x\}} x(e_i) \frac{\gamma v}{2\gamma k} \\
&\leq \frac{v}{2\gamma} \\
&\leq \frac{1}{2\gamma} OPT.
\end{aligned}$$

Rearranging the above inequality, we get

$$f(y) \geq (1 - \frac{1}{2\gamma}) OPT.$$

From the above two cases, we have

$$f(y) \geq \min\{\frac{\lambda \gamma}{2\gamma}, 1 - \frac{1}{2\gamma}\} \cdot OPT.$$

4   Streaming Algorithm with Known $f(\mathcal{X}_e)$

The premise of Algorithm 1 is that we know the value of the optimal solution OPT, which is unrealistic. To address this issue, we present the next algorithm, which is instantiated with different guesses of OPT. Suppose that we know the maximum values $\alpha = \max_{e \in E} f(\mathcal{X}_e)$. Combining with the DR ratio of f, it is easy to see the guesses $v \in V_e$ of OPT are increasing from α to $k\alpha/\gamma$. For each guesses v, Algorithm 2 outputs a feasible solution y^v. Finally, the best of these candidate vectors is returned.

Algorithm 2. Streaming-Know-MAX VAL

Require: function f, cardinality k, ground set E, $\mathbf{B}$, $\alpha = \max_{e \in E} f(\mathcal{X}_e)$.

1: $\mathcal{V}_\epsilon = \{(1+\epsilon)^i | i \in \mathbf{Z}, \alpha/(1+\epsilon) \le (1+\epsilon)^i \le k\alpha/\gamma\}$
2: **for** each $v \in \mathcal{V}_\epsilon$, set $\boldsymbol{y}^v \leftarrow \mathbf{0}$
3: **for** $i = 1, 2, ..., n$ **do**
4:     **for** $v \in \mathcal{V}_\epsilon$ **do**
5:         **if** $\|\boldsymbol{y}^v\|_1 \le k$ **then**
6:             $l_i \leftarrow$ BinarySearchPivot$(f, \boldsymbol{y}^v, B(e_i), e_i, k, \frac{\gamma v}{2^\gamma k})$
7:             $\boldsymbol{y}^v \leftarrow \boldsymbol{y}^v + l_i \mathcal{X}_i$
8:         **end if**
9:     **end for**
10: **end for**
11: **return** $\arg\max_{v \in \mathcal{V}_\epsilon} f(\boldsymbol{y}^v)$

The next lemma is proven in Appendix C.

Lemma 4. *There is a guesses $v \in \mathcal{V}_\epsilon$ such that $(1 - \epsilon) \cdot OPT \le v \le OPT$.*

Theorem 2. *Let f be a non-submodular function with DR ratio $\gamma \in [0,1]$. For any given $\epsilon \in [0,1]$, denote $\boldsymbol{y}$ be the output of Algorithm 2. Then we have*

- *$f(\boldsymbol{y}) \ge (1 - 1/2^\gamma - \epsilon)OPT$, where OPT is the optimal value.*
- *Algorithm 2 requires one pass, at most $O(k\epsilon^{-1} \log k/\gamma)$ space and $O(\epsilon^{-1} \log k/\gamma \log \|\boldsymbol{B}\|_\infty)$ update time per element.*

Proof. By the result of Lemma 4, we know that there must be a v_0 such that $(1 - \epsilon)OPT \le v_0 \le OPT$. Let $\boldsymbol{y}_0$ denote the solution of Algorithm 2 output corresponding to v_0. For the rest of the proof, we suppose that the results of Lemma 1, Lemma 2 and Lemma 3 all hold for the Algorithm 2 with value v_0. Thus, in the same way as Theorem 1, we consider the following two cases

1. When $\|\boldsymbol{y}_0\|_1 = k$, by applying Lemma 3, we have

$$f(\boldsymbol{y}_0) \ge \frac{\gamma v_0 \|\boldsymbol{y}_0\|_1}{2^\gamma k} = \frac{\gamma v_0}{2^\gamma} \ge \frac{(1-\epsilon)\gamma}{2^\gamma} \cdot OPT.$$

2. When $\|\boldsymbol{y}_0\|_1 < k$, we have

$$f(\mathbf{y}_0) \ge (1 - \frac{1}{2^\gamma}) \cdot OPT.$$

Obviously, we can get the following inequality

$$\frac{(1-\epsilon)\gamma}{2^\gamma} \ge 1 - \frac{1}{2^\gamma} - \epsilon, \forall \epsilon \in (0,1), \gamma \in [0,1].$$

From the above two cases, we have

$$f(\boldsymbol{y}_0) \ge \min\{\frac{(1-\epsilon)\gamma}{2^\gamma}, 1 - \frac{1}{2^\gamma}\} \cdot OPT$$

$$\ge \min\{\frac{(1-\epsilon)\gamma}{2^\gamma}, 1 - \frac{1}{2^\gamma} - \epsilon\} \cdot OPT$$

$$= (1 - \frac{1}{2^\gamma} - \epsilon) \cdot OPT.$$

Memory and Query Complexity. We first argue that the amount of $\mathcal{V}_\epsilon$ is $O(\epsilon^{-1} \log k/\gamma)$. For each arriving element e, we need to consider all the parameters in $\mathcal{V}_\epsilon$ to get $O(\epsilon^{-1} \log k/\gamma)$ different solutions. This implies that the memory of Algorithm 2 is $O(k\epsilon^{-1} \log k/\gamma)$. Further, suppose that the element e is fixed, then for each $v \in \mathcal{V}_\epsilon$, we have to call to Algorithm BinarySearchPivot to get a valid pivot. Combine with Lemma 1, we conclude that the query complexity is $O(\epsilon^{-1} \log \|\mathbf{B}\|_\infty \log k/\gamma)$.

5 The One Pass Streaming Algorithm

Algorithm 3. Streaming Algorithm

Require: function f, cardinality k, ground set E, $\mathbf{B}$ and $\epsilon \in (0,1)$.
1: $\mathcal{V}_\epsilon = \{(1+\epsilon)^i | i \in \mathbf{Z}^+\}$
2: for each $v \in \mathcal{V}_\epsilon$, set $\boldsymbol{y}^v \leftarrow \mathbf{0}$
3: $\alpha \leftarrow 0, \beta \leftarrow 0$
4: **for** $i = 1, 2, ..., n$ **do**
5:     $\alpha \leftarrow \max\{\alpha, f(\mathcal{X}_i)\}, \beta \leftarrow \{\beta, f(B(e_i)\mathcal{X}_i)\}$
6:     $\mathcal{V}_\epsilon^i = \{(1+\epsilon)^s | s \in \mathbf{Z}, \alpha/(1+\epsilon) \le (1+\epsilon)^s \le 2^\gamma k\beta/\gamma\}$
7:     Delete all $\boldsymbol{y}^v$, where $v \notin \mathcal{V}_\epsilon^i$
8:     **for** $v \in \mathcal{V}_\epsilon^i$ **do**
9:         **if** $\|y_v\|_1 < k$ **then**
10:            $l_i \leftarrow$BinarySearchPivot$(f, \boldsymbol{y}^v, B(e_i), e_i, k, \frac{\gamma v}{2^\gamma k})$
11:            $\boldsymbol{y}^v \leftarrow \boldsymbol{y}^v + l_i \mathcal{X}_i$
12:        **end if**
13:    **end for**
14: **end for**
15: return $\arg\max_{v \in \mathcal{V}_\epsilon^n} f(\boldsymbol{y}^v)$

Both Algorithm 1 and Algorithm 3 are idealized versions. This is due to the fact that we do not know the exact value of OPT and α defined in Algorithm 2. To solve this problem, we present a new one-pass algorithm in this section, in which we estimate the values of α and a new parameter β. For each element e, denote α and β be the current maximum $f(\mathcal{X}_e)$ and $f(B(e)\mathcal{X}_e)$. This implies that when element e_i and its copies arrive, their corresponding set $\mathcal{V}_\epsilon^i$ will be updated, which is recorded as $\mathcal{V}_\epsilon^i = \{(1+\epsilon)^s | s \in \mathbf{Z}, \alpha/(1+\epsilon) \le (1+\epsilon)^s \le 2^\gamma k\beta/\gamma\}$. When the parameter v first appears in $\mathcal{V}_\epsilon^i$, the value of the vector $\boldsymbol{y}^v$ is $\mathbf{0}$. When the parameter v is not in $\mathcal{V}_\epsilon^i$, we delete the vector $\boldsymbol{y}^v$ to save memory. Finally, the best of $\boldsymbol{y}^v$ is returned.

Lemma 5. *For any $v \in \cup_{i=1}^n \mathcal{V}_\epsilon^i$, denote $\boldsymbol{y}^v$ be the final solution of Algorithm 3 and $\boldsymbol{x}^v = (\boldsymbol{y}^* - \boldsymbol{y}^v) \vee \mathbf{0}$. Suppose that v first appears in $\mathcal{V}_\epsilon^m$. Then for each $e_i, i \in [m-1]$, it always satisfies*

$$f(\boldsymbol{x}^v(e_i)\mathcal{X}_i) < \boldsymbol{x}^v(e_i) \cdot \frac{\gamma v}{2^\gamma k}. \tag{6}$$

Proof. The proof is by contradiction. Suppose that there exists a $i_0 \in [m-1]$ such that

$$f(\boldsymbol{x}^v(e_{i_0})\mathcal{X}_{i_0}) \geq \boldsymbol{x}^v(e_{i_0})\frac{\gamma v}{2^\gamma k}. \tag{7}$$

Let $\alpha_i = \max_{j \in [i]} f(\mathcal{X}_j)$ and $\beta_i = \max_{j \in [i]} f(B(e_j)\mathcal{X}_j)$. Then combine the monotonicity of f and (6), we have

$$v \leq \frac{f(\boldsymbol{x}^v(i_0)\mathcal{X}_{i_0})2^\gamma k}{\boldsymbol{x}^v(e_{i_0})\gamma} \leq \frac{2^\gamma k}{\gamma}f(B(e_{i_0})\mathcal{X}_{i_0}) \leq \frac{2^\gamma k\beta_{i_0}}{\gamma}.$$

Since $v \in \mathcal{V}_\epsilon^m$, we get

$$v \geq \frac{\alpha_m}{1+\epsilon} \geq \frac{\alpha_{i_0}}{1+\epsilon}.$$

Using the above two inequalities, we have $\frac{\alpha_{i_0}}{1+\epsilon} \leq v \leq \frac{2^\gamma k\beta_{i_0}}{\gamma}$. That implies that $v \in \mathcal{V}_{i_0}$, which contradicts the fact v first appears in $\mathcal{V}_\epsilon^m$. Thus we get the desired result.

Theorem 3. *Let f be a non-submodular consider with DR ratio $\gamma \in [0,1]$. For any given $\epsilon \in [0,1]$, denote $\boldsymbol{y}$ be the output of Algorithm 3. Then we have*

- *$f(\boldsymbol{y}) \geq (1 - 1/2^\gamma - \epsilon)OPT$, where OPT is the optimal value.*
- *Algorithm 3 requires one pass, at most $O(k\epsilon^{-1}\log k/\gamma)$ space and $O(\epsilon^{-1}\log k/\gamma \log \|\boldsymbol{B}\|_\infty)$ update time per element.*

Proof. By the result of Lemma 4, we know that there must be a v_0 such that $(1-\epsilon)OPT \leq v_0 \leq OPT$. From Lemma 5, we can assume that v_0 is considered when the first element arriving. For the rest of the proof, we consider the quality of $\boldsymbol{y}^{v_0}$. The analysis is similar as in Theorem 2, thus we have

$$f(\boldsymbol{y}^{v_0}) \geq (1 - 1/2^\gamma - \epsilon)OPT.$$

Since $\boldsymbol{y}$ is the maximum of all solutions, we have

$$f(\boldsymbol{y}) \geq f(\boldsymbol{y}^{v_0}) \geq (1 - 1/2^\gamma - \epsilon)OPT.$$

Thus, we complete the proof.

References

1. Goemans, M.-X., Williamson, D.-P.: Improved approximation algorithms for maximum cut and satisfiability problems using semidefinite programming. J. ACM **42**(6), 1115–1145 (1995)
2. Lin, H., Bilmes, J.: A class of submodular functions for document summarization. In: 49th Annual Meeting of the Association for Computational Linguistics, Portland, Oregon, pp. 510–520. Association for Computational Linguistics (2011)
3. Sipos, R., Swaminathan, A., Shivaswamy, P., Joachims, T.: Temporal corpus summarization using submodular word coverage. In: 21st ACM International Conference on Information and Knowledge Management, Maui, HI, USA, pp. 754–763. Association for Computing Machinery (2012)

4. Calinescu, G., Chekuri, C., Pál, M., Vondrák, J.: Maximizing a submodular set function subject to a matroid constraint (extended abstract). In: Fischetti, M., Williamson, D.P. (eds.) IPCO 2007. LNCS, vol. 4513, pp. 182–196. Springer, Heidelberg (2007). https://doi.org/10.1007/978-3-540-72792-7_15

5. Chen, W., Wang, Y., Yang, S.: Efficient influence maximization in social networks. In: 15th ACM SIGKDD International Conference on Knowledge Discovery and Data Mining, New York, NY, USA, pp. 199–208. Association for Computing Machinery (2009)

6. Seeman, L., Singer, Y.: Adaptive seeding in social networks. In: 54th Annual Symposium on Foundations of Computer Science, Berkeley, CA, USA, pp. 459–468. Institute of Electrical and Electronic Engineers (2013)

7. Chen, W., Wang, C., Wang, Y.: Scalable influence maximization for prevalent viral marketing in large-scale social networks. In: 16th ACM SIGKDD International Conference on Knowledge Discovery and Data Mining, New York, NY, USA, pp. 1029–1038. Association for Computing Machinery (2010)

8. Ageev, A.-A., Sviridenko, M.-I.: An 0.828-approximation algorithm for the uncapacitated facility location problem. Discret. Appl. Math. **93**(2–3), 149–156 (1999)

9. Badanidiyuru, A., Mirzasoleiman, B., Karbasi, A., Krause, A.: Streaming submodular maximization: massive data summarization on the fly. In: 20th ACM SIGKDD International Conference on Knowledge Discovery and Data Mining, New York, NY, USA, pp. 671–680. Association for Computing Machinery (2014)

10. Buchbinder, N., Feldman, M., Schwartz, R.: Online submodular maximization with preemption. In: 26th ACM-SIAM Symposium on Discrete Algorithms, Cambridge, Massachusetts, USA, pp. 1202–1216. Society for Industrial and Applied Mathematics (2014)

11. Norouzi-Fard, A., Tarnawski, J., Mitrovic, S., Zandieh, A., Mousavifar, A., Svensson, O.: Beyond 1/2-approximation for submodular maximization on massive data streams. In: 35th International Conference on Machine Learning, Stockholm, Sweden, pp. 3829–3838. International Machine Learning Society (2018)

12. Kazemi, E., Mitrovic, M., Zadimoghaddam, M., Lattanzi, S., Karbasi, A.: Submodular streaming in all its glory: tight approximation, minimum memory and low adaptive complexity. In: 36th International Conference on Machine Learning, Long Beach, California, pp. 3311–3320. International Machine Learning Society (2019)

13. Soma, T., Yoshida, Y.: Maximizing monotone submodular functions over the integer lattice. Math. Program. 539–563 (2018). https://doi.org/10.1007/s10107-018-1324-y

14. Soma, T., Kakimura, N., Inaba, K., Kawarabayashi, K.-I.: Optimal budget allocation: theoretical guarantee and efficient algorithm. In: 31th International Conference on Machine Learning, Beijing, China, pp. 351–359. International Machine Learning Society (2014)

15. Nong, Q., Fang, J., Gong, S., Du, D., Feng, Y., Qu, X.: A 1/2-approximation algorithm for maximizing a non-monotone weak-submodular function on a bounded integer lattice. J. Comb. Optim. **39**(4), 1208–1220 (2020). https://doi.org/10.1007/s10878-020-00558-4

16. Kuhnle, A., Smith, J.-D., Crawford, V., Thai, M.: Fast maximization of non-submodular, monotonic functions on the integer lattice. In: 35th International Conference on Machine Learning, Stockholm, Sweden, pp. 2786–2795. International Machine Learning Society (2018)

17. Wang, Y., Xu, D., Wang, Y., Zhang, D.: Non-submodular maximization on massive data streams. J. Glob. Optim. **76**(4), 729–743 (2019). https://doi.org/10.1007/s10898-019-00840-8
18. Tan, J., Zhang, D., Zhang, H., Zhang, Z.: Streaming algorithms for monotone DR-submodular maximization under a knapsack constraint on the integer lattice. In: Ning, L., Chau, V., Lau, F. (eds.) PAAP 2020. CCIS, vol. 1362, pp. 58–67. Springer, Singapore (2021). https://doi.org/10.1007/978-981-16-0010-4_6
19. Zhang, Z., Guo, L., Wang, Y., Xu, D., Zhang, D.: Streaming algorithms for maximizing monotone DR-submodular functions with a cardinality constraint on the integer lattice. Asia-Pac. J. Oper. Res. 2140004 (2021)
20. Zhang, Z., Guo, L., Wang, L., Zou, J.: A streaming model for monotone lattice submodular maximization with a cardinality constraint. In: Zhang, Y., Xu, Y., Tian, H. (eds.) PDCAT 2020. LNCS, vol. 12606, pp. 362–370. Springer, Cham (2021). https://doi.org/10.1007/978-3-030-69244-5_32
21. Das, A., Kempe, D.: Submodular meets spectral: greedy algorithms for subset selection, sparse approximation and dictionary selection. In: 28th International Conference on Machine Learning, Bellevue, WA, USA, pp. 1057–1064. International Machine Learning Society (2011)
22. Nong, Q., Sun, T., Gong, S., Fang, Q., Du, D., Shao, X.: Maximize a monotone function with a generic submodularity ratio. In: Du, D.-Z., Li, L., Sun, X., Zhang, J. (eds.) AAIM 2019. LNCS, vol. 11640, pp. 249–260. Springer, Cham (2019). https://doi.org/10.1007/978-3-030-27195-4_23

Causal Inference for Influence Propagation—Identifiability of the Independent Cascade Model

Shi Feng[1]([⊠])[iD] and Wei Chen[2]

[1] IIIS, Tsinghua University, Beijing, China
fengs19@mails.tsinghua.edu.cn
[2] Microsoft Research, Beijing, China
weic@microsoft.com

Abstract. Independent cascade (IC) model is a widely used influence propagation model for social networks. In this paper, we incorporate the concept and techniques from causal inference to study the identifiability of parameters from observational data in extended IC model with unobserved confounding factors, which models more realistic propagation scenarios but is rarely studied in influence propagation modeling before. We provide the conditions for the identifiability or unidentifiability of parameters for several special structures including the Markovian IC model, semi-Markovian IC model, and IC model with a global unobserved variable. Parameter identifiability is important for other tasks such as influence maximization under the diffusion networks with unobserved confounding factors.

Keywords: Influence propagation · Independent cascade model · Identifiability · Causal inference

1 Introduction

Extensive research has been conducted studying the information and influence propagation behavior in social networks, with numerous propagation models and optimization algorithms proposed (cf. [2,16]). Social influence among individuals in a social network is intrinsically a causal behavior—one's action or behavior causes the change of the behavior of his or her friends in the network. Therefore, it is helpful to view influence propagation as a causal phenomenon and apply the tools in causal inference to this domain.

In causal inference, one key consideration is the confounding factors caused by unobserved variables that affect the observed behaviors of individuals in the network. For example, we may observe that user A adopts a new product and a while later her friend B adopts the same new product. This situation could be because A influences B and causes B's adoption, but it could also be caused by an unobserved factor (e.g. an unknown information source) that affects both A and B. Confounding factors are important in understanding the propagation behavior

© Springer Nature Switzerland AG 2021
D. Mohaisen and R. Jin (Eds.): CSoNet 2021, LNCS 13116, pp. 15–26, 2021.
https://doi.org/10.1007/978-3-030-91434-9_2

in networks, but so far the vast majority of influence propagation research does not consider confounders in network propagation modeling. In this paper, we intend to fill this gap by explicitly including unobserved confounders into the model, and we borrow the research methodology from causal inference to carry out our research.

Causal inference research has developed many tools and methodologies to deal with such unobserved confounders, and one important problem in causal inference is to study the identifiability of the causal model, that is, if we can identify the certain effect of an intervention, or identify causal model parameters, from the observational data. In this paper, we introduce the concept of identifiability in causal inference research to influence propagation research and study whether the propagation models can be identified from observational data when there are unobserved factors in the causal propagation model. We propose the extend the classical independent cascade (IC) model to include unobserved causal factors, and consider the parameter identifiability problem for several common causal graph structures. Our main results are as follows. First, for the Markovian IC model, in which each unobserved variable may affect only one observed node in the network, we show that it is fully identifiable. Second, for the semi-Markovian IC model, in which each unobserved variable may affect exactly two observed nodes in the network, we show that as long as a local graph structure exists in the network, then the model is not parameter identifiable. For the special case of a chain graph where all observed nodes form a chain and every unobserved variable affect two neighbors on the chain, the above result implies that we need to know at least $n/2$ parameters to make the rest parameters identifiable, where n is the number of observed nodes in the chain. We then show a positive result that when we know n parameters on the chain, the rest parameters are identifiable. Third, for the global hidden factor model where we have an unobserved variable that affects all observed nodes in the graph, we provide reasonable sufficient conditions so that the parameters are identifiable.

Overall, we view that our work starts a new direction to integrate rich research results from network propagation modeling and causal inference so that we could view influence propagation from the lens of causal inference, and obtain more realistic modeling and algorithmic results in this area. For example, from the causal inference lens, the classical influence maximization problem [16] of finding a set of k nodes to maximize the total influence spread is really a causal intervention problem of forcing an intervention on k nodes for their adoptions, and trying to maximize the causal effect of this intervention. Our study could give a new way of studying influence maximization that works under more realistic network scenarios encompassing unobserved confounders. Due to the limitation of space, the complete proofs of some of the theorems are placed in the full version [8] on arXiv, and only outlines are given in this version.

2   Related Work

Influence Propagation Modeling. As described in [2], the main two models used to describe influence propagation are the independent cascade model and the linear threshold model. Past researches on influence propagation mostly focused on influence maximization problems, such as [16,23]. In these articles, they select seed nodes online, observe the propagation in the network, and optimize the number of activated nodes after propagation by selecting optimal seed nodes. Also, some works are studying the seed-node set minimization problem, such as [12]. However, in our work, we mainly consider restoring the parameters in the independent cascade model by observing the network propagation. After obtaining the parameters in the network, we can then base on this to accomplish downstream tasks including influence maximization and seed-node set minimization.

Causal Inference and Identifiability. For general semi-Markovian Bayesian causal graphs, [14] and [22] have given two different algorithms to determine whether a do effect is identifiable, and these two algorithms have both soundness and correctness. [15] also proves that the ID algorithm and the repeating use of the do calculus are equivalent, so for semi-Markovian Bayesian causal graphs, the do calculus can be used to compute all identifiable do effects.

In addition, for a special type of causal model, the linear causal model, articles [4] and [9] have given some necessary conditions and sufficient conditions on whether the parameters in the graph are identifiable with respect to the structure of the causal graph. However, the necessary and sufficient condition for parameter identifiability problem is not addressed and it remains an open question. In this paper, we study another special causal model derived from the IC model. Since the IC model can be viewed as a Bayesian causal model when the graph structure is a directed acyclic graph and it has some special properties, we try to give some necessary conditions and sufficient conditions for the parameters to be identifiable under some special graph structures.

3   Model and Problem Definitions

Following the convention in causal inference literature (e.g. [20]), we use capital letters $(U, V, X, \ldots)$ to represent variables or a set of variables, and their corresponding lower-case letters to represent their values. For a directed graph, we use U's and V's to represent nodes since each node will also be treated as a random variable in causal inference. For a node V_i, we use $N^+(V_i)$ and $N^-(V_i)$ to represent the set of its out-neighbors and in-neighbors, respectively. When the graph is directed acyclic (DAG), we refer to a node's in-neighbors as its parents and denote the set as $Pa(V_i) = N^-(V_i)$. When we refer to the actual values of the parent nodes of V_i we use $pa(V_i)$. For a positive integer k, we use $[k]$ to denote $\{1, 2, \ldots, k\}$. We use boldface letters to represent vectors, such as $\boldsymbol{r} = (r_1, r_2, \ldots, r_n) = (r_i)_{i \in [n]}$.

The classical *independent cascade model* [16] of influence diffusion in a social network is modeled as follows. The social network is modeled as a directed graph

$G = (V, E)$, where $V = \{V_1, V_2, \cdots, V_n\}$ is the set of nodes representing individuals in the social network, and $E \subseteq V \times V$ is the set of directed edges representing the influence relationship between the individuals. Each edge $(V_i, V_j) \in E$ is associated with an influence probability $p(i, j) \in (0, 1]$ (we assume that $p(i, j) = 0$ if $(V_i, V_j) \notin E$). Each node is either in state 0 or state 1, representing the idle state and the active state, respectively. At time step 0, a *seed set* $S_0 \subseteq V$ of nodes is selected and activated (i.e. their states are set to 1), and all other nodes are in state 0. The propagation proceeds in discrete time steps $t = 1, 2, \ldots$. Let S_t denote the set of nodes that are active by time t, and let $S_{-1} = \emptyset$. At any time $t = 1, 2, \ldots$, the newly activated node $V_i \in S_{t-1} \setminus S_{t-2}$ tries to activate each of its inactive outgoing neighbors $V_j \in N^+(V_i)$, and the activation is successful with probability $p(i, j)$. If successful, V_j is activated at time t and thus $V_j \in S_t$. The activation trial of V_i on its out-neighbor V_j is independent of all other activation trials. Once activated, nodes stay as active, that is, $S_{t-1} \subseteq S_t$. The propagation process ends at a step when there are no new nodes activated. It easy to see that the propagation ends in at most $n - 1$ steps, so we use S_{n-1} to denote the final set of active nodes after the propagation.

Influence propagation is naturally a result of causal effect—one node's activation causes the activation of its outgoing neighbors. If the graph is directed and acyclic, then the IC model on this graph can be equated to a Bayesian causal model. In fact, we can consider each node in the IC model as a variable, and for a node V_i, it takes the value determined by $P(V_i = 1 | pa(V_i)) = 1 - \prod_{j:V_j \in Pa(V_i), v_j = 1 \text{ in } pa(V_i)} (1 - p_{j,i})$. Obviously, this is equivalent to our definition in the IC model. IC model is introduced in [16] to model influence propagation in social networks, but in general, it can model the causal effects among binary random variables. In this paper, we mainly consider the directed acyclic graph (DAG) setting, which is in line with the causal graph setting in the causal inference literature [20]. We discuss the extension to general cyclic graphs or networks in the full version [8].

All variables $V_1, V_2, \ldots, V_n$ are observable, and we call them *observed variables*. They correspond to observed behaviors of individuals in the social network. There are also potentially many *unobserved (or hidden) variables* that affecting individuals' behaviors. We use $U = \{U_1, U_2, \ldots\}$ to represent the set of unobserved variables. In the IC model, we assume each U_i is a binary random variable with probability r_i to be 1 and probability $1 - r_i$ to be 0, and all unobserved variables are mutually independent. We allow unobserved variables U_i's to have directed edges pointing to the observed variables V_j's, but we do not consider directed edges among the unobserved variables in this paper. If U_i has a directed edge pointing to V_j, we usually use $q_{i,j}$ to represent the parameter on this edge. It has the same semantics as the $p_{i,j}$'s in the classical IC model: if $U_i = 1$, then with probability $q_{i,j}$ U_i successfully influence V_j by setting its state to 1, and with probability $1 - q_{i,j}$ V_j's state is not affected by U_i, and this influence or activation effect is independent from all other activation attempts on other edges. Thus, overall, in a network with unobserved or hidden variables, we use $G = (U, V, E)$ to represent the corresponding causal graph, where U is the set of unobserved variables, V is the set of observed variables, and $E \subseteq (V \times V) \cup (U \times V)$ is the set of directed edges. We assume that G is a DAG, and the state of every unobserved variable U_i is sampled from $\{0,1\}$ with parameter r_i, while the state of every observed variable V_j is determined by the states of its parents and the parameters on the incoming edges of V_j following the IC model semantics. In the DAG G, we refer to an observable node V_i as a *root* if it has no observable parents in the graph. Every root V_i has at least one unobserved parent. We use vectors p, q, r to represent parameter vectors associated with edges among observed variables, edges from unobserved to observed variables, and unobserved nodes, respectively. We refer to the model $M = (G = (U, V, E), p, q, r)$ as the *causal IC model*. When the distinction is needed, we use capital letters P, Q, R to represent the parameter names, and lower boldface letters p, q, r to represent the parameter values.

In this paper, we focus on the *parameter identifiability* problem following the causal inference literature. In the context of the IC model, the states of nodes $V = \{V_1, V_2, \ldots, V_n\}$ are observable while the states of $U = \{U_1, U_2, \ldots\}$ are unobservable. We define parameter identifiability as follows.

Definition 1 (Parameter Identifiability). *Given a graph $G = (U, V, E)$, we say that a set of IC model parameters $\Theta \subseteq P \cup Q \cup R$ on G is identifiable if after fixing the values of parameters outside Θ and fixing the observed probability distributions $P(V' = v')$ for all $V' \subseteq V$ and all $v' \in \{0,1\}^{|V'|}$, the values of parameters in Θ are uniquely determined. We say that the graph G is parameter identifiable if $\Theta = P \cup Q \cup R$. Accordingly, the algorithmic problem of parameter identifiability is to derive the unique values of parameters in Θ given graph $G = (U, V, E)$, the values of parameters outside Θ, and the observed probability distributions $P(V' = v')$ for all $V' \subseteq V$ and all $v' \in \{0,1\}^{|V'|}$. Finally, if the algorithm only uses a polynomial number of observed probability values $P(V' = v')$'s and runs in polynomial time, where both polynomials are with respect to the graph size, we say that the parameters in Θ are efficiently identifiable.*

Note that when there are no unobserved variables (except the unique unobserved variables for each root of the graph), the problem is mainly to derive the parameters $p_{i,j}$'s from all observed $P(V' = v')$'s. In this case, the parameter identifiability problem bears similarity with the well-studied network inference problem [1,3,5–7,10,11,13,17–19,21]. The network inference problem focuses on using observed cascade data to derive the network structure and propagation parameters, and it emphasizes on the sample complexity of inferring parameters. Hence, when there are no unobserved variables in the model, we could use the network inference methods to help to solve the parameter identifiability problem. However, in real social influence and network propagation, there are other hidden factors that affect the propagation and the resulting distribution. Such hidden factors are not addressed in the network inference literature. In contrast, our study in this paper is focusing on addressing these hidden factors in network inference, and thus we borrow the ideas from causal inference to study the identifiability problem under the IC model.

In this paper, we study three types of unobserved variables that could commonly occur in network influence propagation. They correspond to three types of IC models with unobserved variables, as summarized below.

Markovian IC Model. In the *Markovian IC model*, each observed variable V_i is associated with a unique unobserved variable U_i, and there is a directed edge from U_i to V_i. This models the scenario where each individual in the social network has some latent and unknown factor that affects its observed behavior. We use q_i to denote the parameter on the edge (U_i, V_i). Note that the effect of U_i on the activation of V_i is determined by probability $r_i \cdot q_i$, and thus we treat $r_i = 1$ for all $i \in [n]$, and focus on identifying parameters q_i's. Thus the graph $G = (U, V, E)$ has parameters $\boldsymbol{q} = (q_i)_{i \in [n]}$, and $\boldsymbol{p} = (p_{i,j})_{(V_i, V_j) \in E}$. Figure 1 shows an example of a Markovian IC model. If some $q_i = 0$, it means that the observed variable V_i has no latent variable influencing it, and it only receives influence from other observed variables.

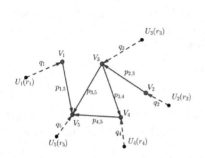

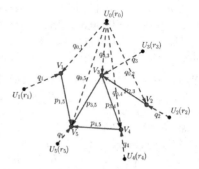

Fig. 1. A Markovian IC model with five nodes.

Fig. 2. A Markovian IC model with five nodes and a global unobserved variable.

Semi-Markovian IC Model. The second type of unobserved variables is the hidden variables connected to exactly two observed variables in the graph. In particular, for every pair of nodes $V_i, V_j \in V$, we allow one unobserved variable $U_{i,j}$ that has two edges, one pointing to V_i and the other pointing to V_j. This models the scenario that two individuals in the social network has a common unobserved confounder that may affect the behavior of two individuals. We call this type of model *semi-Markovian IC model*, following the common terminology of the semi-Markovian model in the literature [20]. In this model, each $U_{i,j}$ has a parameter $r_{i,j}$, and edges $(U_{i,j}, V_i)$ and $(U_{i,j}, V_j)$ have parameters $q_{i,j,1}$ and $q_{i,j,2}$ respectively. Therefore, the graph has parameters $\boldsymbol{r} = (r_{i,j})_{(V_i, V_j) \in E}$, $\boldsymbol{q} = (q_{i,j,1}, q_{i,j,2})_{(V_i, V_j) \in E}$, and $\boldsymbol{p} = (p_{i,j})_{(V_i, V_j) \in E}$.

Within this model, we will pay special attention to a special type of graphs where the observed variables form a chain, i.e. $V_1 \to V_2 \to \cdots \to V_n$, and the unobserved variables always point to the two neighbors on the chain. In this case, we use U_i to denote the unobserved variable associated with edge (V_i, V_{i+1}), and the parameters on the edges (U_i, V_i) and (U_i, V_{i+1}) are denoted as $q_{i,1}$ and $q_{i,2}$, respectively. Figure 3 represents this chain model.

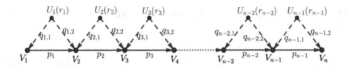

Fig. 3. The semi-Markovian IC chain model.

IC Model with A Global Unobserved Variable. The third type of hidden variables is a global unobserved variable U_0 that points to all observed variables in the network. This naturally models the global causal effect where some common factor affects all or most individuals in the network. For every edge (U_0, V_i), we use $q_{0,i}$ to represent its parameter.

Moreover, we can combine this model with the Markovian IC model, where we allow both unobserved variable U_i for each individual and a global unobserved varoable U_0. Figure 2 represents this model.

4   Parameter Identifiability of the Markovian IC Model

For the Markovian IC model in which every observed variable has its own unobserved variable, we can fully identify the model parameters in most cases, as given by the following theorem.

Theorem 1 (Identifiability of the Markovian IC Model). *For an arbitrary Markovian IC model $G = (U, V, E)$ with parameters $\boldsymbol{q} = (q_i)_{i \in [n]}$ and $\boldsymbol{p} = (p_{i,j})_{(V_i, V_j) \in E}$, all the q_i parameters are efficiently identifiable, and for every $i \in [n]$, if $q_i \neq 1$, then all $p_{j,i}$ parameters for $(V_j, V_i) \in E$ are efficiently identifiable.*

Proof. For an observed variable (node) V_i, suppose that its observed parents are $V_{i_1}, V_{i_2}, \cdots, V_{i_t}$. Therefore, we have

$$P(V_i = 0 | V_{i_1} = 0, \cdots, V_{i_t} = 0) = 1 - q_i, \tag{1}$$

$$P(V_i = 0 | V_{i_j} = 1, V_{i_1} = 0, \cdots, V_{i_{j-1}} = 0, V_{i_{j+1}} = 0, \cdots, V_{i_t} = 0) = (1 - q_i)(1 - p_{i_j, i}). \tag{2}$$

From Eq. (1), we can obtain the value of q_i. Then if $q_i \neq 1$, from Eq. (2), we can derive the value of $p_{i_j, i}$. Moreover, for each root node V_i, we can get q_i by computing $q_i = P(V_i = 1)$. The computational efficiency is obvious.   □

The theorem essentially says that all parameters are identifiable under the Markovian IC model, except for the corner case where some $q_i = 1$. In this case, the observed variable V_i is fully determined by its unobserved parent U_i, so we cannot determine the influence from other observed parents of V_i to V_i. But the influence from the observed parents of V_i to V_i is not useful any way in this case, so the edges from the observed parents of V_i to V_i will not affect the causal inference in the graph and they can be removed.

5   Parameter Identifiability of the Semi-Markovian IC Model

Following the definition in the model section, we then consider the identifiability problem of the semi-Markovian models. We will demonstrate that in most cases, this model is not parameter identifiable. Actually, from [22] we know that the semi-Markovian Bayesian causal model is also not identifiable in general. Essentially, our conclusion is not related to their result. On the other side, we will show that with some parameters known in advance, the semi-Markovian IC chain model will be identifiable.

5.1   Condition on Unidentifiability of the Semi-Markovian IC Model

More specifically, the following theorem shows the unidentifiability of the semi-Markovian IC model with a special structure in it.

Theorem 2 (Unidentifiability of the Semi-Markovian IC Model). *Suppose in a general graph G, we can find the following structure. There are three observable nodes V_1, V_2, V_3 such that $(V_1, V_2) \in E, (V_2, V_3) \in E$ and unobservable U_1, U_2 with $(U_1, V_1), (U_1, V_2), (U_2, V_2), (U_2, V_3) \in E$. Suppose each of U_1, U_2 only has two edges associated to it, the three nodes V_1, V_2, V_3 can be written adjacently in a topological order of nodes in $U \cup V$. Then we can deduce that the graph G is not parameter identifiable.*

Figure 4 is an example of the structure described in the above theorem.

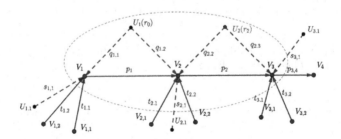

Fig. 4. An example of the structure in Theorem 2.

Proof (Outline). To prove that the parameters in the model with this structure are not identifiable, we give two different sets of parameters directly. We show that these two different sets of parameters produce the same distribution of nodes in V, and thus the set of parameters is not identifiable by observing only the distribution of V. The details of these two sets of parameters and the distributions they produce are included in the full technical report [8]. □

5.2   Identifiability of the Chain Model

We now consider the chain model as described in Sect. 3 and depicted in Fig. 3. In this structure, we present a conclusion of identifiability under the assumption that the valuations of some parameters are our prior knowledge.

We divide the parameters of the graph into four vectors

$$\boldsymbol{q}_1 = (q_{1,1}, q_{2,1}, \cdots, q_{n-1,1}), \boldsymbol{q}_2 = (q_{1,2}, q_{2,2}, \cdots, q_{n-1,2}), \tag{3}$$

$$\boldsymbol{p} = (p_1, p_2, \cdots, p_{n-1}), \boldsymbol{r} = (r_1, r_2, \cdots, r_{n-1}). \tag{4}$$

For the chain model, our theorem below shows that once the parameters p_1 is known, $\boldsymbol{q}_2$ or $\boldsymbol{r}$ is known, the set consists of remaining parameters in the chain is efficiently identifiable.

Theorem 3 (Identifiability of the Semi-Markovian IC Chain Model).
Suppose that we have a semi-Markovian IC chain model with the graph $G = (U, V, E)$ and the IC parameters $\boldsymbol{p} = (p_i)_{i \in [n-1]}$, $\boldsymbol{q}_1 = (q_{i,1})_{i \in [n-1]}$, $\boldsymbol{q}_2 = (q_{i,2})_{i \in [n-1]}$ and $\boldsymbol{r} = (r_i)_{i \in [n-1]}$, and suppose that all parameters are in the range $(0,1)$. If the values of parameter p_1 is known, $\boldsymbol{q}_2$ or $\boldsymbol{r}$ is known, then the remaining parameters are efficiently identifiable.

Proof (Outline). We use induction to prove this theorem. Under the assumption that p_1 is known and $\boldsymbol{q}_2$ or $\boldsymbol{r}$ is known, suppose $p_1, p_2, \cdots, p_{t-2}, r_1, r_2, \cdots, r_{t-2}$, $q_{1,1}, q_{2,1}, \cdots, q_{t-2,1}$ and $q_{1,2}, q_{2,2}, \cdots, q_{t-2,2}, r_{t-1}q_{t-1,1}$ has been determined by us, and we prove that $q_{t-1,1}, r_{t-1}, p_{t-1}, q_{t-1,2}$ and $r_t q_{t,1}$ can also be determined. In fact, by the distribution of the first t nodes on the chain we can obtain three different equations, and after substituting our known parameters, the inductive transition can be completed. It is worthy noting that this inductive process can also be used to compute the unknown parameters efficiently.

The proof is lengthy because of the many corner cases considered and the need to discuss the cases $t = n, t = 2$ and $2 < t < n$.   □

According to Theorem 3 we get that the semi-Markovian chain is parameter identifiable in the case that n particular parameters are known. Simultaneously, by Theorem 2, we can show that if just less than $\lfloor \frac{n+1}{2} \rfloor$ parameters are known, then this semi-Markovian chain will not be parameter identifiable. Actually, if the chain model is parameter identifiable, utilizing Theorem 2, we know that for each $2 \leq t \leq n - 1$, at least one of parameters between $p_{t-1}, p_t, r_{t-1}, r_t, q_{t-1,1}, q_{t-1,2}, q_{t,1}$ and $q_{t,2}$ should be known. Therefore, we let $t = 2, 4, \cdots, 2 \lfloor \frac{n-1}{2} \rfloor$, we can deduce that at least $\lfloor \frac{n-1}{2} \rfloor$ should be known. Formally, we have the following corollary of Theorem 2 and Theorem 3.

Corollary 1. *For a semi-Markovian IC chain model, if no more than $\lfloor \frac{n-1}{2} \rfloor$ parameters are known in advance, the remaining parameters are unidentifiable; if it is allowed to know n parameters in advance, we can choose $p_1, \boldsymbol{q}_2$ or $p_1, \boldsymbol{r}$ to be known, then the remaining parameters are identifiable.*

6   Parameter Identifiability of Model with a Global Hidden Variable

Next, we consider the case where there is a global hidden variable in the causal IC model, defined as those in Sect. 3. If there is only one hidden variable U_0 in the whole model, we prove that the parameters in general in this model are identifiable; if there is not only U_0, the model is also Markovian, that is, there are also n hidden variables $U_1, \cdots, U_n$ corresponding to $V_1, V_2, \cdots, V_n$, then the parameters in this model are identifiable if certain conditions are satisfied.

6.1   Observable IC Model with only a Global Hidden Variable

Suppose the observed variables in the connected DAG graph $G = (U, V, E)$ are $V_1, V_2, \cdots, V_n$ in a topological order and there is a global hidden variable U_0 such that there exists an edge from U_0 to the node for each observable variable V_i. Suppose the activating probability of U_0 is r and the activating probability from U to V_i is $q_i \in [0, 1)$ (naturally, $q_1 \neq 0$ and there are at least 3 of nonzero q_i's). Now we propose a theorem according to these settings.

Theorem 4 (Identifiability of the IC Model with a Global Hidden Variable).   *For an arbitrary IC model with a global hidden variable $G = (U, V, E)$ with parameters $\mathbf{q} = (q_i)_{i \in [n]}$, r and $\mathbf{p} = (p_{i,j})_{(V_i, V_j) \in E}$ such that $q_i \neq 1, p_{i,j} \neq 1$ and $r \neq 1$ for $\forall i, j \in [n]$, all the parameters in $\mathbf{p}, r$ and $\mathbf{q}$ are identifiable.*

Proof (Outline). We discuss this problem in two cases, the first one is the existence of two disconnected points $V_i, V_j, i < j$ in V and $q_i, q_j \neq 0$. At this point we can use $1 - q_j = \frac{P(V_1=0, V_2=0, \cdots, V_i=1, V_{i+1}=0, \cdots, V_j=0)}{P(V_1=0, V_2=0, \cdots, V_i=1, V_{i+1}=0, \cdots, V_{j-1}=0)}$ to solve out q_j, and then use $P(V_1 = 0, V_2 = 0, \cdots, V_j = 0)$ and $P(V_1 = 0, V_2 = 0, \cdots, V_{j-1} = 0)$ to solve out r.

After getting r, by the quotients of probabilities of propagating results, we can get all the parameters.

Another case is that there is no V_i, V_j as described above. At this point there must exist three points V_i, V_j, V_k that are connected with each other and $q_i, q_j, q_k \neq 0$. We observe the probabilities of different possible propagating results of these three points with all other nodes are 0 after the propagation. From these, we can solve out q_i, q_j, q_k, and then solve out all parameters by the same method as in the first case.   □

6.2   Markovian IC Model with a Global Hidden Variable (Mixed Model)

Suppose the model is $G = (U, V, E)$, where $U = \{U_0, U_1, U_2, \cdots, U_n\}$, $V = \{V_1, V_2, \cdots, V_n\}$. Here, $V_1, V_2, \cdots, V_n$ are in a topological order. The parameters are r_0, $\mathbf{q}_0 = (q_{0,i})_{i \in [n]}$, $\mathbf{q} = (q_i)_{i \in [n]}$ and $\mathbf{p} = (p_{i,j})_{(V_i, V_j) \in E}$.

Theorem 5 (Identifiability of Markovian IC Model with a Global Hidden Variable (Mixed Model)). *For an arbitrary Markovian IC Model with a Global Hidden Variable $G = (U, V, E)$ with parameters r_0, $\mathbf{q}_0 = (q_{0,i})_{i \in [n]}$, $\mathbf{q} = (q_i)_{i \in [n]}$ and $\mathbf{p} = (p_{i,j})_{(V_i, V_j) \in E}$, we suppose that all the parameters are not 1. If $\exists i, j, k \in [n], i < j < k$ such that each pair in V_i, V_j, V_k are disconnected and $q_{0,i}, q_{0,j}, q_{0,k} \neq 0$, then the parameters $q_{0,t}, q_t$ and $p_{t,l}, l > t > k$ are identifiable. Moreover, if V_i, V_j, V_k can be adjacently continuous in some topological order, i.e. $j = i + 1, k = i + 2$ without loss of generality, all the parameters are identifiable.*

Proof (Outline). Assuming that there exist V_i, V_j, V_k that satisfy the requirements of the theorem, then we can write expressions for the distribution of these three parameters when all other nodes with subscripts not greater than l are equal to 0. In fact, we can see that with these 8 expressions, we can solve for $P(V_1 = 0, \cdots, V_l = 0, U_0 = 1)$ and $P(V_1 = 0, \cdots, V_l = 0, U_0 = 0)$.

Since we have $P(V_1 = 0, \cdots, V_l = 0, U_0 = 1) = r \prod_{t=1}^{l} (1 - q_t)(1 - q_{0,t})$ and $P(V_1 = 0, \cdots, V_l = 0, U_0 = 0) = (1 - r) \prod_{t=1}^{l} (1 - q_t)$, we will be able to obtain all the parameters very easily by dividing these equations two by two. This proof has some trivial discussion to show that this computational method does not fail due to corner cases. $\qquad \square$

Notice that the parameters in this model are identifiable when and only when a special three-node structure appears in it. Intuitively, this is because through this structure we can more easily obtain some information about the parameters, which does not contradict the intuition of Theorem 2.

7 Conclusion

In this paper, we study the parameter identifiability of the independent cascade model in influence propagation and show conditions on identifiability or unidentifiability for several classes of causal IC model structure. We believe that the incorporation of observed confounding factors and causal inference techniques is important in the next step of influence propagation research and identifiability of the IC model is our first step towards this goal. There are many open problems and directions in combining causal inference and propagation research. For example, seed selection and influence maximization correspond to the intervention (or do effect) in causal inference, and how to compute such intervention effect under the network with unobserved confounders and how to do influence maximization is a very interesting research question. In terms of identifiability, one can also investigate the identifiability of the intervention effect, or whether given some intervention effect one can identify more of such effects.

References

1. Abrahao, B., Chierichetti, F., Kleinberg, R., Panconesi, A.: Trace complexity of network inference. arXiv e-prints, pp. arXiv-1308 (2013)

2. Chen, W., Lakshmanan, L.V., Castillo, C.: Information and Influence Propagation in Social Networks. Morgan & Claypool Publishers (2013)
3. Daneshmand, H., Gomez-Rodriguez, M., Song, L., Schölkopf, B.: Estimating diffusion network structures: recovery conditions, sample complexity & soft-thresholding algorithm. In: ICML (2014)
4. Drton, M., Foygel, R., Sullivant, S.: Global identifiability of linear structural equation models. Ann. Stat. **39**(2), 865–886 (2011)
5. Du, N., Liang, Y., Balcan, M., Song, L.: Influence function learning in information diffusion networks. In: ICML 2014, Beijing, China, 21–26 June 2014 (2014)
6. Du, N., Song, L., Gomez-Rodriguez, M., Zha, H.: Scalable influence estimation in continuous-time diffusion networks. In: NIPS 2013, Lake Tahoe, Nevada, United States, 5–8 December 2013, pp. 3147–3155 (2013)
7. Du, N., Song, L., Smola, A.J., Yuan, M.: Learning networks of heterogeneous influence. In: NIPS 2012, Lake Tahoe, Nevada, United States, 3–6 December 2012, pp. 2789–2797 (2012)
8. Feng, S., Chen, W.: Causal inference for influence propagation - identifiability of the independent cascade model. CoRR abs/2107.04224 (2021). https://arxiv.org/abs/2107.04224
9. Foygel, R., Draisma, J., Drton, M.: Half-trek criterion for generic identifiability of linear structural equation models. Ann. Stat. 1682–1713 (2012)
10. Gomez-Rodriguez, M., Balduzzi, D., Schölkopf, B.: Uncovering the temporal dynamics of diffusion networks. arXiv preprint arXiv:1105.0697 (2011)
11. Gomez-Rodriguez, M., Leskovec, J., Krause, A.: Inferring networks of diffusion and influence. In: KDD 2010 (2010)
12. Goyal, A., Bonchi, F., Lakshmanan, L.V., Venkatasubramanian, S.: On minimizing budget and time in influence propagation over social networks. Soc. Netw. Anal. Min. **3**(2), 179–192 (2013)
13. He, X., Xu, K., Kempe, D., Liu, Y.: Learning influence functions from incomplete observations. arXiv e-prints, pp. arXiv-1611 (2016)
14. Huang, Y., Valtorta, M.: Identifiability in causal Bayesian networks: a sound and complete algorithm. In: AAAI, pp. 1149–1154 (2006)
15. Huang, Y., Valtorta, M.: Pearl's calculus of intervention is complete. arXiv preprint arXiv:1206.6831 (2012)
16. Kempe, D., Kleinberg, J.M., Tardos, É.: Maximizing the spread of influence through a social network. In: KDD (2003)
17. Myers, S.A., Leskovec, J.: On the convexity of latent social network inference. arXiv e-prints, pp. arXiv-1010 (2010)
18. Narasimhan, H., Parkes, D.C., Singer, Y.: Learnability of influence in networks. In: Proceedings of the 29th Annual Conference on Neural Information Processing Systems (2015)
19. Netrapalli, P., Sanghavi, S.: Finding the graph of epidemic cascades. arXiv preprint arXiv:1202.1779 (2012)
20. Pearl, J.: Causality, 2nd edn. Cambridge University Press, Cambridge (2009)
21. Pouget-Abadie, J., Horel, T.: Inferring graphs from cascades: a sparse recovery framework. arXiv e-prints, pp. arXiv-1505 (2015)
22. Shpitser, I., Pearl, J.: Identification of joint interventional distributions in recursive semi-Markovian causal models. In: Proceedings of the 21st National Conference on Artificial Intelligence, pp. 1219–1226 (2006)
23. Tang, Y., Shi, Y., Xiao, X.: Influence maximization in near-linear time: a martingale approach. In: Proceedings of the 2015 ACM SIGMOD International Conference on Management of Data, pp. 1539–1554 (2015)

Streaming Algorithms for Budgeted k-Submodular Maximization Problem

Canh V. Pham[1(✉)], Quang C. Vu[2], Dung K. T. Ha[2,3], and Tai T. Nguyen[3]

[1] ORlab, Faculty of Computer Science, Phenikaa University, Hanoi 12116, Vietnam
`canh.phamvan@phenikaa-uni.edu.vn`
[2] Faculty of Information Security, People's Security Academy, Hanoi, Vietnam
`quangvc.hvan@gmail.com`
[3] Faculty of Information Technology, University of Engineering and Technology,
Vietnam National University, Hanoi, Vietnam
`{20028008,20025032}@vnu.edu.vn`

Abstract. Stimulated by practical applications arising from viral marketing. This paper investigates a novel Budgeted k-Submodular Maximization problem defined as follows: Given a finite set V, a budget B and a k-submodular function $f : (k+1)^V \mapsto \mathbb{R}_+$, the problem asks to find a solution $\mathbf{s} = (S_1, S_2, \ldots, S_k)$, each element $e \in V$ has a cost $c_i(e)$ to be put into i-th set S_i, with the total cost of s does not exceed B so that $f(\mathbf{s})$ is maximized. To address this problem, we propose two streaming algorithms that provide approximation guarantees for the problem. In particular, in the case of each element e has the same cost for all i-th sets, we propose a deterministic streaming algorithm which provides an approximation ratio of $\frac{1}{4} - \epsilon$ when f is monotone and $\frac{1}{5} - \epsilon$ when f is non-monotone. For the general case, we propose a random streaming algorithm that provides an approximation ratio of $\min\{\frac{\alpha}{2}, \frac{(1-\alpha)k}{(1+\beta)k-\beta}\} - \epsilon$ when f is monotone and $\min\{\frac{\alpha}{2}, \frac{(1-\alpha)k}{(1+2\beta)k-2\beta}\} - \epsilon$ when f is non-monotone in expectation, where $\beta = \max_{e \in V, i,j \in [k], i \neq j} \frac{c_i(e)}{c_j(e)}$ and ϵ, α are fixed inputs.

Keywords: k-submodular · Budget constraint · Approximation algorithm · Streaming algorithm

1 Introduction

Maximizing k-submodular functions has attracted a lot of attention because of its potential in solving various combinatorial optimization problems such as influence maximization [8,9,11,12], sensor placement [9,11,12], feature selection [14] and information coverage maximization [11]. Given a finite set V and an integer k, we define $[k] = \{1, 2, \ldots, k\}$ and $(k+1)^V = \{(X_1, X_2, \ldots, X_k) | X_i \subseteq V, \forall i \in [k], X_i \cap X_j = \emptyset, \forall i \neq j\}$ be a family of k disjoint sets. A function $f : (k+1)^V \mapsto \mathbb{R}_+$ is k-submodular iff for any $\mathbf{x} = (X_1, X_2, \ldots, X_k)$ and $\mathbf{y} = (Y_1, Y_2, \ldots, Y_k) \in (k+1)^V$, we have:

$$f(\mathbf{x}) + f(\mathbf{y}) \geq f(\mathbf{x} \sqcap \mathbf{y}) + f(\mathbf{x} \sqcup \mathbf{y}) \tag{1}$$

© Springer Nature Switzerland AG 2021
D. Mohaisen and R. Jin (Eds.): CSoNet 2021, LNCS 13116, pp. 27–38, 2021.
https://doi.org/10.1007/978-3-030-91434-9_3

where
$$\mathbf{x} \sqcap \mathbf{y} = (X_1 \cap Y_1, \ldots, X_k \cap Y_k)$$
and
$$\mathbf{x} \sqcup \mathbf{y} = \left(X_1 \cup Y_1 \setminus (\bigcup_{i \neq 1} X_i \cup Y_i), \ldots, X_k \cup Y_k \setminus (\bigcup_{i \neq k} X_i \cup Y_i) \right)$$

Although there exists a polynomial time to maximize a k-submodular function [16], maximizing a k-submodular function is still NP-hard. Studying on maximizing a k-submodular function was initiated by Singh *et al.* [14] with $k = 2$. Ward *et al.* [17] first studied to maximize an unconstrained k-submodular function for general k and devised a deterministic greedy algorithm which provided an approximation ratio of 1/3. Later on, [6] introduced a random greedy approach which improved the approximation ratio to $\frac{k}{2k-1}$ by applying a probability distribution to select any larger marginal element that has a higher probability. The authors in [10] eliminated the random told above but the number of queries increased to $O(n^2k^2)$. The unconstrained maximizing k-submodular function was further studied in [15] in online settings.

Under the size constraint, Oshaka et al. [9] first proposed 1/2-approximation algorithm by using a greedy approach for maximizing monotone k-submodular maximization functions. [13] showed a greedy selection that could give an approximation ratio of 1/2 under the matroid constraint. The authors in [11] then further proposed multi-objective evolutionary algorithms that provided 1/2-approximation ratio under the size constraint but took $O(kn \log^2 B)$ queries in expectation. Recently, Nguyen [12] *et al.* considered the k-submodular maximization problem subjected to the total size constraint under noises and devised two streaming algorithms which provided the approximation ratio of $O(\epsilon(1 - \epsilon)^{-2} B)$ when f was monotone and $O(\epsilon(1 - \epsilon)^{-3} B)$ when f was non-monotone.

Although there have been many attempts to solve the problem of maximizing a k-submodular function under several kinds of constraints, they did not cover several cases that could happens frequently in reality in which each element could be customized in terms of its private cost or a problem was provided with just limited budgets. Let's consider the following application:

Influence Maximization with k Topics. Given a social network under an information diffusion model and k topics. Each user has a cost to start the influence under a topic which manifests how hard it is to initially influence to a respective person. Given a budget B, we consider the problem of finding a set of users (seed set), each initially adopts a topic, with the total cost is at most B to maximize the expected number of users who are eventually activated by at least one topic. In this application, the expected number of influenced users (objective) function is k-submodular where each user corresponds to each element in the set V [8,9,12].

Motivated by that observation, in this work, we study a novel problem named **Budgeted k-submodular maximization** (BkSM), defined as follows:

Definition 1. *Given a finite set V, a budget B and a k-submodular function $f : (k + 1)^V \mapsto \mathbb{R}_+$. The problem asks to find a solution $\boldsymbol{s} = (S_1, S_2, \ldots, S_k)$,* each element $e \in V$ has a cost $c_i(e) > 0$ to be put in to S_i, with total cost $c(s) = \sum_{i \in [k]} \sum_{e \in S_i} c_i(e) \leq B$ so that $f(s)$ is maximized.

In addition, input data increasing constantly makes it impossible to be stored in computer memory. Therefore it is critical to devise streaming algorithms which not only reduce the requirement of stored memory but also be able to produce guaranteed solutions in a single pass or some passes. Although streaming algorithm is one of efficient methods for solving submodular maximization problems under various kinds of constraints such as cardinality constraint [1,3,7,18], knapsack constraint [5], k-set constraint [4] and matroid constraint [2], it is not potential to directly be applied to our BkSM problem due to intrinsic differences between submodularity and k-submodularity.

Our Contributions. In this paper we propose several algorithms which provide theoretical bounds of BkSM. Overall, our contributions are as follows:

- For a special case when every element has the same cost to be added to any i-th set, we first propose a deterministic streaming algorithm (Algorithm 2) which runs in a single pass, has $O(\frac{kn}{\epsilon} \log B)$ query complexity, $O(\frac{B}{\epsilon} \log B)$ space complexity and returns an approximation ratio of $\frac{1}{4} - \epsilon$ when f is monotone and $\frac{1}{5} - \epsilon$ when f is non-monotone for any input value of $\epsilon > 0$.
- For the general case, we propose a random streaming algorithm (Algorithm 4) which runs in a single pass, has $O(\frac{kn}{\epsilon} \log B)$ query complexity, $O(\frac{B}{\epsilon} \log B)$ space complexity and returns an approximation ratio of $\min\{\frac{\alpha}{2}, \frac{(1-\alpha)k}{(1+\beta)k-\beta}\} - \epsilon$ when f is monotone and $\min\{\frac{\alpha}{2}, \frac{(1-\alpha)k}{(1+2\beta)k-2\beta}\} - \epsilon$ when f is non-monotone in expectation where $\beta = \max_{e \in V, i, j \in [k], i \neq j} \frac{c_i(e)}{c_j(e)}$ and $\alpha \in (0, 1], \epsilon \in (0, 1)$ are inputs.

Our algorithms is an inspired suggestion from [1,5] in which we also sequentially make decision based on the value of incremental objective function per cost of each element and guess the optimal solution through the maximum singleton value. In addition, we introduce a new probability distribution to subsequently select a new element to candidate solutions.

Organization. The rest of the paper is organized as follows: The notations and properties of k-submodular functions are presented in Sect. 2. Section 3 and 4 present our algorithms and theoretical analysis. Finally, we conclude this work in Sect. 5.

2 Preliminaries

Given a finite set V and an integer k, denote $[k] = \{1, 2, \ldots, k\}$, let $(k+1)^V = \{(X_1, X_2, \ldots, X_k) | X_i \subseteq V, \forall i \in [k], X_i \cap X_j = \emptyset, \forall i \neq j\}$ be a family of k disjoint sets, called a k-**set**. We define $supp_i(\mathbf{x}) = X_i$, $supp(\mathbf{x}) = \cup_{i \in [k]} X_i$, X_i is called i-th set of $\mathbf{x}$ and an empty k-set is defined as $\mathbf{0} = (\emptyset, \ldots, \emptyset)$.

For $\mathbf{x} = (X_1, X_2, \ldots X_k)$ and $\mathbf{y} = (Y_1, Y_2, \ldots, Y_k) \in (k+1)^V$, if $e \in X_i$, we write $\mathbf{x}(e) = i$ else if $e \notin \cup_{i \in [k]} X_i$, we write $\mathbf{x}(e) = 0$ and i is called the **position** of e; adding $e \notin supp(\mathbf{x})$ into X_i can be represented by $\mathbf{x} \sqcup (e, i)$. In the case of $X_i = \{e\}$, and $X_j = \emptyset, \forall j \neq i$, we denote $\mathbf{x}$ as (e, i). We denote $\mathbf{x} \sqsubseteq \mathbf{y}$ iff $X_i \subseteq Y_i$ for all $i \in [k]$.

A function $f : (k+1)^V \mapsto \mathbb{R}$ is k-submodular iff for any $\mathbf{x} = (X_1, X_2, \ldots, X_k)$ and $\mathbf{y} = (Y_1, Y_2, \ldots, Y_k) \in (k+1)^V$, we have:

$$f(\mathbf{x}) + f(\mathbf{y}) \geq f(\mathbf{x} \sqcap \mathbf{y}) + f(\mathbf{x} \sqcup \mathbf{y}) \tag{2}$$

where

$$\mathbf{x} \sqcap \mathbf{y} = (X_1 \cap Y_1, \ldots, X_k \cap Y_k)$$

and

$$\mathbf{x} \sqcup \mathbf{y} = \left(X_1 \cup Y_1 \setminus (\bigcup_{i \neq 1} X_i \cup Y_i), \ldots, X_k \cup Y_k \setminus (\bigcup_{i \neq k} X_i \cup Y_i) \right)$$

A function f is monotone iff for any $\mathbf{x} \in (k+1)^V, e \notin supp(\mathbf{x})$ and $i \in [k]$, we have

$$\Delta_{e,i} f(\mathbf{x}) = f(X_1, \ldots, X_{i-1}, X_i \cup \{e\}, X_{i+1}, \ldots, X_k) - f(X_1, \ldots, X_k) \geq 0 \tag{3}$$

From [17], the k-submodularity of f implies the *orthant submodularity*, i.e.,

$$\Delta_{e,i} f(\mathbf{x}) \geq \Delta_{e,i} f(\mathbf{y}) \tag{4}$$

and the *pairwise monotonicity*, i.e.,

$$\Delta_{e,i} f(\mathbf{x}) + \Delta_{e,j} f(\mathbf{x}) \geq 0 \tag{5}$$

for any $\mathbf{x}, \mathbf{y} \in (k+1)^V$ with $\mathbf{x} \sqsubseteq \mathbf{y}$, $e \notin supp(\mathbf{y})$ and $i, j \in [k]$ with $i \neq j$.

In this paper, we assume that f is normalized, i.e., $f(\mathbf{0}) = 0$ and each element e has a cost $c_i(e)$ to be added into i-th set of a solution and the total cost of k-set $\mathbf{x}$ is

$$c(\mathbf{x}) = \sum_{i \in [k], e \in supp_i(\mathbf{x})} c_i(e)$$

We define β as the largest ratio of different costs of an element, i.e.,

$$\beta = \max_{e \in V, i \neq j} \frac{c_i(e)}{c_j(e)}$$

Without loss of generality, throughout this paper, we assume that every element e satisfies $c_i(e) \geq 1, \forall i \in [k]$ and $c_i(e) \leq B$, otherwise we can simply remove it. We only consider $k \geq 2$ because if $k = 1$, the k-submodular becomes to submodular function.

3 Deterministic Streaming Algorithm When $\beta = 1$

In this section, we introduce a deterministic streaming algorithm for the special case when $\beta = 1$, i.e., each element has the same cost for all i-th sets $c_i(e) = c_j(e), \forall e \in V, i \neq j$. For simplicity, we denote $c(e) = c_i(e) = c_j(e)$.

The main idea of our algorithms is that (1) we select each observed element e based on comparing between the ratio of f per total cost at the current solution with a threshold which is set in advance, and (2) we use the maximum singleton value (e_{max}, i_{max}) defined as

$$(e_{max}, i_{max}) = \arg \max_{e \in V, i \in [k]} f((e, i)) \tag{6}$$

to obtain the final solution. We first assume that the optimal solution is known and then remove this assumption by using the method in [1] to approximate the optimal solution.

3.1 Deterministic Streaming Algorithm with Known Optimal Solution

We first present a simplified version of our deterministic streaming algorithm when the optimal solution is known. Denote $\mathbf{o}$ as an optimal solution and $\mathrm{opt} = f(\mathbf{o})$, the algorithm receives v such that $v \leq \mathrm{opt}$ and a parameter $\alpha \in (0, 1]$ as inputs. The role of these parameters are going to be clarified in the main version. The details of the algorithm are fully presented in Algorithm 1. We define the notations as follows:

- (e^j, i^j) as the j-th element and its position added in the main loop of the algorithm;
- $\mathbf{s}^j$ - the solution when adding j elements in the main loop of the algorithm;
- $\mathbf{o}^j = (\mathbf{o} \sqcup \mathbf{s}^j) \sqcup \mathbf{s}^j$;
- $\mathbf{o}^{j-1/2} = (\mathbf{o} \sqcup \mathbf{s}^j) \sqcup \mathbf{s}^{j-1}$;
- $\mathbf{s}^{j-1/2}$: If $e^j \in supp(\mathbf{o})$, then $\mathbf{s}^{j-1/2} = \mathbf{s}^{j-1} \sqcup (e^j, \mathbf{o}(e^j))$. If $e^j \notin supp(\mathbf{o})$, $\mathbf{s}^{j-1/2} = \mathbf{s}^{j-1}$;
- $\mathbf{u}^t = \{(u_1, j_1), (u_2, j_2), \ldots, (u_r, j_r)\}$ - a set of elements that are in $\mathbf{o}^t$ but not in $\mathbf{s}^t$, $r = |supp(\mathbf{u}^t)|$
- $\mathbf{u}_i^t = \mathbf{s}^t \sqcup \{(u_1, j_1), (u_2, j_2), \ldots, (u_i, j_i)\}$

The algorithm initiates a candidate solution $\mathbf{s}^0$ as an empty k-set. For each new incoming element e, the algorithm updates a tuple (e_{max}, i_{max}) to find the maximal singleton then checks that the total cost $c(\mathbf{s}^t) + c(e)$ exceed B or not? If not, it finds a position $i' \in [k]$ that $f(\mathbf{s}^t \sqcup (e, i'))$ is maximal and adds (e, i) into $\mathbf{s}^t$ if $\frac{f(\mathbf{s}^t \sqcup (e, i'))}{c(\mathbf{s}^t) + c(e)} \geq \frac{\alpha v}{B}$. Otherwise, it ignores e and receives the next element. This step helps the algorithm select any element which has high value of marginal value per its cost as well as eliminate bad ones.

After finishing the main loop, the algorithm returns the best solution in $\{\mathbf{s}^t\} \cup \{(e_{max}, i_{max})\}$ when f is monotone or returns the best solution in $\{\mathbf{s}^j : j \leq t\}$

Algorithm 1. Deterministic streaming algorithm with known opt

---

Input: a function $f : (k+1)^V \mapsto \mathbb{R}_+$, $B > 0$, $\alpha \in (0,1]$, v that $v \leq$ opt
Output: a solution s

1. $\mathbf{s}^0 \leftarrow \mathbf{0}$, $t \leftarrow 0$
2. **foreach** $e \in V$ **do**
3. $\quad$ $i_e \leftarrow \arg\max_{i \in [k]} f((e,i))$
4. $\quad$ $(e_{max}, i_{max}) \leftarrow \arg\max_{(e_1,i_1) \in \{(e_{max}, i_{max}),(e,i_e)\}} f((e_1, i_1))$
5. $\quad$ **if** $c(\mathbf{s}^t) + c(e) \leq B$ **then**
6. $\quad\quad$ $i' \leftarrow \arg\max_{i \in [k]} f(\mathbf{s}^t \sqcup (e,i))$
7. $\quad\quad$ **if** $\frac{f(\mathbf{s}^t \sqcup (e,i'))}{c(\mathbf{s}^t) + c(e)} \geq \frac{\alpha v}{B}$ **then**
8. $\quad\quad\quad$ $\mathbf{s}^{t+1} \leftarrow \mathbf{s}^t \sqcup (e,i')$, $t \leftarrow t+1$

9. **return** $\arg\max_{\mathbf{s} \in \{\mathbf{s}^t\} \cup \{(e_{max}, i_{max})\}} f(\mathbf{s})$ if f is monotone,
    $\arg\max_{\mathbf{s} \in \{\mathbf{s}^j : j \leq t\} \cup \{(e_{max}, i_{max})\}} f(\mathbf{s})$ if f is non-monotone.

$\cup \{(e_{max}, i_{max})\}$ when f is non-monotone. We now analysis the approximation guarantee of Algorithm 1. Denote e^t is the last addition of the main loop of the Algorithm 1. By exploiting the relation among $\mathbf{o}$, $\mathbf{o}^j$ and $\mathbf{s}^j$, $j \leq t$, we obtain the following Lemma.

Lemma 1. *If f is monotone then $v - f(\mathbf{o}^t) \leq f(\mathbf{s}^t)$ and if f is non-monotone then $v - f(\mathbf{o}^t) \leq 2f(\mathbf{s}^t)$.*

Due to the space constraint, we omit some proofs and presented them in a full version of this paper. Lemma 1 plays an important role for analyzing approximation ratio of the algorithm, which stated in the following Theorem.

Theorem 1. *Algorithm 1 is a single pass streaming algorithm and returns a solution $\mathbf{s}$ satisfying:*

- *If f is monotone, $f(\mathbf{s}) \geq \min\{\frac{\alpha}{2}, \frac{1-\alpha}{2}\}v$, $f(\mathbf{s})$ is maximized to $\frac{v}{4}$ when $\alpha = \frac{1}{2}$.*
- *If f is non-monotone, $f(\mathbf{s}) \geq \min\{\frac{\alpha}{2}, \frac{1-\alpha}{3}\}v$, $f(\mathbf{s})$ is maximized to $\frac{v}{5}$ when $\alpha = \frac{2}{5}$.*

3.2 Deterministic Streaming Algorithm

We present our deterministic streaming algorithm in the case of $\beta = 1$ which reuses the framework of Algorithm 1 but removes the assumption that $\mathbf{o}$ is known. We use the dynamic update method in [1] to obtain a good approximation of opt.

To specific, denote $m = \max_{e \in V, i \in [k]} f((e,i))$, we have $m \leq$ opt $\leq Bm$. Therefore we use the value $v = (1+\epsilon')^j$ for $\{j | m \leq (1+\epsilon')^j \leq Bm, j \in \mathbb{Z}_+\}$ to guess the value of opt by showing that there exits v such that $(1-\epsilon')$opt $\leq v \leq$ opt. However, in order to find m, we have to require at least one pass over V. Therefore, we adapt the dynamic update method in [1] which updates $m = \max\{m, \max_{i \in [k]} f((e,i))\}$ with an already observed element e to determine the range of guessed optimal values. This method can help algorithm maintain a good estimation of the optimal solution if that range shifts forward when next elements are observed. We implement this method by using variables s^{t_j} and t_j to store a candidate solution and the number of its elements in which $v = (1+\epsilon')^j$ is an guessed value of opt.

We set the value of α by using Theorem 1 which provides the best approximation guarantees. The value of ϵ' is set to several times higher than ϵ to reduce the complexity but still ensure approximation ratios. The detail of our algorithm is presented in Algorithm 2.

Algorithm 2. Deterministic streaming algorithm

Input: a function $f : (k+1)^V \mapsto \mathbb{R}_+$, $B > 0$, $\epsilon > 0$.
Output: a solution s
1. **If** f is monotone $\alpha \leftarrow \frac{1}{2}$, $\epsilon' \leftarrow 4\epsilon$ **else** $\alpha \leftarrow \frac{2}{5}$, $\epsilon' \leftarrow 5\epsilon$
2. **foreach** $e \in V$ **do**
3. $i_e \leftarrow \arg\max_{i \in [k]} f((e, i))$
4. $(e_{max}, i_{max}) \leftarrow \arg\max_{(e_1, i_1) \in \{(e_{max}, i_{max}), (e, i_e)\}} f((e_1, i_1))$
5. $O \leftarrow \{j | f((e_{max}, i_{max})) \leq (1+\epsilon')^j \leq Bf((e_{max}, i_{max})), j \in \mathbb{Z}_+\}$
6.   **for** $j \in O$ **do**
7.   **if** $c(s^{t_j}) + c(e) \leq B$ **then**
8. $i' \leftarrow \arg\max_{i \in [k]} f(s^{t_j} \sqcup (e, i))$ **if** $\frac{f(s^{t_j} \sqcup (e, i'))}{c(s^{t_j}) + c(e)} \geq \frac{\alpha(1+\epsilon')^j}{B}$ **then**
9. $s^{t_j+1} \leftarrow s^{t_j} \sqcup (e, i')$, $t_j \leftarrow t_j + 1$

Lemma 2. *In Algorithm 2, there exists a number $j \in \mathbb{Z}_+$ so that $v = (1+\epsilon')^j \in O$ satisfies $(1 - \epsilon')\text{opt} \leq v \leq \text{opt}$*

Proof. Denote $m = f((e_{max}, i_{max}))$. Duet to k-submodularity of f, we have

$$m \leq \text{opt} = f(\mathbf{o}) \leq \sum_{e \in supp(\mathbf{o})} f(e, \mathbf{o}(e)) \leq Bm$$

Let $j = \lfloor \log_{1+\epsilon'} \text{opt} \rfloor$, we have $v = (1 + \epsilon')^j \leq \text{opt} \leq Bm$ and $v \geq (1 + \epsilon')^{\log_{1+\epsilon'}(\text{opt})-1} = \frac{\text{opt}}{1+\epsilon'} \geq \text{opt}(1 - \epsilon')$.

The performance of Algorithm 2 is claimed in the following Theorem.

Theorem 2. *Algorithm 2 is a single pass streaming algorithm that has $O(\frac{kn}{\epsilon} \log B)$ query complexity, $O(\frac{B}{\epsilon} \log B)$ space complexity and provides an approximation ratio of $\frac{1}{4} - \epsilon$ when f is monotone and $\frac{1}{5} - \epsilon$ when f is non-monotone.*

Proof. The size of O is at most $\frac{1}{\epsilon'} \log B$, finding each s^{t_j} takes at most $O(kn)$ queries and s^{t_j} includes at most B elements. Therefore, the query complexity is $O(\frac{kn}{\epsilon} \log B)$ and total space complexity is $O(\frac{B}{\epsilon} \log B)$.

By Lemma 2, there exists an integer number $j \in \mathbb{Z}_+$ so that $v = (1+\epsilon')^j \in O$ satisfies $(1 - \epsilon')\text{opt} \leq v \leq \text{opt}$. Apply Theorem 1, for the monotone case we have: $f(\mathbf{s}) \geq \frac{1}{4}v \geq \frac{1}{4}(1 - \epsilon')\text{opt} = (\frac{1}{4} - \epsilon)\text{opt}$ and for the non-monotone case: $f(\mathbf{s}) \geq \frac{1}{5}v \geq \frac{1}{5}(1 - \epsilon')\text{opt} = (\frac{1}{5} - \epsilon)\text{opt}$. Hence, the theorem is proved. $\square$

4 Random Streaming Algorithm for General Case

In the case each element e has multiple different cost $c_i(e)$ for each i-th set, we can not apply previous algorithms. Therefore, in this section we introduce one pass streaming which provides approximation ratios in expectation for BkSM problem.

At the core of our algorithm, we introduce a new probability distribution to choose a position for each element to establish the relationship among $\mathbf{o}$, $\mathbf{o}^j$ and $\mathbf{s}^j$ (Lemma 3) and analyze the performance of our algorithm. Besides, we also use a predefined threshold to filter high-value elements into candidate solutions and the maximum singleton value to give the final solution. Similar to the previous section, we first introduce a simplified version of the streaming algorithm when the optimal solution is known in advance.

4.1 Random Algorithm with Known Optimal Solution

This algorithm also receives the inputs $\alpha \in (0, 1)$ and v that $v \leq$ opt. We use the same notations as in Sect. 3. This algorithm also requires one pass over V. The algorithm imitates an empty k-set $\mathbf{s}^0$ and subsequently updates the solution after once passing over V. Be different from Algorithm 1, for each $e \in V$ being observed, the algorithm finds a set collection J that contains positions satisfying the total cost is at most B and the ratio of the increment of the objective function per cost is at least a given threshold, i.e.,

$$J = \left\{ i \in [k] : c(\mathbf{s}^t) + c_i(e) \leq B \text{ and } \frac{f(\mathbf{s}^t \sqcup (e, i)) - f(\mathbf{s}^t)}{c_i(e)} \geq \frac{\alpha v}{B} \right\} \tag{7}$$

These constraints help the algorithm eliminate which position having low increment of the objective function over its cost. If $J \neq \emptyset$, the algorithm puts e into set i of $\mathbf{s}^t$ with a probability:

$$\frac{p_i^{|J|-1}}{T} = \frac{\left(\frac{f(\mathbf{s}^t \sqcup (e,i)) - f(\mathbf{s}^t)}{c_i(e)} \right)^{|J|-1}}{\sum_{i \in J} \left(\frac{f(\mathbf{s}^t \sqcup (e,i)) - f(\mathbf{s}^t)}{c_i(e)} \right)^{|J|-1}} \tag{8}$$

Simultaneously, the algorithm finds the maximum singleton value (e_{max}, i_{max}) by updating the current maximal value from the set of observed elements. As Algorithm 3, the algorithm also uses (e_{max}, i_{max}) as one of candidate solutions and finds the best among them. The full detail of this algorithm is described in Algorithm 3.

Lemma 3 provides the relationship among $\mathbf{o}, \mathbf{o}^j$ and $\mathbf{s}^j, j \leq t$ that play an importance role in analyzing algorithm's performance.

Lemma 3. *In Algorithm 3, if there is no pair $(e, i) \in \mathbf{o}$ satisfying $\exists j \in [t] : e \notin supp(\mathbf{s}^j)$ so that $\frac{f(\mathbf{s}^j \sqcup (e,i))}{c(\mathbf{s}^j) + c_i(e)} \geq \frac{\alpha v}{B}$ and $c(\mathbf{s}^j) + c_i(e) > B$, we have:*

– If f is monotone, then

$$f(\mathbf{o}^{j-1}) - \mathbb{E}[f(\mathbf{o}^j)] \leq \beta\left(1 - \frac{1}{k}\right)(\mathbb{E}[f(\mathbf{s}^j)] - f(\mathbf{s}^{j-1})) + \frac{\alpha v c_{j*}(e^j)}{kB}$$

Algorithm 3. Random streaming algorithm with known opt- RanStreamWithOpt(f, opt, α)

Input: a function $f : (k+1)^V \mapsto \mathbb{R}_+$, $B > 0$, $\alpha \in (0,1]$, $v, v \leq$ opt
Output: a solution **s**

1. $\mathbf{s}_0 \leftarrow \mathbf{0}$, $t \leftarrow 0$
2. **foreach** $e \in V$ **do**
3. $i_e \leftarrow \arg\max_{i \in [k]} f((e,i))$
4. $(e_{max}, i_{max}) \leftarrow \arg\max_{(e_1,i_1) \in \{(e_{max},i_{max}),(e,i_e)\}} f((e_1,i_1))$
5. $J \leftarrow \emptyset$
6.   **foreach** $i \in [k]$ **do**
7.   **if** $c(\mathbf{s}^t) + c_i(e) \leq B$ and $\frac{f(\mathbf{s}^t \sqcup (e,i)) - f(\mathbf{s}^t)}{c_i(e)} \geq \frac{\alpha v}{B}$ **then**
8. $p_i \leftarrow \frac{f(\mathbf{s}^t \sqcup (e,i)) - f(\mathbf{s}^t)}{c_i(e)}$; $J \leftarrow J \cup \{i\}$
9.   **if** $J \neq \emptyset$ **then**
10. $T \leftarrow \sum_{i \in J} p_i^{|J|-1}$
11.   Select a position $i \in J$ with probability $\frac{p_i^{|J|-1}}{T}$
12. $\mathbf{s}^{t+1} \leftarrow \mathbf{s}^t \sqcup (e,i)$; $t \leftarrow t+1$
13. **return** $\arg\max_{\mathbf{s} \in \{\mathbf{s}^t\} \cup \{(e_{max}, i_{max})\}} f(\mathbf{s})$ if f is monotone,
     $\arg\max_{\mathbf{s} \in \{\mathbf{s}^j : j \leq t\} \cup \{(e_{max}, i_{max})\}} f(\mathbf{s})$ if f is non-monotone

---

- *If f is non-monotone, then*

$$f(o^{j-1}) - \mathbb{E}[f(o^j)] \leq 2\beta(1 - \frac{1}{k})(\mathbb{E}[f(\mathbf{s}^j)] - f(\mathbf{s}^{j-1})) + \frac{2\alpha v c_{j^*}(e^j)}{kB}$$

Theorem 3. *Algorithm 3 returns a solution $\mathbf{s}$ satisfying*

- *If f is monotone, $\mathbb{E}[f(\mathbf{s})] \geq \min\{\frac{\alpha}{2}, \frac{(1-\alpha)k}{(1+\beta)k-\beta}\}v$, $f(\mathbf{s})$ is maximized to $\frac{v}{3+\beta-\frac{\beta}{k}}$ when $\alpha = \frac{2}{3+\beta-\frac{\beta}{k}}$.*
- *If f is non-monotone, $\mathbb{E}[f(\mathbf{s})] \geq \min\{\frac{\alpha}{2}, \frac{(1-\alpha)k}{(1+2\beta)k-2\beta}\}v$, $f(\mathbf{s})$ is maximized to $\frac{v}{3+2\beta-\frac{2\beta}{k}}$ when $\alpha = \frac{2}{3+2\beta-\frac{2\beta}{k}}$.*

4.2 Random Streaming Algorithm

In this section we remove the assumption that the optimal solution is known and present the random streaming algorithm which reuses the framework of Algorithm 3.

Similar to the Algorithm 2, we use the method in [1] to estimate opt. We assume that we know β in advance. This is feasible because we can calculate the value of β in $O(kn)$. We set α according to the properties of f to provide the best performance of the algorithm. The algorithm continuously updates $O \leftarrow \{j | f((e_{max}, i_{max})) \leq (1+\epsilon)^j \leq Bf((e_{max}, i_{max})), j \in \mathbb{Z}_+\}$ in order to estimate the value of maximal singleton and uses $\mathbf{s}^{t_j}$ and t_j to save candidate solutions, which is updated by using the probability distribution as in Algorithm 3 with $(1+\epsilon)^j$ is an estimation of optimal solution. The algorithm finally compares all candidate solutions to select the best one. The details of algorithm is presented in Algorithm 4.

Algorithm 4. Random streaming algorithm

Input: a k-submodular function $f : (k+1)^V \mapsto \mathbb{R}_+$, $B > 0$, $\epsilon > 0$, $\alpha \in (0,1]$
Output: a solution $\mathbf{s}$

1. **foreach** $e \in V$ **do**
2. $i_e \leftarrow \arg\max_{i \in [k]} f((e,i))$
3. $(e_{max}, i_{max}) \leftarrow \arg\max_{(e_1,i_1) \in \{(e_{max},i_{max}),(e,i_e)\}} f((e_1,i_1))$
4. $O \leftarrow \{j | f((e_{max}, i_{max})) \leq (1+\epsilon)^j \leq Bf((e_{max}, i_{max})), j \in \mathbb{Z}_+\}$
5.   **foreach** $j \in O$ **do**
6. $J \leftarrow \emptyset$
7.   **foreach** $i \in [k]$ **do**
8.   **if** $c(\mathbf{s}^{t_j}) + c_i(e) \leq B$ **and** $\frac{f(\mathbf{s}^{t_j} \sqcup (e,i)) - f(\mathbf{s}^{t_j})}{c_i(e)} \geq \frac{\alpha(1+\epsilon)^j}{B}$ **then**
9. $p_i \leftarrow \frac{f(\mathbf{s}^{t_j} \sqcup (e,i)) - f(\mathbf{s}^{t_j})}{c_i(e)}$; $J \leftarrow J \cup \{i\}$
10. $T \leftarrow \sum_{i \in J} p_i^{|J|-1}$
11.   Select a position $i \in J$ with probability $\frac{p_i^{|J|-1}}{T}$
12. $\mathbf{s}^{t_j+1} \leftarrow \mathbf{s}^{t_j} \sqcup (e,i)$; $t_j \leftarrow t_j + 1$
13. **return** $\arg\max_{\mathbf{s} \in \{\mathbf{s}^{t_j} : j \in O\} \cup \{(e_{max},i_{max})\}} f(\mathbf{s})$ if f is monotone,
       $\arg\max_{\mathbf{s}\{\mathbf{s}_i^{t_j} : j \in O, i \leq j\} \cup \{(e_{max},i_{max})\}} f(\mathbf{s})$ if f is non-monotone

---

Theorem 4. *Algorithm 4 is one pass streaming algorithm that has* $O(\frac{kn}{\epsilon} \log B)$ *query complexity,* $O(\frac{B}{\epsilon} \log B)$ *space complexity and provides an approximation ratio of* $\min\{\frac{\alpha}{2}, \frac{(1-\alpha)k}{(1+\beta)k-\beta}\} - \epsilon$ *when f is monotone and* $\min\{\frac{\alpha}{2}, \frac{(1-\alpha)k}{(1+2\beta)k-2\beta}\} - \epsilon$ *when f is non-monotone in expectation.*

Proof. By Lemma 2, there exists $j \in \mathbb{Z}_+$ that $v = (1+\epsilon)^j \in O$ satisfies $(1 - \epsilon)\text{opt} \leq v \leq \text{opt}$. Using similar arguments of the proof of Theorem 3, for the monotone case

$$f(\mathbf{s}) \geq \min\{\frac{\alpha}{2}, \frac{(1-\alpha)k}{(1+\beta)k-\beta}\}v \geq \left(\min\{\frac{\alpha}{2}, \frac{(1-\alpha)k}{(1+\beta)k-\beta}\} - \epsilon\right)\text{opt}$$

For the non-monotone case we also obtain the proof by applying the same arguments   □

5   Conclusions

This paper studies the BkSM, a generalized version of maximizing k-submodular functions problem. In order to find the solution, we propose several streaming algorithms with provable guarantees. The core of our algorithms is to exploit the relation between candidate solutions and the optimal solution by analyzing intermediate quantities and applying a new probability distribution to select elements with high contributions to a current solution. In the future we are going to conduct experiments on so some instance of BkSM to show the performance of our algorithms in practice.

Acknowledgements. This work is supported by Vietnam National Foundation for Science and Technology Development (NAFOSTED) under Grant No. 102.01-2020.21.

References

1. Badanidiyuru, A., Mirzasoleiman, B., Karbasi, A., Krause, A.: Streaming submodular maximization: massive data summarization on the fly. In: Macskassy, S.A., Perlich, C., Leskovec, J., Wang, W., Ghani, R. (eds.) The 20th ACM SIGKDD International Conference on Knowledge Discovery and Data Mining, KDD 2014, pp. 671–680. ACM (2014)
2. Chakrabarti, A., Kale, S.: Submodular maximization meets streaming: matchings, matroids, and more. Math. Program. 225–247 (2015). https://doi.org/10.1007/s10107-015-0900-7
3. Gomes, R., Krause, A.: Budgeted nonparametric learning from data streams. In: Fürnkranz, J., Joachims, T. (eds.) Proceedings of the 27th International Conference on Machine Learning (ICML 2010), Haifa, Israel, 21–24 June 2010, pp. 391–398. Omnipress (2010)
4. Haba, R., Kazemi, E., Feldman, M., Karbasi, A.: Streaming submodular maximization under a k-set system constraint. In: Proceedings of the 37th International Conference on Machine Learning, ICML 2020, 13–18 July 2020, Virtual Event. Proceedings of Machine Learning Research, vol. 119, pp. 3939–3949. PMLR (2020)
5. Huang, C., Kakimura, N., Yoshida, Y.: Streaming algorithms for maximizing monotone submodular functions under a Knapsack constraint. Algorithmica **82**(4), 1006–1032 (2020)
6. Iwata, S., Tanigawa, S., Yoshida, Y.: Improved approximation algorithms for k-submodular function maximization. In: Krauthgamer, R. (ed.) Proceedings of the Twenty-Seventh Annual ACM-SIAM Symposium on Discrete Algorithms, SODA 2016, Arlington, VA, USA, 10–12 January 2016, pp. 404–413. SIAM (2016)
7. Kumar, R., Moseley, B., Vassilvitskii, S., Vattani, A.: Fast greedy algorithms in mapreduce and streaming. In: Blelloch, G.E., Vöcking, B. (eds.) 25th ACM Symposium on Parallelism in Algorithms and Architectures, SPAA 2013, pp. 1–10. ACM (2013)
8. Nguyen, L., Thai, M.: Streaming k-submodular maximization under noise subject to size constraint. In: Daumé, H., Singh, A. (eds.) Proceedings of the International Conference on Machine Learning, (ICML-2020), Thirty-Seventh International Conference on Machine Learning (2020)

9. Ohsaka, N., Yoshida, Y.: Monotone k-submodular function maximization with size constraints. In: Cortes, C., Lawrence, N.D., Lee, D.D., Sugiyama, M., Garnett, R. (eds.) Advances in Neural Information Processing Systems 28: Annual Conference on Neural Information Processing Systems 2015, Montreal, Quebec, Canada, 7–12 December 2015, pp. 694–702 (2015)

10. Oshima, H.: Derandomization for k-submodular maximization. In: Brankovic, L., Ryan, J., Smyth, W.F. (eds.) IWOCA 2017. LNCS, vol. 10765, pp. 88–99. Springer, Cham (2018). https://doi.org/10.1007/978-3-319-78825-8_8

11. Qian, C., Shi, J., Tang, K., Zhou, Z.: Constrained monotone k-submodular function maximization using multiobjective evolutionary algorithms with theoretical guarantee. IEEE Trans. Evol. Comput. **22**(4), 595–608 (2018)

12. Rafiey, A., Yoshida, Y.: Fast and private submodular and k-submodular functions maximization with matroid constraints. In: Proceedings of the 37th International Conference on Machine Learning, ICML 2020. Proceedings of Machine Learning Research, vol. 119, pp. 7887–7897. PMLR (2020)

13. Sakaue, S.: On maximizing a monotone k-submodular function subject to a matroid constraint. Discret. Optim. **23**, 105–113 (2017)

14. Singh, A.P., Guillory, A., Bilmes, J.A.: On bisubmodular maximization. In: Lawrence, N.D., Girolami, M.A. (eds.) Proceedings of the Fifteenth International Conference on Artificial Intelligence and Statistics, AISTATS 2012. JMLR Proceedings, vol. 22, pp. 1055–1063. JMLR.org (2012)

15. Soma, T.: No-regret algorithms for online k-submodular maximization. In: Chaudhuri, K., Sugiyama, M. (eds.) The 22nd International Conference on Artificial Intelligence and Statistics, AISTATS 2019, Naha, Okinawa, Japan, 16–18 April 2019. Proceedings of Machine Learning Research, vol. 89, pp. 1205–1214. PMLR (2019)

16. Thapper, J., Zivný, S.: The power of linear programming for valued CSPs. In: 53rd Annual IEEE Symposium on Foundations of Computer Science, FOCS 2012, New Brunswick, NJ, USA, 20–23 October 2012, pp. 669–678. IEEE Computer Society (2012)

17. Ward, J., Zivný, S.: Maximizing bisubmodular and k-submodular functions. In: Chekuri, C. (ed.) Proceedings of the Twenty-Fifth Annual ACM-SIAM Symposium on Discrete Algorithms, SODA 2014, Portland, Oregon, USA, 5–7 January 2014, pp. 1468–1481. SIAM (2014)

18. Yang, R., Xu, D., Cheng, Y., Gao, C., Du, D.: Streaming submodular maximization under noises. In: 39th IEEE International Conference on Distributed Computing Systems, ICDCS 2019, Dallas, TX, USA, 7–10 July 2019, pp. 348–357. IEEE (2019)

Approximation Algorithms for the Lower Bounded Correlation Clustering Problem

Sai Ji[1], Yinhong Dong[2], Donglei Du[3], and Dachuan Xu[4($\boxtimes$)]

[1] Academy of Mathematics and Systems Science, Chinese Academy of Sciences,
Beijing 100190, People's Republic of China
jisai@amss.ac.cn

[2] School of Management, South-Central University for Nationalities,
Wuhan 430074, Hubei, People's Republic of China
3098610@mail.scuec.edu.cn

[3] Faculty of Business Administration, University of New Brunswick,
Fredericton, NB E3B 5A3, Canada
ddu@unb.ca

[4] Beijing Institute for Scientific and Engineering Computing,
Beijing University of Technology, Beijing 100124, People's Republic of China
xudc@bjut.edu.cn

Abstract. Lower bounded correlation clustering problem is a generalization of the classical correlation clustering problem, which has many applications in protein interaction networks, cross-lingual link detection, and communication networks, etc. In the lower bounded correlation clustering problem, we are given a complete graph and an integer L. Each edge is labelled either by a positive sign $+$ or a negative sign $-$ whenever the two endpoints of the edge are similar or dissimilar respectively. The goal of this problem is to partition the vertex set into several clusters, subject to an lower bound L on the sizes of clusters so as to minimize the number of disagreements, which is the total number of the edges with positive labels between clusters and the edges with negative labels within clusters. In this paper, we propose the lower bounded correlation clustering problem and formulate the problem as an integer program. Furthermore, we provide two polynomial time algorithms with constant approximate ratios for the lower bounded correlation clustering problem on some special graphs.

Keywords: Lower bounded · Correlation clustering · Approximation algorithm · Polynomial time

1 Introduction

Correlation clustering problem has numerous applications in the areas of machine learning, computer vision, data mining, social networks and data compression. It has been widely studied in the literature [1,11,13,18–20].

The correlation clustering problem was first introduced by Bansal et al. [4]. In this problem, we are given a complete graph $G = (V, E)$, where each edge

© Springer Nature Switzerland AG 2021
D. Mohaisen and R. Jin (Eds.): CSoNet 2021, LNCS 13116, pp. 39–49, 2021.
https://doi.org/10.1007/978-3-030-91434-9_4

$(u, v) \in E$ is labelled by $+$ or $-$ based on the similarity of vertices u and v. The goal is to find a clustering of vertices V so as to make the edges within clusters are mostly positive and the edges between different clusters are mostly negative. Given a clustering, let each positive edge whose two endpoints lie in different clusters and each negative edge whose endpoints lie in the same cluster be a disagreement. Moreover, let each remaining edge be an agreement.

Based on the goal of the correlation clustering problem, there are two versions of the correlation clustering problem: minimizing disagreements and maximizing agreements. The goal of the former problem is to find a clustering so as to minimize the number of disagreements. The goal of the latter problem is to find a clustering so as to maximize the number of agreements. Given any instance of the correlation clustering problem, the two different versions share the same optimal solution. But the two versions of the problem are essentially different from the point of view of approximation algorithm. In the rest of the paper, we only consider the minimizing disagreements version of correlation clustering problem.

The correlation clustering problem is NP-hard. Bansal et al. [4] give a 17433-approximation algorithm, which is the first constant approximation algorithm for the correlation clustering problem. Charikard et al. [7] first provide a very natural linear programming formulation of the problem and prove that the integrality gap of the linear program is 2. Secondly, they propose a 4-approximation algorithm by using the method of region growth, which significantly improves the approximation ratio of the algorithm provided by Bansal et al. [4]. Finally, for the correlation clustering problem on general graphs, they provided an $O(\log n)$-approximation algorithm. The current best approximation algorithm is provided by Chawla et al. [8], which achieves an approximate ratio of 2.06.

Because of the complexity of the practical applications, the correlation clustering problem has some limitations in modelling real-life situations. In order to adapt to the development of society and guide practice more effectively and realistically, various generalizations of the correlation clustering problem have been proposed and widely studied. Such as the min-max correlation clustering [2], the chromatic correlation clustering [5], the overlapping correlation clustering problem [6], the higher-order correlation clustering [9], the capacitated correlation clustering problem [16], and the correlation clustering with noisy input [14,15], among others.

Lower bound constraint is a natural constraint in combinatorial optimization problems and it has been extensively studied [3,10,12,17]. However, there has been no relevant research on the lower bounded correlation clustering problem. Therefore, this work considers the lower bounded correlation clustering problem, which is a new and natural variant of the correlation clustering problem. In this problem, we are given a labeled complete graph $G = (V, E)$ as well as an integer L. The goal of this problem is to partition set V into several clusters with each size at least L so as to minimize the total number of disagreements.

Note that the lower bounded correlation clustering problem includes the classical correlation clustering problem as a special case by letting $L = 1$. Note the case where $L > |V|/2$ is trivial because only one feasible solution exists for the lower bounded correlation clustering problem. Therefore, we assume that $L \leq |V|/2$ for the rest of the discussion.

There are three main contributions in this paper.

(1) We first propose the lower bounded correlation clustering problem and give an integer programming formulation for the problem.
(2) We provide an algorithm which returns V as the cluster. We show that the algorithm always outputs an optimal solution for the lower bounded correlation clustering problem on $(2|V|/L-1)$-positive edge dominant graphs (Theorem 1). Moreover, we prove that the same algorithm is a 20-approximation algorithm for the lower bounded correlation clustering problem on 4-positive edge dominant graphs (Theorem 2).
(3) We present another algorithm which may return multiple clusters and prove the algorithm is a 20-approximation algorithm for the lower bounded correlation clustering problem on $(5|V|/L - 1)$-positive edge dominant graphs (Theorem 3).

The rest of our paper is structured as follows. Section 2 presents some definitions as well as the formulation of the lower bounded correlation clustering problem. The two algorithms are presented in Sects. 3 and 4, respectively. Some discussions are provided in Sect. 5.

2   Lower Bounded Correlation Clustering Problem

In this section, we give some definitions used in this paper as well as the formulation of the lower bounded correlation clustering problem. Given a complete graph $G = (V, E)$, let E^+ and E^- be the sets of all positive edges and all negative edges, respectively, For each positive integer k, denote set $[k] = \{1, 2, \ldots, k\}$. The lower bounded correlation clustering is defined in Definition 1.

Definition 1 (Lower bounded correlation clustering problem). *Given a labelled complete graph $G = (V, E)$ as well as an integer L, the goal is to find a partition $\mathcal{C} = \{C_1, C_2, \ldots, C_k\}$ of V which satisfies $|C_i| \geq L, i \in [k]$ such that*

$$\sum_{i \in [k]} |(u, v) \in E^- : u, v \in C_i| + \sum_{i,j \in [k]} |(u, v) \in E^+ : u \in C_i, v \in C_j, i \neq j|$$

is minimized.

Definition 2 (M-positive edge dominant graph). *Graph $G = (V, E)$ is an M-positive edge dominant graph if*

$$\inf_{v \in V} \frac{|E_v^+|}{|E_v^-|} \geq M,$$

where $E_v^+ := \{(u, v) \in E^+, u \in V\}$ and $E_v^- := \{(u, v) \in E^-, u \in V\}$.

For each edge (u, v), we introduce a 0-1 variable x_{uv} such that $x_{uv} = 1$ when the two vertices of edge (u, v) lie in different clusters and $x_{uv} = 0$ when the two vertices of edge (u, v) lie in the same cluster. Based on the above variables and Definition 1, we formulate the lower bounded correlation clustering problem as the following integer program (1).

$$
\begin{aligned}
\min \quad & \sum_{(u,v)\in E^+} x_{uv} + \sum_{(u,v)\in E^-} (1 - x_{uv}) \\
\text{s. t.} \quad & x_{uv} + x_{vw} \geq x_{uw}, && \forall u, v, w \in V, \\
& \sum_{v\in V}(1 - x_{uv}) \geq L, && \forall u \in V, \\
& x_{uu} = 0, && \forall u \in V, \\
& x_{uv} \in \{0, 1\}, && \forall u, v \in V.
\end{aligned} \tag{1}
$$

The objective function is the total number of disagreements. The quantity $\sum_{(u,v)\in E^+} x_{uv}$ is the number of disagreements generated by the positive edges, while the quantity $\sum_{(u,v)\in E^-} (1 - x_{uv})$ is the number of disagreements generated by the negative edges. There are three types of constraints in (1). The first one is the triangle inequality, which insures the program returns a feasible cluster of the correlation clustering problem. The second one is a lower bound constraint, which guarantees that there are at least L vertices in each cluster. The third one is the natural binary constraint. By relaxing above 0-1 variables, we obtain the LP relaxation of (1).

$$
\begin{aligned}
\min \quad & \sum_{(u,v)\in E^+} x_{uv} + \sum_{(u,v)\in E^-} (1 - x_{uv}) \\
\text{s. t.} \quad & x_{uv} + x_{vw} \geq x_{uw}, && \forall u, v, w \in V, \\
& \sum_{v\in V}(1 - x_{uv}) \geq L, && \forall u \in V, \\
& x_{uu} = 0, && \forall u \in V, \\
& 0 \leq x_{uv} \leq 1, && \forall u, v \in V.
\end{aligned} \tag{2}
$$

3   A Simple Effecient Algorithm

In this section, we provide Algorithm 1, which returns set V as the only cluster. Furthermore, we prove that Algorithm 1 achieves a constant approximation ration for the lower bounded correlation clustering problem when restricted to two special graphs.

Theorem 1 below can be shown based on the structure of the feasible solutions of the lower bounded correlation clustering problem.

Theorem 1. *Algorithm 1 is an optimal algorithm for the lower bounded correlation clustering problem on $(2|V|/L - 1)$-positive edge dominant graphs.*

Algorithm 1

Input: Integer L, a labelled complete graph $G = (V, E)$.
Output: A partition of vertices.
1: Let V be a cluster.
2: **return** cluster V.

Proof. For each $(2|V|/L - 1)$-positive edge dominant graph $G = (V, E)$, we have $|E_v^-| \leq L/2, \forall v \in V$. Moreover, for each feasible solution $\mathcal{C} = \{C_1, \ldots, C_m\}$ of the instance $I = (G, L)$ which contains more than one cluster. There are at least L cut edges generated by each vertex v. Moreover, there are at least $L/2$ disagreements generated by the positive edges among these cut edges since $|E_v^-| \leq L/2$. Therefore, Algorithm 1 returns an optimal solution since the disagreements generated by each vertex $v \in V$ is no more than $|E_v^-|$ and it can be bounded by $L/2$. $\qquad \square$

For any instance $I = (G, L)$, where $G = (V, E)$ is a complete 4-positive edge dominant graph satisfying $|E_v^-| \leq (|V| - 1)/5, v \in V$, we solve (2) to obtain an optimal fractional solution x^* of I. For each vertex $v \in V$, compute

$$\mathrm{Avg}_v(V) := \frac{\sum_{t \in V} x_{vt}^*}{|V|}.$$

Let

$$cen(V) := \arg \min_{v \in V} \mathrm{Avg}_v(V).$$

be the center vertex of set V. Then we can analyze the upper bound on the number of disagreements based on the value of $\mathrm{Avg}_{cen(V)}(V)$. Specifically, we consider the following two cases:

(1) $\mathrm{Avg}_{cen(V)}(V) \leq 17/80$;
(2) $\mathrm{Avg}_{cen(V)}(V) > 17/80$.

3.1 $\mathrm{Avg}_{cen(V)}(V) \leq 17/80$

Lemma 1. *For each negative edge (u, w) with $x_{ucen(V)}^*, x_{wcen(V)}^* \leq 19/40$. The number of disagreement generated by the negative edge (u, w) is bounded by*

$$20 \left(1 - x_{uw}^* \right).$$

Proof. From the first constraint of (2) and the inequalities $x_{ucen(V)}^*, x_{wcen(V)}^* \leq 19/40$, we have

$$1 - x_{uw}^* \geq 1 - x_{ucen(V)}^* - x_{wcen(V)}^* \geq 1 - \frac{38}{40} = \frac{1}{20}.$$

The number of disagreement generated by edge $(u, w) = 1 \leq 20(1 - x_{uw}^*)$.

We conclude the lemma. $\qquad \square$

Lemma 2. *For each vertex $w \in V$ with $x^*_{wcen(V)} > 19/40$, the number of disagreements generated by the negative edges (u, w) with $x^*_{ucen(V)} \leq x^*_{wcen(V)}$ can be bounded by*

$$20 \left[\sum_{(u,w)\in E^+, x^*_{ucen(V)} \leq x^*_{wcen(V)}} x^*_{uw} + \sum_{(u,w)\in E^-, x^*_{ucen(V)} \leq x^*_{wcen(V)}} (1 - x^*_{uw}) \right].$$

Proof. For each vertex $w \in V$, denote

$$P_w(V) := \left\{ u \in V : (u, w) \in E^+, x^*_{ucen(V)} \leq x^*_{wcen(V)} \right\},$$

$$N_w(V) := \left\{ u \in V : (u, w) \in E^-, x^*_{ucen(V)} \leq x^*_{wcen(V)} \right\}.$$

Recall $\mathrm{Avg}_{cen(V)}(V) \leq 17/80$ and $x^*_{wcen(V)} > 19/40$. We obtain that there are at least $|V|/2$ vertices in $P_w(V) \cup N_w(V)$. Moreover we have $|P_w(V)| \geq |N_w(V)|$ since

$$|N_w(V)| \leq |E_w^-| \leq \frac{|V| - 1}{5}.$$

Then, we get

$$\sum_{u \in P_w(V)} x^*_{uw} + \sum_{u \in N_w(V)} (1 - x^*_{uw})$$

$$\geq \sum_{u \in P_w(V)} \left(x^*_{wcen(V)} - x^*_{ucen(V)} \right) + \sum_{u \in N_w(V)} \left(1 - x^*_{wcen(V)} - x^*_{ucen(V)} \right)$$

$$= x^*_{wcen(V)} |P_w(V)| + \left(1 - x^*_{wcen(V)} \right) |N_w(V)| - \sum_{u \in P_w(V) \cup N_w(V)} x^*_{ucen(V)}$$

$$\geq \frac{19}{40} |P_w(V)| - \frac{17}{80} (|P_w(V)| + |N_w(V)|)$$

$$\geq \frac{1}{20} |P_w(V)|.$$

Therefore, the number of disagreements generated by the negative edges (u, w) with $x^*_{ucen(V)} \leq x^*_{wcen(V)}$ is bounded by

$$20 \left[\sum_{u \in P_w(V)} x^*_{uw} + \sum_{u \in N_w(V)} (1 - x^*_{uw}) \right].$$

We conclude the lemma. □

3.2 $\mathrm{Avg}_{cen(V)}(V) > 17/80$

Lemma 3. *For each vertex $w \in V$, the number of disagreements generated by the negative edges (u, w) is bounded by*

$$20 \sum_{(u,w)\in E^+, u \in V} x^*_{uw}.$$

Proof. From the definition of $cen(V)$, we obtain that if $\text{Avg}_{cen(V)}(V) > 17/80$, then for each vertex $w \in V$

$$\frac{\sum_{u \in V} x_{uw}^*}{|V|} > \frac{17}{80} \tag{3}$$

holds. Moreover, for each vertex $w \in V$, we have

$$|E_w^-| \leq \frac{(|V| - 1)}{5} \text{ and } |E_w^+| \geq \frac{4(|V| - 1)}{5}. \tag{4}$$

Combining (3) and (4), we obtain that for each vertex $w \in V$,

$$\frac{\sum_{u \in E_w^+} x_{uw}^*}{|E_w^+|} \geq \frac{17}{80} - \frac{1}{5} = \frac{1}{80}$$

holds. Therefore, for each vertex $w \in V$, the number of disagreements generated by the negative edges (u, w) is no more than $|E_w^+|/4$ and it is bounded by

$$20 \sum_{u \in E_w^+} x_{uw}^*.$$

$\square$.

Combining Lemma 1–3, we obtain Theorem 2.

Theorem 2. *Algorithm 1 is a 20-approximation algorithm for the lower bounded correlation clustering problem on 4-positive edge dominant graphs.*

4   A Complex Algorithm May Outputs Multiple Clusters

In Sect. 3, we give a algorithm which only return one cluster. However, in some applications, we may need to output more than one clusters. Therefore, we provide Algorithm 2 for some special graphs in this section which may output multiple clusters. The detailed algorithm is shown in Algorithm 2.

We assume without loss of generality that the solution returned by Algorithm 2 contains exactly k clusters. The center set of vertices of the solution is $C := \{v_1, v_2, \ldots, v_k\}$. The corresponding clusters are $C_{v_1}, C_{v_2}, \ldots, C_{v_k}$. The number of the disagreements generated by partition $\mathcal{C}$ is

$$\sum_{i \in [k-1]} |(u, v) \in E^+ : u \in C_{v_i}, v \in \cup_{t \in [k] \setminus [i]} C_{v_t}| + \sum_{i \in [k]} |(u, v) \in E^- : u, v \in C_{v_i}|.$$

The first part is the number of disagreements generated by the positive edges and the upper bound on the number of these disagreements is analyzed in Subsect. 4.1. The second part is the number of disagreements generated by the negative edges and the upper bound on the number of these disagreements is analyzed in Subsect. 4.2.

Algorithm 2

Input: Integer L, and a labelled complete graph $G = (V, E)$ with $|E_v^-| \leq L/5, v \in V$.
Output: A partition of vertices.
1: Solve (2) to obtain an optimal fractional solution x^*.
2: Initialize the un-cluster set $S := V$, and the center set of vertices $C := \emptyset$.
3: **while** $S \neq \emptyset$ **do**
4:     **for** each vertex $v \in S$ **do**
5:         Order the vertices in S in nondecreasing value of x^* from v. Let set T_v^1 be the
           first L vertices in S according to above order and $T_v^2 := \{t \in S : x_{vt}^* \leq 1/2\}$.
           Set $T_v = T_v^1 \cup T_v^2$ and compute

$$\mathrm{Avg}_v(T_v) := \frac{\sum_{t \in T_v} x_{vt}^*}{|T_v|}.$$

6:     **end for**
7:     Choose vertex v with minimum $\mathrm{Avg}_v(T_v)$.
8:     **if** $|S \backslash T_v| \geq L$ and $\mathrm{Avg}_v(T_v) \leq 17/80$ **then**
9:         Let $C_v := T_v$ be a cluster.
10:        Update $S := S \backslash C_v$ and $C := C \cup \{v\}$.
11:    **else**
12:        Select

$$v := \arg\min_{s \in S} \frac{\sum_{t \in S} x_{st}^*}{|S|}.$$

13:        Let $C_v := S$ be a cluster.
14:        Update $S := \emptyset$ and $C := C \cup \{v\}$.
15:    **end if**
16: **end while**
17: **return** Set C and $\mathcal{C} = \{C_v : v \in C\}$.

4.1   Disagreements Generated by Positive Edges

Recall Algorithm 2. For each $i \in [k-1]$, we have $\mathrm{Avg}_{v_i}(C_{v_i}) \leq 17/80$. We analyze the upper bound on the number of disagreements generated by the positive edges in the following lemma.

Lemma 4. *For each $i \in [k-1]$ and vertex $v \in V \backslash \cup_{t \in [k] \backslash [i]} C_{v_t}$, the number of disagreements generated by the positive edges $(q, v), q \in C_{v_i}$ is bounded by*

$$\frac{64}{17} \left[\sum_{(q,v) \in E^+, q \in C_{v_i}} x_{qv}^* + \sum_{(q,v) \in E^-, q \in C_{v_i}} (1 - x_{qv}^*) \right].$$

Proof. For each $v \in V \backslash \cup_{t \in [k] \backslash [i]} C_{v_t}$, denote

$$E_v^+(C_{v_i}) := \left\{ (q, v) \in E^+ : q \in C_{v_i} \right\},$$
$$E_v^-(C_{v_i}) := \left\{ (q, v) \in E^- : q \in C_{v_i} \right\}.$$

From $|E_v^-| \leq L/5$ and $|E_v^-(C_{v_i})| + |E_v^+(C_{v_i})| \geq L$, we have

$$|E_v^-(C_{v_i})| \leq |E_v^+(C_{v_i})|/4. \tag{5}$$

Combining (5) and Step 5 of Algorithm 2, we have

$$\sum_{q \in E_v^+(C_{v_i})} x_{qv}^* + \sum_{q \in E_v^-(C_{v_i})} \left(1 - x_{qv}^*\right)$$

$$\geq \sum_{q \in E_v^+(C_{v_i})} \left(x_{vv_i}^* - x_{qv_i}^*\right) + \sum_{q \in E_v^-(C_{v_i})} \left(1 - x_{vv_i}^* - x_{qv_i}^*\right)$$

$$= x_{vv_i}^* |E_v^+(C_{v_i})| + \left(1 - x_{vv_i}^*\right)|E_v^-(C_{v_i})| - \sum_{q \in C_{v_i}} x_{qv_i}^*$$

$$\geq \frac{1}{2}|E_v^+(C_{v_i})| - \frac{17}{80}(|E_v^+(C_{v_i})| + |E_v^-(C_{v_i})|)$$

$$\geq \frac{17}{64}|E_v^+(C_{v_i})|.$$

Therefore, the number of disagreements generated by the positive edges $(q, v), q \in C_{v_i}$ equals $|E_v^+(C_{v_i})|$ and it is bounded by

$$\frac{64}{17}\left[\sum_{(q,v) \in E^+, q \in C_{v_i}} x_{qv}^* + \sum_{(q,v) \in E^-, q \in C_{v_i}} \left(1 - x_{qv}^*\right)\right].$$

We conclude the lemma.   □

4.2   Disagreements Generated by Negative Edges

In this subsection, we consider the disagreements generated by the negative edges. Similar to Lemmas 2 and 3, we obtain the following two lemmas.

Lemma 5. *For each cluster C_{v_i} with $Avg_{v_i}(C_{v_i}) \leq 17/80$ and vertex $w \in C_{v_i}$, if $x_{wv_i}^* > 19/40$, then the number of disagreements generated by the negative edges (u, w) with $x_{uv_i}^* \leq x_{wv_i}^*$ is bounded by*

$$20\left[\sum_{u \in C_{v_i}:(u,w) \in E^+, x_{uv_i}^* \leq x_{wv_i}^*} x_{uw}^* + \sum_{u \in C_{v_i}:(u,w) \in E^+, x_{uv_i}^* \leq x_{wv_i}^*} \left(1 - x_{uw}^*\right)\right].$$

Lemma 6. *If $Avg_{v_k}(C_{v_k}) > 17/80$, then for each vertex $w \in C_{v_k}$, the number of disagreements generated by the negative edges $(u, w), u \in C_{v_k}$ is bounded by*

$$20\sum_{(u,w) \in E^+, u \in C_{v_k}} x_{uw}^*.$$

Combining Lemmas 1, 4–Lemma 6, we obtain Theorem 3.

Theorem 3. *Algorithm 2 is a 20-approximation algorithm for the lower bounded correlation clustering problem on $(5|V|/L - 1)$-positive edge dominant graphs.*

5   Discussions

In this paper, we first study the lower bounded correlation clustering problem and give an integer programming formulation for the problem. We provide two polynomial time approximation algorithms and prove that the algorithms in this paper achieve constant approximate ratios for the lower bounded correlation clustering problem on some special graphs. About the lower bounded correlation clustering problem, we propose the following future research questions:

- In this paper, we prove that our algorithms achieve constant ratio for the correlation clustering problem on some special graphs. It will be interesting to design a polynomial time constant approximation algorithm for the lower bounded correlation clustering problem on general complete graphs.
- In this paper, we study the minimizing disagreements version of the lower bounded correlation clustering problem. There is no relevant research on the maximizing agreements version of the lower bounded correlation clustering problem. Therefore, another interesting future work is to study the maximizing agreements version of the lower bounded correlation clustering problem.

Acknowledgement. The first author is supported by National Natural Science Foundation of China (No. 12101594) and the Project funded by China Postdoctoral Science Foundation (No. 2021M693337). The second author is supported by National Social Science Foundation (NO. 20BGL259). The third author is supported by the NSERC (06446) and National Natural Science Foundation of China (Nos. 11771386, 11971349, 71771117). The fourth author is supported by National Natural Science Foundation of China (No. 12131003) and Beijing Natural Science Foundation Project No. Z200002.

References

1. Ailon, N., Avigdor-Elgrabli, N., Liberty, E., Zuylen, A.V.: Improved approximation algorithms for bipartite correlation clustering. SIAM J. Comput. **41**(5), 1110–1121 (2012)
2. Ahmadi, S., Khuller, S., Saha, B.: Min-max correlation clustering via MultiCut. In: Lodi, A., Nagarajan, V. (eds.) IPCO 2019. LNCS, vol. 11480, pp. 13–26. Springer, Cham (2019). https://doi.org/10.1007/978-3-030-17953-3_2
3. Ahmadian, S., Swamy, C.: Improved approximation guarantees for lower-bounded facility location. In: Erlebach, T., Persiano, G. (eds.) WAOA 2012. LNCS, vol. 7846, pp. 257–271. Springer, Heidelberg (2013). https://doi.org/10.1007/978-3-642-38016-7_21
4. Bansal, N., Blum, A., Chawla, S.: Correlation clustering. Mach. Learn. **56**(1–3), 89–113 (2004)
5. Bonchi, F., Gionis, A., Gullo, F., Tsourakakis, C.E., Ukkonen, A.: Chromatic correlation clustering. ACM Trans. Knowl. Discov. Data **9**(4), 1–24 (2015)
6. Bonchi, F., Gionis, A., Ukkonen, A.: Overlapping correlation clustering. Knowl. Inf. Syst. **35**(1), 1–32 (2013)
7. Charikar, M., Guruswami, V., Wirth, A.: Clustering with qualitative information. J. Comput. Syst. Sci. **71**(3), 360–383 (2005)

8. Chawla, S., Makarychev, K., Schramm, T., Yaroslavtsev, G.: Near optimal LP rounding algorithm for correlation clustering on complete and complete k-partite graphs. In: Proceedings of the 47th ACM Symposium on Theory of Computing, pp. 219–228 (2015)

9. Fukunaga, T.: LP-based pivoting algorithm for higher-order correlation clustering. J. Comb. Optim. **37**(4), 1312–1326 (2018). https://doi.org/10.1007/s10878-018-0354-y

10. Han, L., Hao, C., Wu, C., Zhang, Z.: Approximation algorithms for the lower-bounded k-median and its generalizations. In: Kim, D., Uma, R.N., Cai, Z., Lee, D.H. (eds.) COCOON 2020. LNCS, vol. 12273, pp. 627–639. Springer, Cham (2020). https://doi.org/10.1007/978-3-030-58150-3_51

11. Jafarov, J., Kalhan, S., Makarychev, K., Makarychev, Y.: Correlation clustering with asymmetric classification errors. In: Proceedings of the 37th International Conference on Machine Learning, pp. 4641–4650 (2020)

12. Li, S.: On facility location with general lower bounds. In: Proceedings of the 30th Annual ACM-SIAM Symposium on Discrete Algorithms, pp. 2279–2290 (2019)

13. Li, P., Puleo, G.J., Milenkovic, O.: Motif and hypergraph correlation clustering. IEEE Trans. Inf. Theory **66**(5), 3065–3078 (2019)

14. Makarychev, K., Makarychev, Y., Vijayaraghavan, A.: Correlation clustering with noisy partial information. In: Proceedings of the 28th Annual Conference Computational Learning Theory, pp. 1321–1342 (2015)

15. Mathieu, C., Schudy, W.: Correlation clustering with noisy input. In: Proceedings of the 21th Annual ACM-SIAM Symposium on Discrete Algorithms, pp. 712–728 (2010)

16. Puleo, G.J., Milenkovic, O.: Correlation clustering with constrained cluster sizes and extended weights bounds. SIAM J. Optim. **25**(3), 1857–1872 (2015)

17. Svitkina, Z.: Lower-bounded facility location. ACM Trans. Algorithms **6**(4), 1–16 (2010)

18. Swamy, C.: Correlation clustering: maximizing agreements via semidefinite programming. In: Proceedings of the 15th Annual ACM-SIAM Symposium on Discrete Algorithms, pp. 526–527 (2004)

19. Saha, B., Subramanian, S.: Correlation clustering with same-cluster queries bounded by optimal cost. In: Proceedings of the 27th Annual European Symposium on Algorithms, pp. 81:1–81:17 (2019)

20. Veldt, N., Gleich, D.F., Wirth, A.: A correlation clustering framework for community detection. In: Proceedings of the 27th World Wide Web Conference, pp. 439–448 (2018)

Approximation Algorithm for Maximizing Nonnegative Weakly Monotonic Set Functions

Min Cui[1], Donglei Du[2], Dachuan Xu[3], and Ruiqi Yang[4(✉)]

[1] Department of Operations Research and Information Engineering,
Beijing University of Technology, Beijing 100124, People's Republic of China
B201840005@emails.bjut.edu.cn
[2] Faculty of Management, University of New Brunswick,
Fredericton, NB E3B 5A3, Canada
ddu@unb.ca
[3] Beijing Institute for Scientific and Engineering Computing,
Beijing University of Technology, Beijing 100124, People's Republic of China
xudc@bjut.edu.cn
[4] School of Mathematical Sciences, University of Chinese Academy Sciences,
Beijing 100049, People's Republic of China
yangruiqi@ucas.ac.cn

Abstract. In recent years, submodularity has been found in a wide range of connections and applications with different scientific fields. However, many applications in practice do not fully meet the characteristics of diminishing returns. In this paper, we consider the problem of maximizing unconstrained non-negative weakly-monotone non-submodular set function. The generic submodularity ratio γ is a bridge connecting the non-negative monotone functions and the submodular functions, and no longer applicable to the non-monotone functions. We study a class of non-monotone functions, define as the weakly-monotone function, redefine the submodular ratio related to it and name it weakly-monotone submodularity ratio $\widehat{\gamma}$, propose a deterministic double greedy algorithm, which implements the $\frac{\widehat{\gamma}}{\widehat{\gamma}+2}$ approximation of the maximizing unconstrained non-negative weakly-monotone function problem. When $\widehat{\gamma} = 1$, the algorithm achieves an approximate guarantee of $1/3$, achieving the same ratio as the deterministic algorithm for the unconstrained submodular maximization problem.

Keywords: Non-submodular optimization · Unconstrained · Submodularity ratio

1 Introduction

The research on combination problems with submodular property has received extensive attention in recent years. We say the function $f : 2^N \rightarrow R^+$ is submodular on the finite ground set N if and only if for any subsets S, T of N, we have:

© Springer Nature Switzerland AG 2021
D. Mohaisen and R. Jin (Eds.): CSoNet 2021, LNCS 13116, pp. 50–58, 2021.
https://doi.org/10.1007/978-3-030-91434-9_5

$$f(S) + f(T) \geq f(S \cup T) + f(S \cap T) \tag{1}$$

i.e. for any two subsets $S \subseteq T \subseteq N$ and $e \in N \setminus T$

$$f(S \cup \{e\}) - f(S) \geq f(T \cup \{e\}) - f(T).$$

We say the function f is monotone if for any two subsets S, T of N such that $S \subseteq T$, we have

$$f(S) \leq f(T) \tag{2}$$

Submodularity has a very intuitive interpretation in economics, which called the diminishing marginal utility. The diminishing marginal utility enables submodular functions to accurately simulate diversity and information gain in practical applications. At the same time, submodular functions can be solved accurately to minimize and approximately maximize in polynomial time [16]. These make submodular functions getting increasing attention in the field of artificial intelligence [24] and data mining [1], such as: social network influence [19], deep compressed sensing [22], sensor placement [21], targeted marketing [6], to name a few.

Unconstrained submodular problems is one of the most basic problem in submodular optimization. The factor of the approximation algorithm for the unconstrained submodular maximization problem is hardly better than $1/2$ in polynomial time [8]. In fact, many basic NP-hard problems are special cases of unconstrained submodular maximization, including undirected cut problems [11], directed cut problems [14], the maximum facility location problems [18], and some limited satisfiability problems. In addition, the approximation algorithms of the unconstrained submodular maximization problem have been used as a subroutine of many other algorithms, such as social network marketing [15], and so on.

The study of unconstrained submodular maximization problems began in the 1960s [5]. Obviously, there are not many results. Feige et al. [9] were the first team to rigorously study general unconstrained submodular maximization issues: they proposed an uniform random subset algorithm and a local search algorithm, then increased the approximation guarantee to $2/5$ by adding noise to the local search algorithm; they also showed that it may require exponential query to achieve an approximate ratio of $1/2+\epsilon$ in the value oracle model. Gharan et al. [12] and Feldman et al. [10] used methods such as simulated annealing to further improve the noisy local search technique. Buchbinder et al. [3] showed that a simple random algorithm strategy can be used to achieve the tight $1/2$-approximate ratio. Later, their team [4] gave the de-randomized algorithm with the same approximate ratio. Roughgarden et al. [20] studied online unconstrained submodular maximization problem, provided a polynomial-time no-$1/2$-regret algorithm for this problem.

The applicability of submodular function is quite convenient and extensive. Is it possible to apply skills to connect the general function problem with the submodular problem, and then use the algorithms for submodular problem to solve the general problem effectively and securely? Based on one of the equivalent definitions of submodular functions, Das and Kempe [7] proposed the submodularity ratio $\gamma_{N,k}$ with respect to the ground set U and the parameter k, which is a quantity characterizing how close a general set function is to being submodular. Bian et al. [2] combined and generalized the ideas of curvature α and the submodularity ratio $\gamma_{N,k}$. Gong et al. [13] provided a more practical measurement γ which is called generic submodularity ratio which is depend on the monotonicity of the function.

In application, the problem with monotonic and submodularity is an idealized situation. For social network marketing, participants' decision-making are affected by the following conditions in actual social networks [17]: the social trust between participants, the social relationship between participants, and the preference similarity between participants. An appropriate number of "Big V", "Opinion Leader", "KOL" and "Online Celebrity" recommendations could make things widely spread without causing too much disgust, so as to ensure the positive growth of marketing effect. For the facility location problems, suppose the objective function is to evaluate the overall income of supermarkets in a city. The scope of the city would not change in a short time, and the construction and daily operation of the supermarket is a fixed cost. When the supermarket supply meets the urban demand [23], the newly opened supermarkets and previous supermarkets have to carry out price reduction and promotion in order to survive, which reduces the overall income.

Our Results. In this paper, we study a class of non-monotone functions which is a relaxed version of the monotonicity called the weakly-monotone, and discuss the problem of maximizing unconstrained weakly-monotone functions.

- We first give the definition of weakly-monotone function, then define the weakly-monotone submodularity ratio $\widehat{\gamma}$.
- Second, we present a deterministic double greedy algorithm for the unconstrained weakly-monotone maximization problem, and prove the algorithm achieves $\frac{\widehat{\gamma}}{\widehat{\gamma}+2}$-approximation ratio in $2n+2$ times query and in $O(n)$ computing time.

In addition, when the $\widehat{\gamma}$ reaches 1 (i.e. the function is submodular), the approximation guarantee of the algorithm recovers the tight ratio as deterministic algorithm for the unconstrained submodular maximization problem.

Organization. The rest of this paper is structured as follows. In Sect. 2, we introduce the basic definitions and symbols used throughout this article, and give new definitions. We provide a deterministic algorithm for the unconstrained weakly-monotone functions maximization problem in Sect. 3, and give the approximate ratio analysis. Section 4 offers direction of future work.

2   Preliminaries

In this paper, we consider the problem of maximizing an unconstrained nonnegative weakly-monotone function, the objective is to select a subset S of the ground set N to maximize $f(S)$. The set function $f : 2^N \rightarrow R^+$ is a non-negative weakly-monotone set function with $f(\emptyset) = 0$. This problem can be stated as:

$$\max_{S \subseteq N} f(S). \tag{3}$$

$f(S \cup T) - f(S)$ denotes the marginal gain of adding the set $T \subseteq N$ to the set $S \subseteq N$. Specially, when the set $T = \{e\} \in N \setminus S$, the marginal gain of adding the single element e to the set S is defined as $f(S \cup \{e\}) - f(S)$.

Then, we defined the weakly-monotone function in following:

Definition 1. Weakly-monotone: *Let $f : 2^N \rightarrow R^+$ be a non-negative set function with for any two subsets of different sizes S, T of N that $f(S) \neq f(T)$. We say f is weakly-monotone if for any subset $A \subseteq N$ and any $e \in N \setminus A$,*

1. if $f(A \cup \{e\}) > f(A)$, for any $S \subseteq N$ such that $A \subsetneqq S$, $f(A) \neq f(S)$;
2. if $f(A \cup \{e\}) < f(A)$, for any $S \subseteq N$ such that $A \subsetneqq S$, $f(A) > f(S)$.

Example 1. $N = \{1, 2, 3, 4, 5\}$, $f(S) = \min\{|S|, |N| - |S| + 0.5\}$.

When $A = \emptyset$, $e \in N \setminus A$, $f(A \cup \{e\}) = 1 > 0 = f(A)$: for any $A \subsetneqq S \subseteq \{1, 2, 3, 4, 5\}$, $f(S) \in \{1, 2, 2.5, 1.5, 0.5\} \neq 0 = f(A)$;

When $A \in \{\{1\}, \{2\}, \{3\}, \{4\}, \{5\}\}$, $e \in N \setminus A$, $f(A \cup \{e\}) = 2 > 1 = f(A)$: for any $A \subsetneqq S \subseteq \{1, 2, 3, 4, 5\}$, $f(S) \in \{2, 2.5, 1.5, 0.5\} \neq 1 = f(A)$;

When $A \in \{\{1, 2\}, \{1, 3\}, \{1, 4\}, \{1, 5\}, \{2, 3\}, \{2, 4\}, \{2, 5\}, \{3, 4\}, \{3, 5\}, \{4, 5\}\}$, $e \in N \setminus A$, $f(A \cup \{e\}) = 2.5 > 2 = f(A)$: for any $A \subsetneqq S \subseteq \{1, 2, 3, 4, 5\}$, $f(S) \in \{2.5, 1.5, 0.5\} \neq 2 = f(A)$;

When $A \in \{\{1, 2, 3\}, \{1, 2, 4\}, \{1, 2, 5\}, \{1, 3, 4\}, \{1, 3, 5\}, \{1, 4, 5\}, \{2, 3, 4\}, \{2, 3, 5\}, \{2, 4, 5\}, \{3, 4, 5\}\}$, $e \in N \setminus A$, $f(A \cup \{e\}) = 1.5 < 2.5 = f(A)$: for any $A \subsetneqq S \subseteq \{1, 2, 3, 4, 5\}$, $f(S) \in \{1.5, 0.5\} < 2.5 = f(A)$;

When $A \in \{\{1, 2, 3, 4\}, \{1, 2, 3, 5\}, \{1, 2, 4, 5\}, \{1, 3, 4, 5\}, \{2, 3, 4, 5\}\}$, $e \in N \setminus A$, $f(A \cup \{e\}) = 0.5 < 1.5 = f(A)$: for any $A \subsetneqq S \subseteq \{1, 2, 3, 4, 5\}$, $f(S) = 0.5 < 1.5 = f(A)$.

For the function that does not conform to the strictly property, we only need to add a small disturbance to the value of the function. For example, if there are two equal function values Q, we could add $\frac{Q}{n \cdot 10^k}$ to the second function value, and so on.

Then, we give the definition of the weakly-monotone submodularity ratio $\hat{\gamma}$.

Definition 2. Weakly-monotone submodularity ratio: *For a non-negative weakly-monotone set function $f : 2^N \rightarrow R^+$, the weakly-monotone submodularity ratio of f is the largest scalar $\hat{\gamma}$ satisfied the corresponding inequailities under these two cases, for any $S \subseteq N$ and any $e \in N \setminus S$:*

1. When $f(S \cup \{e\}) - f(S) > 0$, for any $T \subseteq N \setminus \{e\}$ such that $S \subseteq T$:

$$f(S \cup \{e\}) - f(S) \geqslant \hat{\gamma} \mid f(T \cup \{e\}) - f(T) \mid;$$

2. When $f(S \cup \{e\}) - f(S) < 0$, for any $T \subseteq N \setminus \{e\}$ such that $S \subseteq T$:

$$f(S \cup \{e\}) - f(S) \geqslant \frac{1}{\widehat{\gamma}}[f(T \cup \{e\}) - f(T)].$$

Note 1. The weakly-monotone submodularity ratio $\widehat{\gamma}$ of f in Example 1 is 0.5.

For non-negative weakly-monotone set functions, the following lemma always holds.

Lemma 1. *Given a non-negative weakly-monotone set function* $f : 2^N \to R^+$ *with weakly-monotone submodularity ratio* $\widehat{\gamma}$, *it always holds that*

a. $\widehat{\gamma} \in (0, 1]$.
b. If the function f *is submodular, then* $\widehat{\gamma} = 1$.

Throughout this paper, denotes S as the output solution by the algorithm; denote O and OPT (i.e. $f(O) = OPT$) as the optimal set and the value of the optimal sets respectively. We assume that a single query on the oracle value requires $O(1)$ time.

3   The Deterministic-Greedy Algorithm

In this section, we propose a deterministic algorithm for maximizing unconstrained weakly-monotone functions. The algorithm runs in n iterations. In the i-th iteration, we only consider whether to keep the element e_i in the solution. The algorithm always maintains two feasible solutions S and T. The initial setting of S is an empty set and the T is the ground set N. The algorithm in arbitrary sequence checks each element $e_i \in N$ one by one to decide on adding it to S or deleting it from T. The decision is greedy depend on the size of marginal gain a_i of adding e_i to S and marginal gain b_i of abandoning e from S. If a_i is not less than b_i, adding e_i to S; otherwise, removing e_i from T. After traversing all the elements in N, we get $S = T$ which as the output of the algorithm. The principle of double greedy is intuitive as the operation for the element that we decide brings greater marginal benefits. A formal description of the algorithm appears as Algorithm 1.

In order to prove the approximate ratio of the algorithm, we give some additional notations. According to the construction of S and T, S staring with the empty set denotes S_0, T staring with the ground set N denotes T_0; in the i-th iteration, the algorithm either adds e_i to S_{i-1} or removes e_i from T_{i-1}. Record the S as S_i and record T as T_i when we have finished the operation.

First, we introduce an intermediate function $f((O \cup S_i) \cap T_i)$, using the change of intermediate function to bound the loss of $f(S)$ and $f(T)$ in each iteration.

Lemma 2. *For all* $i = 1, 2, \cdots, n$:

$$f((O \cup S_{i-1}) \cap T_{i-1}) - f((O \cup S_i) \cap T_i) \leq \frac{1}{\widehat{\gamma}}[f(S_i) - f(S_{i-1}) + f(T_i) - f(T_{i-1})]$$

Algorithm 1. Deterministic-Greedy

Input: evaluation oracle $f : 2^N \to R^+$
Output: the set S

1: Initialize $S \leftarrow \emptyset$, $T \leftarrow N$
2: **for** $i = 1$ to n **do**
3:     Initialize $a_i \leftarrow f(S \cup \{e_i\}) - f(S)$
4:     Initialize $b_i \leftarrow f(T \setminus \{e_i\}) - f(T)$
5:     **if** $a_i \geq b_i$ **then**
6:         $S \leftarrow S \cup \{e_i\}$
7:         $T \leftarrow T$
8:     **else**
9:         $S \leftarrow S$
10:        $T \leftarrow T \setminus \{e_i\}$
11: **return** S

Due to the length limitation, we only give the main idea of proof: from the relationship between a_i and b_i, the proof is divided into two cases: $a_i \geq b_i$ or $a_i < b_i$. In each case, we need to find out the relationship between S_{i-1}, T_{i-1} and S_i, T_i, and show one of a_i and b_i must more than 0; then we discuss whether e_i is in the optimal solution O and whether the value of $f((O \cup S_{i-1}) \cap T_{i-1})$ increases; in the end, using the definition of $\hat{\gamma}$ to find the relationship between the $f((O \cup S_{i-1}) \cap T_{i-1}) - f((O \cup S_i) \cap T_i)$ and $f(S_i) - f(S_{i-1}) + f(T_i) - f(T_{i-1})$.

Then, we use the total change of intermediate function $f((O \cup S_i) \cap T_i)$ to bound the total loss of $f(S)$ and $f(T)$ at the algorithm.

Lemma 3.

$$f((O \cup S_0) \cap T_0) - f((O \cup S_n) \cap T_n)) \leq \frac{1}{\hat{\gamma}}[f(S_n) + f(T_n) - f(S_0) - f(T_0)]$$

Proof. Summing up the inequalities in Lemma 2 for all $i = 1, 2, \cdots, n$, we get

$$\sum_{i=1}^{n}(f((O \cup S_{i-1}) \cap T_{i-1}) - f(O \cup S_i) \cap T_i)))$$
$$\leq \frac{1}{\hat{\gamma}}\sum_{i=1}^{n}[f(S_i) - f(S_{i-1}) + f(T_i) - f(T_{i-1})] \tag{4}$$

Combine the similar items, we have

$$f((O \cup S_0) \cap T_0) - f((O \cup S_n) \cap T_n))$$
$$\leq \frac{1}{\hat{\gamma}}[f(S_n) + f(T_n) - f(T_0)] \tag{5}$$

Notice, at the begining of the algorithm the intermediate function $f((O \cup S_0) \cap T_0)$ is $f(O)$.

Theorem 1. *For any non-negative weakly-monotone function $f : 2^N \to R^+$, Algorithm Deterministic-Greedy is a $\frac{\widehat{\gamma}}{\widehat{\gamma}+2}$-approximation algorithm, the query complexity is $2n + 2$, the computing time is $O(n)$.*

Proof. From the setting of S_i, T_i ($i = 0, 1, \cdots, n$), we can easily get $S_n = T_n = S$; $(O \cup S_n) \cap T_n = S_n = S$. So

$$f(O) - f(S) \leq \frac{2}{\widehat{\gamma}} f(S)$$

Thus,

$$f(S) \geq \frac{\widehat{\gamma}}{\widehat{\gamma} + 2} OPT.$$

Then, we consider the queries of the algorithm by two parts. The first part is to compute the value of a_i when every element e_i arrive, the number of queries in this part is $n + 1$. The second part is to compute the value of b_i whenevery element e_i arrive, the number of queries in this part is also $n + 1$. Consequently, the number of queries is $2n + 2$. Thus, the computing time of the algorithm is $O(n)$ oracle queries plus $O(n)$ other operations.

From [3], we know that for any $\epsilon > 0$, $(1/3 + \epsilon)$-approximation is tight for deterministic algorithm for unconstrained submodular maximization problem. When $\widehat{\gamma} = 1$, the algorithm has an approximation ratio of $1/3$. Therefore, we can say our algorithm is tight.

4 Discussion

Today, we would face the increasingly large data sets that are ubiquitous in modern machine learning and data mining applications. Greedy algorithm is highly continuous and would no longer have advantages in large-scale data sets. Various algorithms have been proposed to solve numerous submodular problems including large-scale problems according to application requirements, which can be roughly divided into centralized algorithms, streaming algorithms, distributed algorithms and decentralized framework. My recent interest is parallel algorithms for submodular problems. One future work is the research of parallel algorithms for maximizing an unconstrained non-negative weakly-monotone function on large-scale data sets or streaming setting.

Acknowledgements. The first and third authors are supported by National Natural Science Foundation of China (No. 12131003) and Beijing Natural Science Foundation Project No. Z200002. The second author is supported by the Natural Sciences and Engineering Research Council of Canada (NSERC) grant 06446, and Natural Science Foundation of China (Nos. 11771386, 11728104). The fourth author is supported by the Fundamental Research Funds for the Central Universities (No. E1E40108).

References

1. Badanidiyuru, A., Mirzasoleiman, B., Karbasi, A., Krause, A.: Streaming submodular maximization: massive data summarization on the fly. In: 20th International Proceedings on Knowledge Discovery and Data Mining, New York, NY, USA, pp. 671–680. Association for Computing Machinery (2014)
2. Bian, A.A., Buhmann, J.M., Krause, A., Tschiatschek, S.: Guarantees for Greedy maximization of non-submodular functions with applications. In: 35th International Conference on Machine Learning, NSW, Australia, vol. 70, pp. 498–507. Proceedings of Machine Learning Research (2017)
3. Buchbinder, N., Feldman, M., Naor, J.S., Schwartz, R.: A tight linear time (1/2)-approximation for unconstrained submodular maximization. SIAM J. Comput. **44**(5), 1384–1402 (2015)
4. Buchbinder, N., Feldman, M.: Deterministic algorithms for submodular maximization problems. In: 27th International Symposium on Discrete Algorithms, Arlington, VA, USA, pp. 392–403. Society for Industrial and Applied Mathematics (2016)
5. Cherenin, V.: Solving some combinatorial problems of optimal planning by the method of successive calculations. In: Conference of Experiences and Perspectives of the Applications of Mathematical Methods and Electronic Computers in Planning, Russian, Mimeograph, Novosibirsk (1962)
6. Coelho, V.N., et al.: Generic Pareto local search metaheuristic for optimization of targeted offers in a bi-objective direct marketing campaign. Comput. Oper. Res. **78**, 578–587 (2017)
7. Das, A., Kempe, D.: Submodular meets spectral: Greedy algorithms for subset selection, sparse approximation and dictionary selection. In: 28th International Conference on Machine Learning, Bellevue, Washington, USA, pp. 1057–1064. Omnipress (2011)
8. Dobzinski, S., Vondrak, J.: From query complexity to computational complexity. In: 44th International Symposium on Theory of Computing, New York, NY, USA, pp. 1107–1116. Society for Industrial and Applied Mathematics (2012)
9. Feige, U., Mirrokni, V.S., Vondrák, J.: Maximizing non-monotone submodular functions. SIAM J. Comput. **40**(4), 1133–1153 (2011)
10. Feldman, M., Naor, J.S., Schwartz, R.: Nonmonotone submodular maximization via a structural continuous greedy algorithm. In: Aceto, L., Henzinger, M., Sgall, J. (eds.) ICALP 2011. LNCS, vol. 6755, pp. 342–353. Springer, Heidelberg (2011). https://doi.org/10.1007/978-3-642-22006-7_29
11. Galbiati, G., Maffioli, F.: Approximation algorithms for maximum cut with limited unbalance. Theor. Comput. Sci. **385**(1), 78–87 (2007)
12. Gharan, S.O., Vondrák, J.: Submodular maximization by simulated annealing. In: 22nd International Symposium on Discrete Algorithms, San Francisco, California, USA, pp. 1098–1116. Society for Industrial and Applied Mathematics (2011)
13. Gong, S., Nong, Q., Liu, W., Fang, Q.: Parametric monotone function maximization with matroid constraints. J. Global Optim. **75**(3), 833–849 (2019). https://doi.org/10.1007/s10898-019-00800-2
14. Halperin, E., Zwick, U.: Combinatorial approximation algorithms for the maximum directed cut problem. In: 20th International Symposium on Discrete Algorithms, Washington, DC, USA, pp. 1–7. Association for Computing Machinery/Society for Industrial and Applied Mathematics (2001)
15. Hartline, J., Mirrokni, V., Sundararajan, M.: Optimal marketing strategies over social networks. In: 17th International Proceedings on World Wide Web, Beijing, China, pp. 189–198. Association for Computing Machinery (2008)

16. Iwata, S., Fleischer, L., Fujishige, S.: A combinatorial strongly polynomial algorithm for minimizing submodular functions. J. Assoc. Comput. Mach. **48**(4), 761–777 (2001)
17. Mansell, R., Collins, B.S.: Introduction: Trust and Crime in Information Societies. Edward Elgar (2005)
18. Melo, M.T., Nickel, S., Saldanha-da-Gama, F.: Facility location and supply chain management - A review. Eur. J. Oper. Res. **196**(2), 401–412 (2009)
19. Mossel, E., Roch, S.: On the submodularity of influence in social networks. In: 30th International Symposium on Theory of Computing, San Diego, California, USA, pp. 128–134. Association for Computing Machinery (2007)
20. Roughgarden, T., Wang, J.: An optimal learning algorithm for online unconstrained submodular maximization. In: 31st International Conference on Learning Theory, Stockholm, Sweden, pp. 1307–1325. Proceedings of Machine Learning Research (2018)
21. Tran, A.K., Piran, M.J., Pham, C.: SDN controller placement in IoT networks: an optimized submodularity-based approach. Sensors **19**(24), 5474 (2019)
22. Tsai, Y.-C., Tseng, K.-S.: Deep compressed sensing for learning submodular functions. Sensors **20**(9), 2591 (2020)
23. Tsiang, S.C.: A note on speculation and income stability. Economica **10**(40), 27–47 (1943)
24. Zhang, H., Vorobeychik, Y.: Submodular optimization with routing constraints. In: 30th International Proceedings on Artificial Intelligence, Phoenix, Arizona, USA, pp. 819–825. AAAI Press (2016)

Differentially Private Submodular Maximization over Integer Lattice

Jiaming Hu[1], Dachuan Xu[2], Donglei Du[3], and Cuixia Miao[4(✉)]

[1] Department of Operations Research and Information Engineering,
Beijing University of Technology, Beijing 100124, People's Republic of China
hujiaming89@emails.bjut.edu.cn
[2] Beijing Institute for Scientific and Engineering Computing, Beijing University
of Technology, Beijing 100124, People's Republic of China
xudc@bjut.edu.cn
[3] Faculty of Management, University of New Brunswick,
Fredericton, NB E3B 5A3, Canada
ddu@unb.ca
[4] School of Mathematical Sciences, Qufu Normal University,
Qufu 273165, People's Republic of China
miaocuixia@126.com

Abstract. Many machine learning problems, such as medical data summarization and social welfare maximization, can be modeled as the problems of maximizing monotone submodular functions. Differentially private submodular functions under cardinality constraints are first proposed and studied to solve the Combinatorial Public Projects (CPP) problem, in order to protect personal data privacy while processing sensitive data. However, the research of these functions for privacy protection has received little attention so far. In this paper, we propose to study the differentially private submodular maximization problem over the integer lattice. Our main contributions are to present differentially private approximation algorithms for both DR-submodular and integer submodular function maximization problems under cardinality constraints and analyze the sensitivity of our algorithms.

Keywords: Integer submodular · Differentially private · DR-submodular

1 Introduction

Submodular set functions defined on a ground set V containing n elements (or equivalently the Boolean lattice $\mathbb{B}^n = \{0,1\}^n$) have been extended to the integer lattice $\mathbb{Z}^n$ (cf. Soma et al. [16]; Soma and Yoshida [19]). Many combinatorial optimization and machine learning problems can be unified under the more general integer lattice $\mathbb{Z}$, including budget allocation with a competitor (Soma et al. [16]), causal structure discovery (Agrawal et al. [1]), sensor placement with different power levels (Soma and Yoshida [17]), pareto optimization problem (Qian et al. [11]) and social welfare maximization, etc.

© Springer Nature Switzerland AG 2021
D. Mohaisen and R. Jin (Eds.): CSoNet 2021, LNCS 13116, pp. 59–67, 2021.
https://doi.org/10.1007/978-3-030-91434-9_6

Submodular function maximization problems on the integer lattice $\mathbb{Z}^n$ are different from the traditional set submodular function on the Boolean lattice $\mathbb{B}^n$ maximization, where we choose each element only once. The objective function is defined over multisets instead of subsets. Note that multiset theory is a generalization of the set theory, whose multiple elements need to be calculated according to the number of times they occur.

It is well-known that submodularity is equivalent to the diminishing return property on the Boolean lattice. However, over the integer lattice, submodularity is characterized of each individual dimension. If the function satisfies submodularity in all dimensions, we call it a DR-submodular function. We note that functions satisfying lattice submodularity do not necessarily satisfy DR-submodularity. The problem of maximizing such DR-submodular functions subject to different constraints has been extensively studied (Soma and Yoshida [18,19]; Bian et al. [2,3]; Chen et al. [5]).

When sensitive information about individuals are involved, such as medical data, sensor placement data and social welfare data, it is of great importance to protect users' privacy. Some of the most compelling use cases for these applications concern sensitive data about individuals. Effective optimization methods are generally needed to ensure the privacy of individuals. Informally, a randomized algorithm is differentially private if changing a single entry in the input database only results in a small distributional change in the outputs. Therefore an adversary cannot information-theoretically infer whether or not a single individual participated in the database.

Given a dataset to record some information from one domain, and if two datasets differ by a single point, then they are *neighboring*. Formally, for $\epsilon, \delta > 0$, a randomized computation A is said to be $(\epsilon, \delta)-$differentially private if for any set of outputs $V \subseteq range(A)$ and neighboring datasets $D \sim D'$,

$$Pr[A(D) \in V] \leq \exp(\epsilon)Pr[A(D') \in V] + \delta.$$

When $\delta = 0$, we say A is $\epsilon-$differentially private. For small changes in the input dataset, differentially private algorithms adjust the sensitivity of the function to ensure the performance of the algorithm.

In this paper, we consider differentially private integer submodular maximization under cardinality constraints. Our main contribution is to provide an algorithm to maximize differentially private DR-submodular function and lattice submodular function subject to cardinality constraint by using the exponential mechanism. The problem we are interested in are described as follows.

Problem 1. Given a sensitive dataset D associated with a monotone integer submodular function $f_D : \mathbb{Z}_+^V \to \mathbb{R}$ and a cardinality parameter r, the objective is to find a multiset $\mathbf{y}$ of carnality no more than r to maximize $f_D(\mathbf{y})$ while guaranteeing differential privacy with respect to the input dataset D.

1.1 Our Contributions

Our main contributions in this paper are listed as follows.

1. To our knowledge, we give the first differentially private algorithm for a submodular maximization problem under cardinality constraint over the integer lattice.
2. We also provide a differentially private algorithm for DR-submodular function over the integer lattice. And we provide the sensitivity proof for our algorithm.
3. We provide a simple proof that the sum of submodular functions on integer lattices satisfies submodular property.

In order to devise polynomial-time algorithms to ensure differential privacy, we adopt the decreasing-threshold greedy framework of the monotone lattice submodular function and combine it with the Diff. Private Greedy algorithm recently introduced by Mitrovic et al. [9], and find a solution in the following way. We start from $\mathbf{y} = \mathbf{0}$, and greedily increase a multiset $\{\mathbf{x}_t\}$ if the expectation of the average gain in the increase is above a threshold in each iteration t. The multiplicity of $\{\mathbf{x}_t\}$ can be obtained by the binary search. While decreasing the threshold, we repeat the greedy inner loop until the algorithm outputs a solution.

1.2 Related Work

Integer Submodular Maximization. For maximizing non-negative monotone submodular functions over the integer lattice, Soma and Yoshida [19] provide polynomial-time $(1 - 1/e - \epsilon)$-approximation algorithms for various constraints. Kuhnle et al. [7] provide a fast threshold greedy algorithm for maximizing monotone non-submodular functions on the integer lattice. Qian et al. [12] consider integer submodular maximization problem with a size constraint and propose a randomized iterative approach POMS. Sahin et al. [14] introduce a probabilistic integer submodular model and present generalized multilinear extensions (GME) for integer submodular functions. For maximizing lattice submodular functions over discrete polymatroid constraints, Sahin et al. [15] use the GME tool to get an $(1 - 1/e - \epsilon)$-approximation guarantee with the monotone condition and $(1/e - \epsilon)$-approximation guarantee for the non-monotone condition. Recently, submodular functions with various constraints on integer lattices have been studied as well, and approximation algorithms for maximizing these functions can be found in [10, 20, 21].

Private Submodular Maximization. Under differential privacy constraints and sensitive data sets, Gupta et al. [6] study a variety of combinatorial optimization problems and give a differentially private algorithm for the combinatorial public project problem (CPP). For submodular functions under certain constraints, Mitrovic et al. [9] provide differentially private greedy algorithms and solve the facility location problem by using a dataset of Uber pickup locations in Manhattan. Chaturvedi et al. [4] provide two private continuous greedy algorithms for maximizing monotone and non-monotone decomposable submodular functions. Recently, Rafiey and Yoshida [13] provide a $(1 - 1/e)-$approximation differentially private algorithm for a monotone k-submodular function maximization problem under matroid constraints and the first 1/2-approximation differentially private algorithm for monotone k-submodular function maximization problem under matroid constraints.

1.3   Organizations

The rest of this paper is organized as follows. Section 2 provides the preliminaries for submodularity and differential privacy. Section 3 presents the main differentially private algorithms for DR-submodular and integer submodular function maximization problems subject to cardinality constraints. Section 4 provides the concluding remarks for our work.

2   Preliminaries

The Problem. We denote the sets of integers and non-negative integers by $\mathbb{Z}$ and $\mathbb{Z}_+$, and the sets of reals and non-negative reals by $\mathbb{R}$ and $\mathbb{R}_+$, respectively. Given a finite set V which we refer to as the *groundset* and a finite set X which we refer to as the *data universe*. A dataset is an n-tuple $D = \{1, \ldots, n\} \in X^n$. Assume that each dataset D is associated with a function $f_D : \mathbb{Z}_+^V \to \mathbb{R}$ over the integer lattice $\mathbb{Z}^V$, or equivalently over $\mathcal{M}$, the multiset of V. For a multiset $\{\mathbf{x}\}$, we denote the multiplicity of one element v by $x(v)$, and define $|\{\mathbf{x}\}| := x(V)$.

Definition 1. *(Submodular function (Soma et al. 2014)) A function $f_D : \mathbb{Z}^V \to \mathbb{R}$ is* submodular *if for any $\mathbf{x}, \mathbf{y} \in \mathbb{Z}_+^V$,*

$$f_D(\mathbf{x}) + f_D(\mathbf{y}) \geq f_D(\mathbf{x} \wedge \mathbf{y}) + f_D(\mathbf{x} \vee \mathbf{y}).$$

Definition 2. *(DR-Submodular function (Soma and Yoshida 2016)) A function $f_D : \mathbb{Z}^V \to \mathbb{R}$ is* DR-submodular *if for any $\mathbf{x} \leq \mathbf{y}$ and $i \in [m]$,*

$$f_D(\mathbf{x} + \chi_i) - f_D(\mathbf{x}) \geq f_D(\mathbf{y} + \chi_i) - f_D(\mathbf{y}).$$

In the above, $\mathbf{x} \wedge \mathbf{y}$ and $\mathbf{x} \vee \mathbf{y}$ denote the coordinate-wise minimum and maximum, i.e., $\mathbf{x} \wedge \mathbf{y} = (\min\{\mathbf{x}_1, \mathbf{y}_1\}, \ldots, \min\{\mathbf{x}_m, \mathbf{y}_m\})$ and $\mathbf{x} \vee \mathbf{y} = (\max\{\mathbf{x}_1, \mathbf{y}_1\}, \ldots, \max\{\mathbf{x}_m, \mathbf{y}_m\})$. Let χ_i donate the i-th unit vector, i.e., the i-th entry of χ_i is 1 and others are 0. We donate the all-zeros and all-ones vectors by $\mathbf{0}$ and $\mathbf{1}$.

A function $f_D : \mathbb{Z}^V \to \mathbb{R}$ is monotone if for any $\mathbf{x} \leq \mathbf{y}$, $f_D(\mathbf{x}) \leq f_D(\mathbf{y})$. Assume that monotone functions are normalized, i.e., $f_D(\mathbf{0}) = 0$. Note that a function $f_D : \mathbb{Z}^V \to \mathbb{R}$ is diminishing return submodular (DR-submodularity) implies that it must be submodular, but not vice versa.

Differential Privacy. In our setting, a dataset D consists of private integer submodular functions $f_1, \ldots, f_n : \mathbb{Z}^V \to \mathbb{R}$. For given neighboring datasets D and D', differentially private algorithms adjust the sensitivity of the function to ensure the performance of the algorithm, defined formally as follows.

Definition 3. *For a given dataset D, the sensitivity of a function $f_D : V \to U$ is defined as*

$$\max_{D' : D' \sim D} \max_{v \in V} |f_D(V) - f_{D'}(V)|.$$

If a function has sensitivity σ, it is called $\sigma-$ sensitive.

2.1 Composition of Differential Privacy

The composition theorem of differential privacy is used to analyze the approximation ratio performance of our algorithms. Let $\{(\epsilon_t, \delta_t)\}_{t=1}^T$ be an ordered set of privacy parameters and A^* be a mechanism that behaves as follows on an input D. In each iteration t, the algorithm A^* adopts an $(\epsilon_t, \delta_t)-$differentially private algorithm A_t depending upon the preceding outputs $A_1(D), \ldots, A_{t-1}(D)$ and obtains $A_t(D)$.

2.2 Exponential Mechanism

The key of our framework is the *exponential mechanism* of McSherry and Talwar [8]. Given a dataset D and a family of candidate result $\mathcal{R}$, we can define the exponential mechanism to a *quality function* $q_D : \mathcal{R} \to \mathbb{R}$.

Definition 4. *(McSherry and Talwar (2007)). For $\epsilon, \sigma > 0$, let $q_D : \mathcal{R} \to \mathbb{R}$ be a quality score function. Then, the exponential mechanism $EM(\epsilon, \sigma, q_D)$ outputs $R \in \mathcal{R}$ with probability proportional to $\exp\left(\frac{\epsilon}{2\sigma} \cdot q_D(R)\right)$.*

3 Algorithm

In this section, we provide a variant of decreasing-threshold greedy algorithm which enable monotone submodular function maximization problems over the integer lattice. Mitrovic et al. (2017) provide a differentially private greedy algorithm for the class of low-sensitivity monotone submodular functions given in Algorithm 1.

Algorithm 1: Diff. Private Greedy (Cardinality)

Input: Score function $f_D : 2_+^V \to \mathbb{R}_+$, sensitive dataset D, cardinality constraint k, and the value of privacy parameters, ϵ_0, δ_0.
Output: Size k subset of V
Initialize $S_0 = \emptyset$;
for $i = 1, \ldots, k$ **do**
    Define $q_i : (V \backslash S_{i-1}) \times X^n \to \mathbb{R}$ via $q_i(v, \tilde{D}) = f_{\tilde{D}}(S_{i-1} \cup \{v\}) - f_{\tilde{D}}(S_{i-1})$
    Compute $v_i \leftarrow_R \mathcal{O}(q_i; D; \epsilon_0; \delta_0)$
    Update $S_i \leftarrow (S_{i-1} \cup \{v_i\})$
end
Return S_k

3.1 Diff. Private DR-Submodular Algorithm

We consider a Diff. Private DR-Submodular algorithm for DR-submodular function with differential privacy. In our setting, we choose one dimension with a differential privacy parameter. Firstly, we find the maximum marginal gain of a single point $v \in V$. In each iteration, we select the direction of iteration by the exponential mechanism, and then use binary search to determine the step size of the greedy phase. As the algorithm iteration threshold keeps dropping, the final algorithm outputs solution $\mathbf{y}$. It means that the expectation of the average marginal benefit obtained by our algorithm can largely guarantee that the approximate optimal solution can be obtained while protecting privacy.

Algorithm 2: Diff. Private DR-Submodular

Input: Submodular function $f_D : \mathbb{Z}_+^V \to \mathbb{R}_+$, dataset D, $\mathbf{c} \in \mathbb{Z}_+^V$, $r \in \mathbb{Z}_+$, and the value of privacy parameters, ϵ_0, δ_0, $\rho > 0$.

Output: $\mathbf{y} \leq \mathbf{c}$

Initialize $r_0 = 0$, $\mathbf{y} = \mathbf{0}$ and $d \leftarrow \max_{v \in V} f(\chi_v)$;

for $(\theta = d; \theta \geq \frac{\rho}{r}d; \theta \leftarrow \theta(1 - \rho))$ **do**

  for $t = 0$ to T **do**

    Define $q_t : (V \backslash V_{t-1}) \times X^n \to \mathbb{R}$ via $q(\chi_v, \tilde{D}) = f_{\tilde{D}}(\mathbf{y} + \{\chi_v\}) - f_{\tilde{D}}(\mathbf{y})$.

    Compute $\chi_v \leftarrow_R \mathcal{O}(q_i, D; \epsilon_0, \delta_0)$.

    Compute the maximum integer $k \leq \min\{c(v) - y(v), r - y(V)\}$ with $\frac{q(k\chi_v, \tilde{D})}{k}$ by binary search.

    if *such k exists* **then**

      | $y_t \leftarrow y_{t-1} + k\chi_v, r_t \leftarrow r_{t-1} + k$

    end

  end

end

Return $\mathbf{y}_T$

Theorem 1. *Suppose DR-submodular function $f_D : \mathbb{Z}_+^V \to \mathbb{R}_+$ is monotone and has sensitivity σ over the integer lattice. Then Alg. 2 with $\mathcal{O} = EM$ and parameter $\epsilon' > 0$ provides $(\epsilon = r\epsilon', \delta = 0)$−differential privacy. It also provides (ϵ, δ)−differential privacy for every $\delta > 0$ with ϵ. Moreover, for every $D \in X^n$,*

$$\mathbb{E}[f_D(\mathbf{y})] \geq \left(1 - \frac{1}{e} - \rho\right) OPT - \frac{2r\sigma \ln |V|}{\epsilon}.$$

where $\mathbf{y} \leftarrow_R \mathcal{G}(D)$. Moreover, Alg. 2 evaluates f_D at most $O(\frac{n}{\rho} \log ||\mathbf{c}||_\infty \log \frac{r}{\rho})$ times.

We notice that integer submodular functions are different from set submodular functions, because they add k_t unit vectors χ of the same dimension at the same time in each iteration t. Hence we need to consider the average marginal gain of each unit vector. We can prove that Algorithm 2 is ϵ−differentially private by the basic composition theorem.

Theorem 2. *Algorithm 2 preserves $O(\epsilon k^2)$−differential privacy.*

3.2   Diff. Private Integer Submodular Algorithm

Our algorithm for maximizing a differentially private integer submodular function is based on the decreasing threshold greedy framework and the exponential mechanism and uses subroutine $BinarySearchLattice$ to find points whose expectation of the average marginal gain is similar to that of the current threshold. The proof of theorem 2 is similar to theorem 1, but due to the weak submodularity, we need to find the gap between the optimal solution and the actual solution in the iterative process.

Algorithm 3: Diff. Private Integer Submodular

Input: Submodular function $f_D : \mathbb{Z}_+^V \to \mathbb{R}_+$, dataset D, $\mathbf{c} \in \mathbb{Z}_+^V$, $r \in \mathbb{Z}_+$, and the value of privacy parameters, ϵ_0, δ_0, $\rho > 0$.

Output: $\mathbf{y} \leq \mathbf{c}$

Initialize $r_0 = 0$ $\mathbf{y} = \mathbf{0}$ and $d \leftarrow \max_{v \in V} f(c(v)\chi_v)$;

for $(\theta = d; \theta \geq \frac{\rho}{r} d; \theta \leftarrow \theta(1 - \rho))$ **do**

    for $t = 0$ *to* T **do**

        Define $q_t : (V \backslash V_{t-1}) \times X^n \to \mathbb{R}$ via $q(\chi_v, \tilde{D}) = f_{\tilde{D}}(\mathbf{y} \vee \{\chi_v\}) - f_{\tilde{D}}(\mathbf{y})$.

        Compute $\chi_v \leftarrow_R \mathcal{O}(q_t, D; \epsilon_0, \delta_0)$.

        Compute integer k by invoking

        BinarySearchLattice($q(k\chi_v, \tilde{D})$, v, θ, $\min\{c(v) - y(v), r - y(V)\}$, ρ) .

        if *BinarySearchLattice returned* $k \in \mathbb{N}$ **then**

            | $\mathbf{y} \leftarrow \mathbf{y} \vee k\chi_v$, $r_t \leftarrow r_{t-1} + k$

        end

    end

end

Return $\mathbf{y}$

Theorem 3. *Suppose the integer submodular function $f_D : \mathbb{Z}_+^E \to \mathbb{R}_+$ is monotone and has sensitivity σ. Then Alg. 4 with $\mathcal{O} = EM$ and parameter $\epsilon' > 0$ provides $(\epsilon = r\epsilon', \delta = 0)-$differential privacy. It also provides $(\epsilon, \delta)-$differential privacy for every $\delta > 0$ with ϵ. Moreover, for every $D \in X^n$,*

$$\mathbb{E}[f_D(\mathbf{y})] \geq \left(1 - \frac{1}{e} - O(\rho)\right) OPT - \frac{2r\sigma \ln |V|}{\epsilon}.$$

where $\mathbf{y} \leftarrow_R \mathcal{G}(D)$.

4   Conclusion

In this paper, we consider the problem of maximizing integer submodular functions under cardinality constraints and differential privacy. We provide differentially private algorithms for DR-submodular and integer submodular function maximization problems, respectively. One future direction of research is to consider an integer submodular function under polymatroid constraint and differential privacy.

Acknowledgements. The first two authors are supported by National Natural Science Foundation of China (No. 12131003) and Beijing Natural Science Foundation Project No. Z200002. The third author is supported by National Natural Sciences and Engineering Research Council of Canada (NSERC) grant 06446, and National Natural Science Foundation of China (Nos. 11771386, 11728104). The fourth author is supported by the Province Natural Science Foundation of Shandong (No. ZR2017MA031) and the National Natural Science Foundation of China (No. 11801310).

References

1. Agrawal, R., Squires, C., Yang, K., Shanmugam, K., Uhler, C.: Abcd-strategy: budgeted experimental design for targeted causal structure discovery. In: Proceedings of the 22nd International Conference on Artificial Intelligence and Statistics, pp. 3400–3409 (2019)
2. Bian, Y., Buhmann, J., Krause, A.: Optimal continuous DR-submodular maximization and applications to provable mean field inference. In: Proceedings of the 36th International Conference on Machine Learning, pp. 644–653 (2020)
3. Bian, A., Levy, K., Krause, A., Buhmann, J. M.: Non-monotone continuous DR-submodular maximization: Structure and algorithms. In: Proceedings of the 30th Annual Conference on Neural Information Processing Systems, pp. 486–496 (2017)
4. Chaturvedi, A., Nguyễn, H. L., Zakynthinou, L.: Differentially private decomposable submodular maximization. In: Proceedings of the 35th AAAI Conference on Artificial Intelligence, pp. 6984–6992 (2021)
5. Chen, L., Hassani, H., Karbasi, A.: Online continuous submodular maximization. In: Proceedings of the 21st International Conference on Artificial Intelligence and Statistics, pp. 1896–1905 (2018)
6. Gupta, A., Ligett, K., McSherry, F., Roth, A., Talwar, K.: Differentially private combinatorial optimization. In: Proceedings of the 21st Annual ACM-SIAM Symposium on Discrete Algorithms, pp. 1106–1125 (2010)
7. Krause, A., Smith, D., Crawford, V.G., Thai, M.T.: Fast maximization of non-submodular, monotonic functions on integer lattice. In: Proceedings of the 35th International Conference on Machine Learning, pp. 2791–2800 (2018)
8. McSherry, F., Talwar, K.: Mechanism design via differential privacy. In: Proceedings of the 48th Annual IEEE Symposium on Foundations of Computer Science, pp. 94–103 (2007)
9. Mitrovic, M., Bun, M., Krause, A., Karbasi, A.: Differentially private submodular maximization: Data summarization in disguise. In: Proceedings of the 34th International Conference on Machine Learning, pp. 2478–2487 (2017)
10. Nong, Q., Fang, J., Gong, S., Du, D., Feng, Y., Qu, X.: A 1/2-approximation algorithm for maximizing a non-monotone weak-submodular function on a bounded integer lattice. J. Comb. Optim. **39**(4), 1208–1220 (2020). https://doi.org/10.1007/s10878-020-00558-4
11. Qian, C., Yu, Y., Zhou, Z.H.: Subset selection by pareto optimization. In: Proceedings of the 28th Annual Conference on Neural Information Processing Systems, pp. 1765–1773 (2015)
12. Qian, C., Zhang, Y., Tang, K., Yao, X.: On multiset selection with size constraints. In: Proceedings of the 32nd AAAI Conference on Artificial Intelligence, pp. 1395–1402 (2018)

13. Rafiey, A., Yoshida, Y.: Fast and private submodular and k-submodular functions maximization with matroid constraints. In: Proceedings of the 37th International Conference on Machine Learning, pp. 7887–7897 (2020)

14. Sahin, A., Buhmann, J.M., Krause, A.: Constrained maximization of lattice submodular functions. IN: ICML 2020 workshop on Negative Dependence and Submodularity for ML, Vienna, Austria, PMLR 119 (2020)

15. Sahin, A., Bian, Y., Buhmann, J.M., Krause, A.: From sets to multisets: provable variational inference for probabilistic integer submodular models. In: Proceedings of the 37th International Conference on Machine Learning, pp. 8388–8397 (2020)

16. Soma, T., Kakimura, N., Inaba, K., Kawarabayashi, K.: Optimal budget allocation: theoretical guarantee and efficient algorithm. In: Proceedings of the 31st International Conference on Machine Learning, pp. 351–359 (2014)

17. Soma, T., Yoshida, Y.: A generalization of submodular cover via the diminishing return property on the integer lattice. In: Proceedings of Advances in Neural Information Processing Systems, pp. 847–855 (2015)

18. Soma, T., Yoshida, Y.: Non-monotone DR-submodular function maximization. In: Proceedings of the 31st AAAI conference on Artificial Intelligence, pp. 898–904 (2017)

19. Soma, T., Yoshida, Y.: Maximizing monotone submodular functions over the integer lattice. Math. Program. **172**(4), 539–563 (2018). https://doi.org/10.1007/s10107-018-1324-y

20. Tan, J., Zhang, D., Zhang, H., Zhang, Z.: Streaming algorithms for monotone DR-submodular maximization under a knapsack constraint on the integer lattice. In: Ning, L., Chau, V., Lau, F. (eds.) PAAP 2020. CCIS, vol. 1362, pp. 58–67. Springer, Singapore (2021). https://doi.org/10.1007/978-981-16-0010-4_6

21. Zhang, Z., Guo, L., Wang, L., Zou, J.: A streaming model for monotone lattice submodular maximization with a cardinality constraint. In: Proceedings of the 21st International Conference on Parallel and Distributed Computing, Applications and Technlolgies, pp. 362–370 (2020)

Maximizing the Sum of a Supermodular Function and a Monotone DR-submodular Function Subject to a Knapsack Constraint on the Integer Lattice

Jingjing Tan[1], Yicheng Xu[2,3(✉)], Dongmei Zhang[4], and Xiaoqing Zhang[5]

[1] School of Mathematics and Information Science, Weifang University, Weifang 261061, People's Republic of China
[2] Shenzhen Institute of Advanced Technology, Chinese Academy of Sciences, Shenzhen 518055, People's Republic of China
yc.xu@siat.ac.cn
[3] Guangxi Key Laboratory of Cryptography and Information Security, Guilin 541004, People's Republic of China
[4] School of Computer Science and Technology, Shandong Jianzhu University, Jinan 250101, People's Republic of China
zhangdongmei@sdjzu.edu.cn
[5] Department of Operations Research and Information Engineering, Beijing University of Technology, Beijing 100124, People's Republic of China

Abstract. Submodular maximization is of great significance as it has included many classical combinatorial problems. In this paper, we consider the maximization of the sum of a supermodular function and a monotone DR-submodular function on the integer lattice. As our main contribution, we present a streaming algorithm under the assumption that the optimum is known, and a two-pass streaming algorithm in general case. The proposed algorithms are proved to have polynomial time and space complexity, and a performance guarantee dependent on the curvature of the supermodular function.

Keywords: Submodular maximization · Supermodular · DR-submodular · Knapsack constraint · Integer lattice

1 Introduction

Submodular maximization problem has a widespread application in practice and thus attracts a lot of research interest in the past decades, such as [1,4,5,13,14]. Many theoretically beautiful techniques like greedy and local search has been investigated to solve this problem [7–9,11,12,16].

Cardinality and knapsack constraint are two natural constraints in the submodular maximization literature. Badanidiyuru et al. [2] propose a so-called

© Springer Nature Switzerland AG 2021
D. Mohaisen and R. Jin (Eds.): CSoNet 2021, LNCS 13116, pp. 68–75, 2021.
https://doi.org/10.1007/978-3-030-91434-9_7

Sieve-Streaming algorithm that achieves a $(1/2 - \varepsilon)$-approximation for the submodular maximization subject to a cardinality constraint with $O(K \log K)$ memory. Buchbinder et al. [3] propose a 1/4-approximation for the same problem, with improved memory of $O(K)$. Norouzi-Fard et al. [17] prove that unless P = NP, one cannot obtain an approximation ratio better than 1/2 with $O(n/K)$ memory. For the submodular maximization subject to a knapsack constraint, the first approximation result is a 0.35-approximation Wolsey [20]. Yu et al. [24] propose a one pass streaming algorithm with approximation ratio $(1/3 - \varepsilon)$. Huang et al. [13] then improve this result to $(0.4 - \varepsilon)$ for the problem.

In many cases, we also in face of the maximization of the sum of, say, the submodular and supermodular functions. Sviridenko et al. [18] give a $(1 - e - c)$-approximation algorithm, where c is the curvature, for maximization the sum of a submodular and supermodular function subject to a cardinality constraint. In addition to the above, there are many impressive results in the literature ([3,6, 10,15,21–23]). Moreover, submodular maximization problem over a multiset or integer lattices is introduced. Soma et al. [19] give a polynomial-time streaming algorithm for this problem subject to a cardinality constraint, a polymatroid constraint and a knapsack constraint.

2 Preliminaries

Let $E = \{e_1, e_2, \cdots, e_n\}$ be the ground set and $k \in \mathbb{N}$ is a positive integer. $[k]$ denotes the set of all integers from 1 to k. Given $\boldsymbol{x}$, an n-dimensional vector in $\mathbb{N}^E$, let $\boldsymbol{x}(e)$ be the e_{th} component of $\boldsymbol{x}$. $\boldsymbol{0}$ is the zero vector and χ_{e_i} is the standard unit vector. Denote $\boldsymbol{x}(X) := \sum_{e_i \in X} \boldsymbol{x}(e_i)$. The support of $\boldsymbol{x}$ is $supp^+(\boldsymbol{x}) = \{e \in E | \boldsymbol{x}(e) > 0\}$. A multi-set of $\boldsymbol{x}$ is a set where each element e_i can appears at most $\boldsymbol{x}(e_i)$ times. Denote $|\{\boldsymbol{x}\}| := \boldsymbol{x}(E)$.

For arbitrary multi-sets $\{\boldsymbol{x}\}$ and $\{\boldsymbol{y}\}$, define $\{\boldsymbol{x}\} \setminus \{\boldsymbol{y}\} := \{(\boldsymbol{x} \setminus \boldsymbol{y}) \vee \boldsymbol{0}\}$. We call a function $f(\mathbb{N}^E \rightarrow \mathbb{R}_+)$ monotone if $f(\boldsymbol{x}) \leq f(\boldsymbol{y})$ for any $\boldsymbol{x} \leq \boldsymbol{y}$. We call a function f nonnegative if $f(\boldsymbol{x}) \geq 0$ for all $\boldsymbol{x} \in \mathbb{N}^E$. We call f normalized if $f(\boldsymbol{0}) = 0$.

Definition 1. *We call a function f DR-submodular if it holds for any $e \in E, \boldsymbol{x}, \boldsymbol{y} \in \mathbb{N}^E$ with $\boldsymbol{x} \leq \boldsymbol{y}$ that*

$$f(\boldsymbol{y} + \chi_e) - f(\boldsymbol{y}) \leq f(\boldsymbol{x} + \chi_e) - f(\boldsymbol{x}).$$

Similarly, we give the definition of supermodular function and lattice supermodular over the integer lattice.

Definition 2. *For any $e \in E$, we call a function g supermodular if it holds that*

$$g(\boldsymbol{x} + \chi_e) - g(\boldsymbol{x}) \leq g(\boldsymbol{y} + \chi_e) - g(\boldsymbol{y})$$

where $\boldsymbol{x}, \boldsymbol{y} \in \mathbb{N}^E$ with $\boldsymbol{x} \leq \boldsymbol{y}$.

Throughout this paper, the set consisting of all monotone non-negative DR-submodular functions is denoted as $\mathcal{F}_b$, where $f(\mathbf{0}) = 0$. Also, the set of all increasing non-negative supermodular functions is denoted as $\mathcal{G}_b$, where $g(\mathbf{0}) = 0$. The domain of f and g is $\{\mathbf{x} \in \mathbb{N}^E : \mathbf{x} \leq \mathbf{b}\}$. For $f \in \mathcal{F}_b$, we denote the marginal increasement of $\mathbf{x}$ as $f(\mathbf{x}|\mathbf{y}) = f(\mathbf{x} + \mathbf{y}) - f(\mathbf{y})$, where $\mathbf{x}, \mathbf{y} \in \mathbb{N}^E$. Herein, we aim to maximize the sum of $f \in \mathcal{F}_b$ and $g \in \mathcal{G}_b$ subject to a knapsack constraint. Denote $h(\mathbf{x}) = f(\mathbf{x}) + g(\mathbf{x})$. The studied problem can be described as

$$\text{maximize } h(\mathbf{x}) \text{ subject to } \mathbf{c}^T \mathbf{x} \leq K, \tag{1}$$

where $K \in \mathbb{N}$ is the total budget, and $\mathbf{c}$ is the weight function.

Let OPT and $\mathbf{x}^*$ be the optimal value and the optimal solution of the problem 1, respectively. On the integer lattice, we give the definition of the curvature as follows.

Definition 3. *For any* $e \in E$, *if* $f(\chi_e) \neq 0$,

$$\alpha_f = 1 - min_{e \in E} \frac{f(\mathbf{b}) - f(\mathbf{b} - \chi_e)}{f(\chi_e)}$$

is defined as the curvature of a non-negative non-decreasing DR-submodular function on the integer lattices.

Definition 4. *For any* $e \in E$,

$$c^g = \alpha_{g(b)-g(b-x)} = 1 - min_{e \in E} \frac{g(\chi_e)}{g(\mathbf{b}) - g(\mathbf{b} - \chi_e)}$$

is the curvature of a non-negative supermodular function on the integer lattices.

Let $h(\mathbf{x}) = f(\mathbf{x}) + g(\mathbf{x})$, we conclude the follows.

Remark 1. *If* $f \in \mathcal{F}_b$, $f \in \mathcal{G}_b$, *then for any* $\mathbf{x} \leq \mathbf{y}$ *and* $\mathbf{y} + \chi_e \leq \mathbf{b}$, *we have*

$$(1 - c^g)[h(\mathbf{y} + \chi_e) - h(\mathbf{y})] \leq h(\mathbf{x} + \chi_e) - h(\mathbf{x}).$$

3   The Streaming Algorithms

We give two streaming algorithms for maximizing the sum of a DR-submodular function and a supermodular function.

3.1   Algorithms with Known Optimum

Suppose that the optimal value of the problem is known. The idea of Algorithm 1 is that, we calculate a threshold value $l \in [b(e_i)]$ to help us to make decision on whether to keep current arrival element or not. Towards this end, we design Algorithm 2, a binary search for the value of l. Algorithm 1 and 2 are presented as follows.

Algorithm 1

Input: $f \in \mathcal{F}_b$, $g \in \mathcal{G}_b$, weight function c, E, $\gamma \in (0,1]$, $v : \gamma OPT \leq v \leq OPT$.

Output: a vector $x \in \mathbb{N}^E$.

1: $h \leftarrow f + g$, $x \leftarrow 0$;

2: **for** $i \in [1, n]$, **do**

3:    **if** $c^T x$ is less than K, **then**

4:        $l \leftarrow$ **Half Search** $\left(h, \frac{v - h(x)}{2K - c^T x}\right)$;

5:        **if** $c^T(x + l\chi_{e_i})$ satisfies the constraint condition, **then**

6:            update x to $x + l\chi_{e_i}$;

7:        **end if**

8:    **end if**

9: **end for**

10: **return** x

Lemma 1. *We have for the i_{th} of Algorithm 1 that*

$$h(x_{i-1} + l_i \chi_{e_i}) \geq \frac{vc^T x_i}{2K}. \tag{2}$$

Let x^* be the optimal solution for the same instance the algorithms are running on.

Lemma 2. *If $c^T \tilde{x} < K$ holds for any $e \in \{x^*\} \setminus \{\tilde{x}\}$, then it must be that $h(\chi_e | \tilde{x}) < \frac{vc(e)}{2(1 - c^g)K}$.*

Definition 5. *We call $e \in \{x^*\}$ a bad item if it satisfies the threshold condition or exceeds knapsack constraint when we add it to the current solution, i.e., $h(l_e \chi_e | x) \geq \frac{v - h(x)}{2K - c^T x}$, $c^T(x + l_e \chi_e) > K$ but $c^T x \leq K$.*

Lemma 3. *If there is no bad item and $v \leq h(x^*)$ holds, it must be that*

$$h(\tilde{x}) \geq (1 - \frac{1}{1 - c^g})v.$$

Denote $\hat{x}$ be the output of Algorithm 3. Combine with Algorithm 1, we can obtain the following conclusion.

Theorem 1. *Running Algorithm 1 and Algorithm 3 in parallel. We conclude that the proposed algorithm is a $\min\left\{(1 - \frac{1}{2(1 - c^g)})\gamma, \frac{(1 - c^g)\gamma}{4}\right\}$-approximation with $O(K)$ space complexity and $O(\log K)$ time complexity.*

Algorithm 2. Half Search(h, τ)

Input: $h : \mathbb{N}^E \to \mathbb{R}_+$, $e \in E$, weight function c, $x \in \mathbb{N}^E$, and $\tau \in \mathbb{R}_+$.
Output: $l_s \in \mathbb{R}^+$.

1: $l_s \leftarrow 1, l_t \leftarrow \min \left\{ b(e), \left\lfloor \frac{k - c^T x}{c(e)} \right\rfloor \right\}$;

2: **if** $\frac{h(l_t \chi_e | x)}{l_t c(e)}$ is greater than or equal to τ, **then**

3:    return l_t

4: **end if**

5: **if** $\frac{h(\chi_e | x)}{c(e)}$ is less than τ, **then**

6:    return 0.

7: **end if**

8: **while** $l_s < l_t + 1, a \leftarrow \lfloor \frac{l_s + l_t}{2} \rfloor$ **do**

9:    **if** $\frac{h(a \chi_e | x)}{a c(e)}$ is greater than or equal to τ **then**

10:       $l_s = a,$

11:    **else**

12:       $l_t = a,$

13:    **end if**

14: **end while**

15: **return** l_s

Algorithm 3

Input: stream of data E, $e \in E$, $h : \mathbb{N}^E \to \mathbb{R}_+$, b, $x \in \mathbb{N}^E$.

1: $x \leftarrow 0$;

2: **while** item e is arrive, **do**

3:    **if** $h(b(e)\chi_e) > h(x)$, **then**

4:       $x \leftarrow b(e)\chi_e$;

5:    **end if**

6: **end while**

7: **return** x

3.2   Algorithms with Unknown Optimum

Obviously, a shortcoming of Algorithm 1 is the assumption of known OPT. However, it is not always the case. A modified algorithm for the case that the OPT is unknown is presented in Algorithm 4 and we conclude the follows.

Algorithm 4

Input: h, weight function c, E, $\varepsilon > 0$.

Output: a vector $\boldsymbol{x} \in \mathbb{N}^E$.

1: $m \leftarrow \max\limits_{e \in E} h(\chi_e)$;

2: guess set $\mathcal{I} = \{(1+\varepsilon)^s | \frac{m}{1+\varepsilon} \leq (1+\varepsilon)^s \leq \frac{Km}{1-c^g}\}$, for each $v \in \mathcal{I}_\varepsilon$, set $\boldsymbol{x}^v \leftarrow \boldsymbol{0}$;

3: **for** $i \in [1, n]$ **do**

4:     **if** $c^T \boldsymbol{x}$ is less than K, **then**

5:         find l by **Half Search** $\left(h, \frac{v - h(\boldsymbol{x})}{2K - cT\boldsymbol{x}} \right)$;

6:         **if** $c^T (\boldsymbol{x} + l\chi_{e_i})$ is less than or equal to K, **then**

7:             update $\boldsymbol{x}$ to $\boldsymbol{x} + l\chi_{e_i}$;

8:         **end if**

9:     **end if**

10: **end for**

11: **return** $\boldsymbol{x}$

Lemma 4. *Let m be the maximum of the unit vector, there is a v within set $\mathcal{I}$ such that $v \leq OPT \leq (1+\varepsilon)v$.*

Theorem 2. *The Algorithm 4 is a two pass* $\min\left\{ 1 - \frac{1}{2(1-c^g)} - \varepsilon, \frac{(1-c^g)}{4} - \varepsilon \right\}$-*approximation with $O(K \log K/\varepsilon)$ space complexity and $O(\log^2 K/\varepsilon)$ time complexity.*

4   Conclusions

In this work, we give two streaming algorithms for maximizing a DR-submodular function plus a supermodular function with a knapsack constraint on integer lattices. Assume that the optimum is known, we prove that the proposed algorithm based on the above threshold inequality rule a $\min\left\{ (1 - \frac{1}{2(1-c^g)})\gamma, \frac{(1-c^g)\gamma}{4} \right\}$-approximation algorithm. Moreover, in the general case that the optimum is unknown, we obtain a $\min\left\{ 1 - \frac{1}{2(1-c^g)} - \varepsilon, \frac{(1-c^g)}{4} - \varepsilon \right\}$-approximation streaming algorithm.

Acknowledgememts. Jingjing Tan is supported by Natural Science Foundation of Shandong Province (Nos. ZR2017LA002, ZR2019MA022) and Doctoral Research Foundation of Weifang University (No. 2017BS02). Yicheng Xu is supported by Guangxi Key Laboratory of Cryptography and Information Security (No. GCIS202116). Dongmei Zhang is supported by National Natural Science Foundation of China (No. 11871081).

References

1. Bai, W., Bilmes, J.A.: Greed is still good: maximizing monotone submodular+supermodular functions. In: Proceedings of ICML, pp. 1–10 (2018)
2. Badanidiyuru, A., Mirzasoleiman, B., Karbasi, A., Krause, A.: Streaming submodular maximization: massive data summarization on the fly. In: Proceedings of KDD, pp. 671–680 (2014)
3. Buchbinder, N., Feldman, M., Schwartz, R.: Onling submodular maximization with preemption. In: Proceedings of SODA, pp. 1202–1216 (2015)
4. Balkanski, E., Rubinstein, A., Singer, Y.: An exponential speedup in parallel running time for submodular maximization without loss in approximation. In: Proceedings of SODA, pp. 283–302 (2019)
5. Calinescu, G., Chekuri, C., Pal, M., Vondrak, J.: Maximizing a momotone submodular function subject to a matroid constraint. SIAM J. Comput. **40**, 1740–1766 (2011)
6. Chakrabarti, A., Kale, S.: Submodular maximization meets streaming: matchings, matroids, and more. Math. Program. **154**, 225–247 (2015). https://doi.org/10.1007/s10107-015-0900-7
7. Chekuri, C., Quanrud, K.: Submodular function maximization in parallel via the multilinear relaxation. In: Proceedings of SODA, pp. 303–322 (2019)
8. Das, A., Kempe, D.: Algorithms for subset selection in linear regression. In: Proceedings of STC, pp. 45–54 (2008)
9. Das, A., Kempe, D.: Submodular meets spectral: greedy algorithms for subset selection, sparse approximation and dictionary selection. In: Proceedings of ICML, pp. 1057–1064 (2011)
10. EI-Arini, K., Guestrin, C.: Beyond keyword search: discovering relevant scientific literature. In: Proceedings of ICKDDM, pp. 439–447 (2011)
11. Ene, A., Nguyen, H.L.: Submodular maximization with nearly-optimal approximation and adaptivity in nearly-linear time. In: Proceedings of SODA, pp. 274–282 (2019)
12. Gong, S., Nong, Q., Liu, W., Fang, Q.: Parametric monotone function maximization with matroid constraints. J. Glob. Optim. **75**(3), 833–849 (2019). https://doi.org/10.1007/s10898-019-00800-2
13. Huang, C., Kakimura, N.: Improved streaming algorithms for maximising monotone submodular functions under a knapsack constraint. In: Proceedings of WADS, pp. 438–451 (2019)
14. Ji, S., Xu, D., Li, M., Wang, Y., Zhang, D.: Stochastic greedy algorithm is still good: maximizing submodular + supermodular functions. In: Le Thi, H.A., Le, H.M., Pham Dinh, T. (eds.) WCGO 2019. AISC, vol. 991, pp. 488–497. Springer, Cham (2020). https://doi.org/10.1007/978-3-030-21803-4_49
15. Jiang, Y.-J., Wang, Y.-S., Xu, D.-C., Yang, R.-Q., Zhang, Y.: Streaming algorithm for maximizing a monotone non-submodular function under d-knapsack constraint. Optim. Lett. **14**, 1235–1248 (2020). https://doi.org/10.1007/s11590-019-01430-z
16. Khanna, R., Elenberg, E.-R., Dimakis, A.-G., Negahban S., Ghosh J.: Scalable greedy feature selection via weak submodularity. In: Proceedings of ICAIS, pp. 1560–1568 (2017)
17. Norouzi-Fard, A., Tarnawski, J., Mitrovic, S., Zandieh, A., Mousavifar, A., Svensson, O.: Beyond 1/2-approximation for submodular maximization on massive data streams. In: Proceedings of ICML, pp. 3829–3838 (2018)

18. Sviridenko, M., Vondrak, J., Ward, J.: Optimal approximation for submodular and supermodular optimization with bounded curvature. In: Proceedings of SODA, pp. 1134–1148 (2015)
19. Soma, T., Yoshida, Y.: Maximization monotone submodular functions over the integer lattice. Math. Program. **172**, 539–563 (2018). https://doi.org/10.1007/s10107-018-1324-y
20. Wolsey, L.: Maximising real-valued submodular set function: primal and dual heuristics for location problems. Math. Oper. Res. **7**, 410–425 (1982)
21. Wang, Y.-J., Xu, D.-C., Wang, Y.-S., Zhang, D.-M.: Non-submodular maximization on massive data streams. J. Glob. Optim. **76**, 729–743 (2020). https://doi.org/10.1007/s10898-019-00840-8
22. Yang, R., Xu, D., Li, M., Xu, Y.: Thresholding methods for streaming submodular maximization with a cardinality constraint and its variants. In: Du, D.-Z., Pardalos, P.M., Zhang, Z. (eds.) Nonlinear Combinatorial Optimization. SOIA, vol. 147, pp. 123–140. Springer, Cham (2019). https://doi.org/10.1007/978-3-030-16194-1_5
23. Yang, R., Xu, D., Du, D., Xu, Y., Yan, X.: Maximization of constrained non-submodular functions. In: Du, D.-Z., Duan, Z., Tian, C. (eds.) COCOON 2019. LNCS, vol. 11653, pp. 615–626. Springer, Cham (2019). https://doi.org/10.1007/978-3-030-26176-4_51
24. Yu, Q.-L., Xu, E.-L., Cui, S.-G.: Streaming algorithms for news and scientific literature recommendation: submodular maximization with a d-knapsack constraint. In: Proceedings of IEEE GCSI (2016). https://arxiv.org/abs/1603.05614

Deep Learning and Applications to Complex and Social Systems

A Framework for Accelerating Graph Convolutional Networks on Massive Datasets

Xiang Li[1]([⊠]), Ruoming Jin[2]([⊠]), Rajiv Ramnath[1]([⊠]), and Gagan Agrawal[3]([⊠])

[1] Ohio State University, Columbus, OH 43210, USA
{li.3880,ramnath.6}@osu.edu
[2] Kent State University, Kent, OH, USA
jin@cs.kent.edu
[3] Augusta University, Augusta, GA, USA
gagrawal@augusta.edu

Abstract. In recent years, there has been much interest in Graph Convolutional Networks (GCNs). There are several challenges associated with training GCNs. Particularly among them, because of massive scale of graphs, there is not only a large computation time, but also the need for partitioning and loading data multiple times. This paper presents a different framework in which existing GCN methods can be accelerated for execution on large graphs. Building on top of ideas from meta-learning we present an optimization strategy. This strategy is applied to three existing frameworks, resulting in new methods that we refer to as GraphSage++, ClusterGCN++, and GraphSaint++. Using graphs with order of 100 million edges, we demonstrate that we reduce the overall training time by up to 30%, while not having a noticeable reduction in F1 scores in most cases.

1 Introduction

In recent years, there has been much interest in Graph Convolutional Networks (GCNs) [2,15]. There are several challenges associated with training GCNs. One of them is the *neighborhood explosion* when training a k-deep GCN, where the value at each node needs to be computed as an aggregation from its k-hop neighborhood. For graphs with large degrees, this computation is often intractable. To address this challenge, a number of sampling methods have been developed [4,5,9,10,16,18].

Another problem in GCN when applied to very large graphs is the need for partitioning the data and loading them multiple times because all of the data may not fit in the memory of the GPU. To explain the issue, consider the following summary of a typical training process [18]. *"1. Construct a complete GCN on the full training graph. 2. Sample nodes or edges of each layer to form mini-batches. 3. Perform forward and backward propagation among the sampled GCN. Steps (2) and (3) proceed iteratively"*. Now, if the training graph in the

© Springer Nature Switzerland AG 2021
D. Mohaisen and R. Jin (Eds.): CSoNet 2021, LNCS 13116, pp. 79–92, 2021.
https://doi.org/10.1007/978-3-030-91434-9_8

step 1 can fit into the memory of a single GPU, steps (2) and (3) can be applied without the need for loading or unloading the data. However, GPUs often do not have sufficient memory to allow a full training graph to be loaded. Only a limited amount of work to date has considered this problem [6,17]. These works build minibatches from subgraphs, and thus do not require the entire training graph to fit into the GPU memory. However, the problem with this approach is the high cost of frequently loading subgraphs into the GPU during each iteration.

This paper presents a different framework in which different GCN methods can be accelerated for execution on large graphs. Our work draws its inspiration from the idea of *meta-learning* [14], In meta-learning, we assume there is a large number of tasks over the same dataset, and our goal is to optimize these tasks all together. The correspondence we can draw is that training of a GCN using a single large graph can be viewed as a collection of training tasks over subgraphs or partitions, each of which fits into GPU memory.

Based on this idea, we develop an overall framework for accelerating GCN training over large graphs. The main idea is that by focusing on training of GCN over each subgraph that has already been loaded into memory, we can reduce the data loading times as compared to a normal implementation. We apply this idea to three recent algorithms for GCN training, GraphSaint [18], GraphSage [9], and ClusterGCN [6], resulting in new algorithms GraphSaint++, GraphSage++, and ClusterGCN++, respectively. We show mathematical analysis denoting why these methods are still able to converge, while reducing data loading costs.

We have carried out a detailed experimental evaluation of our three new algorithms using four graph datasets. We demonstrate how we are able to obtain comparable convergence and final F1 scores while reducing the data loading time by up to 90% and total training time up to 30%.

2   Technical Details

This section provides important backgrounds on GCNs, followed by discussion of existing methods, with an emphasis on the memory requirements and associated data loading costs.

2.1   Background

Consider a graph $G = (V, E)$ where V is the set of vertices and E is the set of edges $E \subseteq V \times V$ represented by an adjacency matrix M, where an entry $M(i, j)$ denotes an edge between nodes i and j. Associated with every node in the graph are F features. Thus $X \in R^{|V| \times F}$ captures the F features for all nodes in the graph.

A GCN framework is composed of a number of layers (say, L). At each layer, the GCN computes a latent representation, using representation from the previous layer. For simplicity of presentation, we assume that the latent representation has the same dimension F as that of node feature. Thus, we denote the representation computed at the layer l by X^l, and $X^0 = X$. Now, the computation at each layer can be denoted as $X^{l+1} = A \times X^l \times W^l$. Here, W^l is a *feature transformation matrix*, $W^l \in R^{F \times F}$. The goal of the training process is to learn these matrices. In an inductive supervised learning based on GCNs, the goal is to learn the L weight transformation matrices while minimizing a *loss function*.

2.2   Existing Methods, Memory Requirements, and Data Loading Costs

Consider the original GCN method [13]. This method evaluates embeddings for each node in the graph for each layer yielding memory requirement as $|V| \times F \times L$. In addition, the process needs to maintain the matrix A and the current values of W^l for each l. Thus, the total memory requirement will be $|V| \times F \times L + |A| + L \times F^2$. For large graphs, this can easily exceed the available memory on a single GPU. In this work, the authors have not discussed any method for partitioning the problem that will allow us to work on different parts of the graph. Besides large memory requirements, this method also suffers from large computational time cost.

Since the original GCN method was presented, several researchers have developed methods for improving the efficiency of the process [4–6,9,18] Most of these approaches involve the use of *mini-batches*, possibly together with sampling of the neighborhood. Unfortunately, these approaches do not sufficiently reduce memory requirements for most graphs, especially when the number of layers is large. In the mini-batch approach, consider a batch size of b. If the average degree of a node is d, then with L layers, there are $b \times d^L$ nodes for which embeddings need to be computed. Depending upon the value of b, d, and L, this number can easily approach $|V|$, resulting in memory requirements comparable to the original GCN method. Some reduction in the exponential growth of the number of layers can be achieved with sampling of the neighbors. For example, Graph-SAGE [9] takes a fixed number of neighbors for each node. If this number is s ($s < d$), then the number of nodes for which embeddings need to be calculated reduces to $b \times s^L$. Note, however, that, as different nodes are selected to be part of the mini-batch for each epoch, we have one of the two possibilities. First, we store all nodes and their features on the device (such as the GPU). This limits the size of the graph that can be processed. The second possibility is to load the set of nodes that are part of the mini-batch and their neighborhood for each epoch. This, however, means high cost of reloading data for each epoch.

Two new efforts have specifically focused on the need of processing large graphs – ClusterGCN [6] and GraphSAINT [18]. We now describe these approaches with an emphasis of examining the data loading costs associated with them. ClusterGCN is an approach based on partitioning the graph, and subsequently, choosing a mini-batch from within a partition. The advantage of this approach is that nodes within a mini-batch are more likely to have common neighbors, thus allowing greater reuse of computations done on some of the nodes. Because of partitioning, this approach can also deal with very large graphs, which others approaches may not be able to handle. However, a hidden cost associated with this method in dealing with large graphs is that of data loading. If n partitions are created from the graph and the training is conducted over m epochs, each partition needs to be loaded m times during training.

GraphSAINT [18] can also handle very large graphs, but takes a different approach. Instead of choosing nodes that form a given mini-batch, it samples a smaller graph from the larger graph. Each epoch of the method works with one such sampled graph. Because the size of the sampled graph can be quite small, this method can also train very large graphs. However, there is a cost of sampling and loading the sampled graph for each epoch.

3   Overall Approach and Implementations

We discussed how the cost of loading either a partition or a k-step neighborhood of mini-batch vertices, or sampled subgraphs, can be quite high. To address this problem, we draw motivation from the previous work on *meta-learning* [14].

3.1   Background: Meta-learning Approach

In meta-learning, we assume there is a large number of tasks over the same dataset, and our goal is to optimize these tasks all together. In [14], a remarkably simple algorithm *Reptile* is proposed. We summarize the approach as Algorithm 1. Here τ denotes a task (line 2) and $U_\tau^k(\phi)$ (line 3) denotes the function that performs k gradient updates from the training algorithm on sampled (mini-batched) data starting with ϕ. This training is performed using Stochastic Gradient Descent (SGD) or Adam [12]. In line 4, we update ϕ, treating $\phi - \tilde{\phi}$ as the gradient. For this update, a parameter ϵ is used as the step size.

Algorithm 1. Reptile (serial version)

   Initialize ϕ (the vector of initial weights)
1: **for** iteration $i = 1, 2, \ldots$ **do**
2:     Sample task τ with loss $\tilde{\phi}$
3:     Compute $\tilde{\phi} = U_\tau^k(\phi)$, denoting k steps (SGD or Adam)
4:     Update $\phi \leftarrow \phi + \epsilon(\tilde{\phi} - \phi)$
5: **end for**

In [14], it has been argued that the Reptile converges towards a solution ϕ that is close (in Euclidean distance) to each task τ's manifold of the optimal solutions. As stated above, meta-learning is concerned with a large number of tasks that are being optimized together.

3.2 Our Approach

Algorithm 2. Large-Scale GCN training framework

Input: GCN model, Graph $G(V, E)$, feature X, label $\bar{Y}$
Output: GCN model with trained weights;

1: Φ: initialization parameters
2: **for** $macro\_epoch = 1, 2, \ldots, T_{macro}$ **do**
3:      Shuffle training nodes
4:      **for** $s = 1, 2, \ldots$ **do**
5:          Load or Generate $G_s(V_s, E_s)$
6:          **for** $mini\_epoch = 1, 2, \cdots T_{micro}$ **do**
7:              $\tilde{\phi} = U_s(\phi)$, denoting a full-batch train on G_s
8:          **end for**
9:      **end for**
10: **end for**

Based on the discussion above, we can consider training a GCN on each subgraph as a learning task (denoted as τ_1, τ_2, ...) and then training the GCN on the original graph (denoted as G) as the meta-learning task. By doing this, we can perform more computation/training using one subgraph that is already loaded into the GPU memory, and thus saving the loading cost from CPU main memory, or disk, or even remote storage (through network). For each training task, we go through a specific number of training epochs T_{Total} before the convergence is reached. The total training epoch duration is divided into two parameters, $macro\_epoch$ and $micro\_epoch$, such that

$$T_{Total} = T_{macro} \times T_{micro} \tag{1}$$

During each macro_epoch, one subgraph will be generated, uploaded on to GPU and trained for T_{micro} epochs. In this way, each subgraph data only need to be generated and uploaded on to GPU for a total of T_{macro} times. The overall framework is shown as Algorithm 2. This method involves loops over $macro\_epochs$. Each iteration starts with loading or generating a subgraph G_s (line 3). This subgraph can be pre-computed or be constructed on the fly, as we will explain later. The size of the each subgraph will be adjusted to be able to fit GPU memory – since the GCN model is trained by aggregating node features from a subgraph, a larger subgraph is expected to provide more information for the learning task. Therefore, during each micro_epoch, each subgraph will be full-batch trained to utilize all of its information for a better prediction performance and $U_s(\phi)$ (line 7) denotes the function that performs one-step gradient full-batch training. Overall, the update in line 6–8 corresponds to T_{micro} steps full-batch training on the entire subgraph G_s. This is also the reason why we do not use the parameter ϵ (line 4 from Algorithm 1) in line 7 of Algorithm 2.

Our training technique can be applied to multiple GCN learning frameworks since it is orthogonal to both graph sampling/partition methods and

GCN architecture. Three different GCN training algorithm have been adopted under our framework to illustrate the effectiveness of our proposed training strategy: GraphSaint [18], GraphSage [9] and ClusterGCN [6]. The main difference between these GCN training algorithm is their distinct strategies of constructing subgraphs. We mainly applied GCN architecture from the previous GraphSaint work with all the hyper-parameters carefully tuned for each benchmark dataset [18]. We refer to the resulting algorithms as GraphSaint++, GraphSage++, and ClusterGCN++, respectively.

The GraphSaint is a graph sampling based algorithm. To generate representative subgraphs for efficient information aggregation during training, it uses samplers that aggregate nodes with high influence on each other and also sample edges [18]. Several samplers have been used, such as random node sampler, random edge sampler, and random walk based sampler[18]. In applying our approach here, in each macro_epoch, we first shuffle all the train nodes. A user selected sampler will be executed to sample train nodes to construct each subgraph. The data that is required to be uploaded on GPU for the training process includes the adjacency matrix of subgraph, node features, edge weights, and the node labels. A series of T_{micro} epochs full-batch training is performed to update the model weights. The data uploading and full-batch training in the $micro\_epoch$ phase will be similar in all three GCN training algorithms.

The GraphSage is an inductive framework that learns node embeddings with good generalization performance by utilizing a node's neighborhood information. More specifically, features of a node's neighborhood will be sampled and aggregated [9]. The topological structure of each node's neighborhood and the node features distribution will be learnt simultaneously [9]. Nodes can get information from its neighbors at multiple hops. The number of hops and the number of neighbor nodes on each hop are user defined parameters and can be specified by the Neighbor Number List $L = [S_1, S_2, \ldots]$. For example, $L = [10, 5]$ means we include two-hop neighbors with 10 neighbor train nodes on the first hop and 5 on the second. In applying our approach, we first randomly sample a specific number of train nodes at the beginning of each $macro\_epoch$. We further expand the sampled nodes by further including their neighbor train nodes based on the array L to construct the subgraph. The other details are identical to the previous method.

The ClusterGCN framework exploits the graph clustering structure for SGD-based training for an improved memory and computation efficiency [6]. A graph clustering algorithm will be applied to partition the whole graph into disjoint clusters. These clusters will later be randomly recombined into multiple isolated subgraphs. During training, each train node can only utilize features of nodes which locate in the same isolated subgraph [6]. We first partition the training Graph into isolated clusters using METIS [11]. During each macro_epoch, clusters will be shuffled to reduce bias and a subgraph will be constructed by combining a specific number of clusters. Both the number of clusters and subgraphs are user defined hyper-parameters and the subgraph size should be chosen so it fits in GPU memory. The other details are the same as in other methods.

4   Experimental Results

Table 1. Dataset Details ("s" for single class and "m" for multiclass classification)

Dataset	Nodes	Edges	Feature	Classes	Train/Val/Test
Flickr	$89,250$	$899,756$	500	7 (s)	0.50/0.25/0.25
Reddit	$232,965$	$11,606,919$	602	41 (s)	0.66/0.10/0.24
Yelp	$716,847$	$6,977,410$	300	100 (m)	0.75/0.10/0.15
Amazon	$1,598,960$	$132,169,734$	200	107 (m)	0.85/0.05/0.10

4.1   Implementations and Setup

Our framework implementations are based on the GraphSaint architecture [18] with three major components: sampler, GCN model and subgraph generator. Inside each part, we incorporate implementation for three GCN training methods (GraphSaint [18], ClusterGCN [6] and GraphSage [9]) and we apply our strategy by separating the training process into two phases as described in Sect. 3.2, resulting in GraphSaint++, ClusterGCN++, and GraphSage++, respectively. The hyper-parameters obtained after the tuning process are listed in the Table 2. We use Adam [12] optimizer with learning rates carefully tuned for all our experiments. Dropout regularization is applied. GCN architecture is specified as $L \times F$, where L is the GCN depth and F is the *hidden dimension*, i.e. the dimension of latent representation in the GCN model.

The last column of Table 2 depends on a specific GCN method as explained below. For GraphSaint, we use the edge sampler and each edge will be sampled into a subgraph based on an independent decision [18]. The Edge budget in Table 2 is given as a sampling parameter to specify the expected number of edges in each subgraph [18]. For GraphSage neighborhood sampling, the previous work [9] shows that high learning performance can be obtained by including a neighbor number list $L = [S_1, S_2](S_1 \cdot S_2 \leq 500)$ and we are using $L = [S_1, S_2](S_1 \cdot S_2 \leq 500)$ in our work. The Node budget in Table 2 is to enforce a pre-defined budget on the subgraph size [9]. For ClusterGCN, we applied the strategy of stochastic multiple partitions [6] to reduce the bias and the diagonal enhancement technique to further improve the performance. In the subgraph generation procedure, isolated clusters are formed by using METIS clustering algorithm [11] and later recombined into subgraphs randomly without replacement. The number of isolated clusters N and the number of subgraphs K are user-defined sampling parameters as given in the table.

Our framework is implemented in Pytorch on CUDA 10.1. The sampling part for the GraphSaint++ and GraphSage++ is implemented in Cython 0.29.21. Our experiments are performed on nodes with Dual Intel Xeon8268s @2.9 GHz CPU and NVIDIA Volta V100 w/32 GB memory GPU and 384 GB DDR4 memory on OSC cluster [3]. Generation of subgraphs is performed in serial with 1 CPU core.

Table 2. Training configuration

GraphSaint++

Dataset	Learning rate	Dropout	T_{Total}	GCN architecture	Edge budget
Flickr	2×10^{-4}	0.2	30	3×256	6000
Reddit	1×10^{-3}	0.1	100	4×128	6000
Yelp	1×10^{-3}	0.1	100	3×512	2500
Amazon	1×10^{-2}	0.1	30	3×512	2000

GraphSage++

Dataset	Learning rate	Dropout	T_{Total}	GCN architecture	Node budget
Flickr	5×10^{-5}	0.2	15	2×256	8000
Reddit	1×10^{-3}	0.1	100	2×128	8000
Yelp	1×10^{-3}	0.1	100	2×512	5000
Amazon	2×10^{-4}	0.1	80	2×512	4500

ClusterGCN++

Dataset	Learning rate	Dropout	T_{Total}	GCN architecture	$K(N)$
Flickr	1×10^{-3}	0.2	15	3×256	16 (64)
Reddit	2×10^{-3}	0.1	100	4×128	64 (256)
Yelp	2×10^{-3}	0.1	100	3×512	64 (256)
Amazon	2×10^{-3}	0.2	100	3×512	64 (256)

4.2   Datasets

We use four benchmark datasets, which have also been used in other recent efforts (for example, GraphSaint [18]). Detailed statistics of all datasets are listed in Table 1. Flickr and Reddit are used for single-class classification task, i.e., each node can only belong to a single class while Yelp and Amazon are for multi-class classification. Each dataset has a specific fraction for the split of training/validation/test data, which is shown in Table 1.

Flickr aims at classifying images based on descriptions and common properties of online images. A node in this graph stands for one image uploaded to Flickr. An edge between two nodes will be established if comment properties exist between two images such as geographic location and comments from the same users. **Reddit** utilize users' comments to generate predictions about online posts communities. Each node is one user and edges will be established based on the friendship between users. This dataset has more than 10 million edges. **Yelp** is about the categorization of business according to customer's reviews and friendship in the open challenge website. One node represents one user and an edge will be created between two nodes if two corresponding users are friends. **Amazon** categorizes types of products by referring to buyers' reviews and activities. A node corresponds to one product on the Amazon website. If two products share the same customer, then an edge will be created between them. This dataset has more than 100 million edges.

Table 3. Test F1 score for different T_{micro} (%)

T_{micro}	Flickr	Reddit	Yelp	Amazon
GraphSaint++				
1	50.19 ± 0.52	96.52 ± 0.03	65.10 ± 0.06	80.70 ± 0.04
5	47.39 ± 0.39	96.50 ± 0.03	64.81 ± 0.07	79.50 ± 0.09
10	–	96.43 ± 0.01	64.43 ± 0.01	78.76 ± 0.13
ClusterGCN++				
1	50.78 ± 0.15	96.01 ± 0.05	64.31 ± 0.09	80.97 ± 0.02
5	49.49 ± 0.42	95.89 ± 0.04	64.47 ± 0.08	80.85 ± 0.03
10	–	95.67 ± 0.11	64.34 ± 0.06	80.70 ± 0.02
GraphSage++				
1	50.53 ± 0.35	96.58 ± 0.04	64.61 ± 0.06	78.12 ± 0.03
5	50.81 ± 0.03	96.47 ± 0.04	64.41 ± 0.04	78.16 ± 0.05
10	–	96.34 ± 0.04	64.20 ± 0.06	78.09 ± 0.07

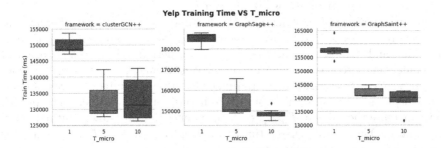

Fig. 1. Training time with different T_{micro} across different frameworks on Yelp

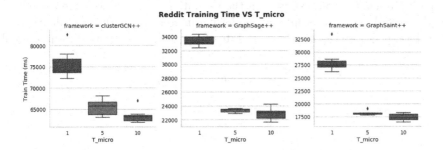

Fig. 2. Training time with different T_{micro} across different frameworks on Reddit

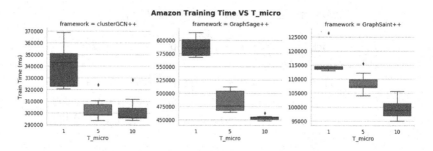

Fig. 3. Training time with different T_{micro} across different frameworks on Amazon

4.3 F1-Score

Our first experiment focused on evaluating the impact of T_{micro} on convergence. It should be noted that when T_{micro} is 1, the computations performed are identical to the original framework, i.e., GraphSage++ is same as GraphSage, and so on (though implementations are different).

In our experiments, as different GCN training algorithms have different convergence rates, we set a different value of T_{Total} for each combination of GCN algorithm and dataset. During the training, we periodically take a snapshot of the model and perform an evaluation on the validation dataset to record the convergence curve. As experiments with different datasets resulted in a very similar behavior, we show results only from the Yelp dataset in this paper. It turns out, as we increase the value of T_{micro}, the convergence did slow down, but only very marginally. Overall, we can see that use of higher values of T_{micro} parameter remains a feasible approach for training GCNs.

We compare the F1-score performance on test data among distinct values of T_{micro} in Table 3. Each data point is generated by 7 runs under the same hyper-parameter settings. Based on the validation F1-score, we choose the best model snapshot, i.e. the optimal model parameters, to perform evaluation on test data. First, we can see that the state-of-art results as reported from original

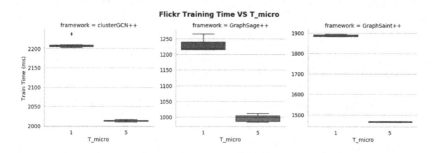

Fig. 4. Training time with different T_{micro} across different frameworks on Flickr

publications of these frameworks have been reproduced when T_{micro} is 1, i.e., when our implementation is simply reproducing original algorithm. Next, as reflected in Table 3, the influence of T_{micro} on test F1 score can vary through different frameworks. There are some cases where $T_{micro} > 1$ can outperform the baseline case ($T_{micro} = 1$). For example, ClusterGCN++ on Yelp obtains its best F1-score with $T_{micro} = 5$. Similar things happen to GraphSage++ on Flickr and Amazon. For the GraphSaint++, a larger T_{micro} does result in some loss of F1 score. Overall, across different combination of GCN training methods and datasets, decrease in F1 score is limited to at most 1–3%, again establishing that use of higher values of T_{micro} parameter remains a feasible approach for training GCNs.

A larger value of T_{micro} implies that we are more focusing on training each subgraph and we will iterate through different subgraphs less frequently. However, compared with the baseline case ($T_{micro} = 1$), we can still achieve good Test F1 score while maintaining a fast convergence speed with a larger T_{micro}. As long as the subgraph is well sampled to be representative of the target graph, we are able to obtain both fast convergence speed as reflected by the validation F1-score convergence curve and good generalization performance as indicated by the test F1-scores.

4.4 Training Times

Finally, we focus on the gains from training times.

The training time excludes all the data pre-processing such as sampling as in GraphSaint++, METIS partitioning as in ClusterGCN++ or neighbor nodes generation as in GraphSage++. It includes time cost of data uploading to the GPU device and the on-GPU computation of training loss and parameters updating. We investigated training time for each dataset using different training methods with multiple T_{micro} values as shown in Figs. 1, 2, 3 and 4. Seven runs are performed for each experimental task to include variations for the training time. The data loading time will shrink T_{micro} times with $T_{micro} > 1$. Significant savings on training time can be achieved especially when the data loading takes a relatively bigger portion in the training time as in Table 4. We see a proportional decrease of training time in Fig. 2. Relatively less train time saving have been achieved for Yelp in Fig. 1 and Amazon (Fig. 3). That is because they only have data loading time with a fraction around $10 - 15\%$ of their training time and GCN computation is their major time cost. That also explains why Flickr in Fig. 4 achieves more time saving for GraphSage++ and GraphSaint++ than it does for ClusterGCN++. Overall, the improvement from $T_{micro} = 5$ to $T_{micro} = 10$ remains limited and only the GraphSaint++ on Amazon as in Fig. 3 shows a relatively obvious difference. That is because a value as $T_{micro} = 5$ will reduce data loading time into a small enough fraction of training time so that further optimization with $T_{micro} = 10$ will not make a substantial difference.

Table 4. Data load time fraction of train time (%)

Framework	Flickr	Reddit	Yelp	Amazon
GraphSaint	27.41 ± 0.19	32.32 ± 3.63	11.76 ± 1.89	10.04 ± 1.58
ClusterGCN	10.34 ± 0.04	11.81 ± 0.95	14.94 ± 3.05	13.59 ± 0.34
GraphSage	23.06 ± 0.77	30.02 ± 1.10	14.58 ± 2.10	14.91 ± 2.39

5 Related Work

Besides GraphSage [9], ClusterGCN [6] and GraphSaint [18] that we have extensively discussed and built on, other prominent efforts are as follows. FastGCN [5]
interprets the graph convolution as integral transforms of embedding functions
under probability measures and applies importance sampling among graph vertices, though significant overhead could be induced by its sampling algorithm.
S-GCN [4] reduces the neighborhood sampling size using a variance reduction
technique. These both methods still have scalability issues due to the requirement of keeping all nodes' intermediate embeddings in memory. There has been
previous work on developing an efficient out-of-core implementation of Convolution Neural Networks (CNNs) [1]. However, the computation and data access
patterns for a CNN is very different from GCNs.

In one aspect, the idea of our work is similar to and related to the recent
work of data echoing [7] and minibatch persistency [8]. However, these works
are based on training using mini-batches, whereas we consider subgraph reuse.
Subgraphs are typically much larger and have an internal structures, whereas
the standard minibatch consists of randomized data points. In fact, the data
echoing and minibatch persistency are mainly used in settings like CNN; to
our best of knowledge, it has never been attempt in GCN. Also, we have given
mathematical analysis based on the meta-learning, whereas data echoing [7] and
minibatch persistency [8] never did.

6 Conclusions

In this paper, we have focused on the problem of training large-scale Graph Convolutional Networks (GCNs), which is becoming increasingly important. Drawing inspiration from the idea of *meta-learning*, we observe that training of a GCN
using a single large graph can be viewed as a collection of training tasks over subgraphs or partitions, each of which fits into GPU memory. Based on this idea, we
developed both an overall framework as well as three instantiations – converting
three recent methods GraphSaint, GraphSage, and ClusterGCN, resulting into
new algorithms GraphSaint++, GraphSage++, and ClusterGCN++, respectively. We have also shown mathematical analysis denoting why these methods
are still able to converge, while reducing data loading costs.

We have carried out a detailed experimental evaluation of our three new algorithms using four graph datasets. We demonstrate how we are able to obtain comparable convergence and final F1 scores while reducing the data loading time by up to 90% and total training time up to 30%.

References

1. Awan, A.A., Hamidouche, K., Hashmi, J.M., Panda, D.K.: S-caffe: co-designing MPI runtimes and Caffe for scalable deep learning on modern GPU clusters. In: Proceedings of the 22nd ACM SIGPLAN Symposium on Principles and Practice of Parallel Programming, SIGPLAN Notices, vol. 52, no. 8, pp. 193–205, January 2017
2. Cai, H., Zheng, V.W., Chang, K.C.C.: A comprehensive survey of graph embedding: Problems, techniques and applications. CoRR, abs/1709.07604 (2017)
3. Ohio Supercomputer Center. Ohio supercomputer center (1987)
4. Chen, J., Zhu, J., Song, L.: Stochastic training of graph convolutional networks with variance reduction. In: ICML, pp. 941–949 (2018)
5. Chen, J., Ma, T., Xiao, C.: FastGCN: fast learning with graph convolutional networks via importance sampling. In: International Conference on Learning Representations (ICLR) (2018)
6. Chiang, W.L., Liu, X., Si, S., Li, Y., Bengio, S., Hsieh, C.J.: Cluster-GCN. In: Proceedings of the 25th ACM SIGKDD International Conference on Knowledge Discovery & Data Mining (KDD), July 2019
7. Choi, D., Passos, A., Shallue, C.J., Dahl, G.E.: Faster neural network training with data echoing. CoRR, abs/1907.05550 (2019)
8. Fischetti, M., Mandatelli, I., Salvagnin, D.: Faster SGD training by minibatch persistency. CoRR, abs/1806.07353 (2018)
9. Hamilton, W., Ying, Z., Leskovec, J.: Inductive representation learning on large graphs. In: Advances in Neural Information Processing Systems, vol. 30, pp. 1024–1034 (2017)
10. Huang, W., Zhang, T., Rong, Y., Huang, J.: Adaptive sampling towards fast graph representation learning. In: Advances in Neural Information Processing Systems, pp. 4558–4567 (2018)
11. Karypis, G., Kumar, V.: A fast and high quality multilevel scheme for partitioning irregular graphs. SIAM J. Sci. Comput. 20(1), 359–392 (1998)
12. Kingma, D.P., Ba, J.: Adam: a method for stochastic optimization. In: 3rd International Conference for Learning Representations (2015)
13. Kipf, T.N., Welling, M.: Semi-supervised classification with graph convolutional networks. In: International Conference on Learning Representations (ICLR), abs/1609.02907 (2017)
14. Nichol, A., Schulman, J.: Reptile: a scalable metalearning algorithm. arXiv preprint arXiv:1803.02999, vol. 2, no. 3, p. 4 (2018)
15. Wu, Z., Pan, S., Chen, F., Long, G., Zhang, C., Philip, S.Y.: A comprehensive survey on graph neural networks. CoRR, abs/1901.00596 (2019)
16. Ying, R., He, R., Chen, K., Eksombatchai, P., Hamilton, W.L., Leskovec, J.: Graph convolutional neural networks for web-scale recommender systems. In: Proceedings of the 24th ACM SIGKDD International Conference on Knowledge Discovery & Data Mining, KDD 2018 (2018)

17. Zeng, H., Zhou, H., Srivastava, A., Kannan, R., Prasanna, V.: Accurate, efficient and scalable graph embedding. In: 2019 IEEE International Parallel and Distributed Processing Symposium (IPDPS), May 2019
18. Zeng, H., Zhou, H., Srivastava, A., Kannan, R., Prasanna, V.: Graphsaint: graph sampling based inductive learning method. In: International Conference on Learning Representations (ICLR) abs/1907.04931 (2020)

AdvEdge: Optimizing Adversarial Perturbations Against Interpretable Deep Learning

Eldor Abdukhamidov[1], Mohammed Abuhamad[2], Firuz Juraev[1],
Eric Chan-Tin[2], and Tamer AbuHmed[1(✉)]

[1] Sungkyunkwan University, Seoul, South Korea
tamer@skku.edu
[2] Loyola University Chicago, Chicago, IL 60660, USA

Abstract. Deep Neural Networks (DNNs) have achieved state-of-the-art performance in various applications. It is crucial to verify that the high accuracy prediction for a given task is derived from the correct problem representation and not from the misuse of artifacts in the data. Hence, interpretation models have become a key ingredient in developing deep learning models. Utilizing interpretation models enables a better understanding of how DNN models work, and offers a sense of security. However, interpretations are also vulnerable to malicious manipulation. We present AdvEdge and AdvEdge$^+$, two attacks to mislead the target DNNs and deceive their combined interpretation models. We evaluate the proposed attacks against two DNN model architectures coupled with four representatives of different categories of interpretation models. The experimental results demonstrate our attacks' effectiveness in deceiving the DNN models and their interpreters.

Keywords: Adversarial image · Deep learning · Interpretability

1 Introduction

Due to the complex architecture of Deep Neural Networks (DNNs), it is still not explicit how a DNN proposes a certain decision. This is the drawback of black-box models for applications in which explainability is required. Being inherently vulnerable to crafted adversarial inputs is another drawback of DNN models, which leads to unexpected model behaviors in the decision-making process.

To represent the behavior of the DNN models in an understandable form to humans, interpretability would be an indispensable tool. For example, in Fig. 1 (a), based on the prediction, an attribution map emphasizes the most informative regions of the image, showing the causal relationship. Using interpretability helps understand the inner workings of DNNs (to debug models, conduct security analysis, and detect adversarial inputs). Figure 1 (b) shows that an adversarial input causes the target DNN to misclassify, making an attribution map highly distinguishable from its original attribution map, and is therefore detectable.

© Springer Nature Switzerland AG 2021
D. Mohaisen and R. Jin (Eds.): CSoNet 2021, LNCS 13116, pp. 93–105, 2021.
https://doi.org/10.1007/978-3-030-91434-9_9

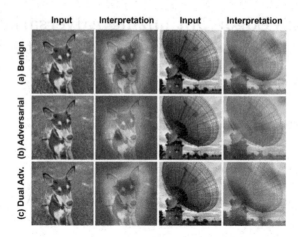

Fig. 1. Example images for (a) benign, (b) regular adversarial and (c) dual adversarial and interpretations on ResNet (classifier) and CAM (interpreter).

Classifiers with their interpreters (IDLSes) provide a sense of security in the decision-making process with human involvement, as experts can distinguish whether an attribution map matches the models' prediction. However, interpretability is sensitive to malicious manipulations, and this expands the vulnerability of DNN models to the interpretability models against adversarial attacks [17]. Crafting adversarial input is both valid and practical to mislead the target DNN and deceive its corresponding interpreters simultaneously. Figure 1 (c) shows an example of these dual adversarial inputs that are misclassified by the target DNNs and interpreted highly similar to the interpretation of benign inputs. Therefore, IDLSes offer limited security in the decision-making process.

This paper proposes AdvEdge and AdvEdge$^+$, which are optimized versions of an adversarial attack that deceive the target DNN model and its corresponding interpreter. AdvEdge and AdvEdge$^+$ take advantage of the edge information of the image to allow perturbation to be added to the edges in regions highlighted by the interpreters' attribution map. This enables a much stealthier attack as the generated adversarial samples are challenging to detect even with interpretation and human involvement. Moreover, the proposed attacks generate effective adversarial samples with less perturbation size.

Our Contribution. Firstly, we indicate that the existing IDLSes can be manipulated by adversarial inputs. We present two attack approaches that generate adversarial inputs to mislead the target DNN and deceive its interpreter. We evaluate our attacks against four major types of IDLSes on a dataset and compare them with the existing attack ADV^2 [17]. We summarize our contributions as follows:

– We propose AdvEdge and AdvEdge$^+$ attacks that incorporate edge information to enhance interpretation-derived attacks. We show that even restricting the perturbation to edges in regions spotted by interpretation models, the adversarial input can be very effective. We evaluate our attacks against two common DNNs architectures accompanied by four interpretation models that represent different categories.

– Our evaluation includes measuring the effectiveness of the attacks in terms of success rate and deceiving the coupled interpreters, and comparing to the existing attack (ADV^2). The results show that the proposed attacks are as effective as ADV^2 in terms of misclassification and outperform ADV^2 in generating adversarial inputs with highly similar interpretations to their benign cases. Moreover, this level of effectiveness is maintained with a smaller amount of noise as compared with ADV^2.

Organization. The rest of the paper is organized as follows: Sect. 2 highlights the relevant literature; Sect. 3 presents the fundamental concepts; Sect. 4 and Sect. 5 describe AdvEdge and AdvEdge+ attacks and their implementation against four major interpreter types; Sect. 6 shows the evaluation of the attacks' effectiveness; and Sect. 7 offers the conclusion.

2   Related Work

In this section, we provide different categories of research work that are relevant to our work: adversarial attacks and interpretability.

Attacks. Basically, there are two main threats for machine learning models: infecting the training data to weaken the target models (poisoning attack [5]) and manipulating the input data to make the target model misbehaves (evasion attack [5]). Attacking deep neural networks (DNNs) has been more challenging due to their high complexity in model architecture. Our work explores attacks against DNNs with interpretability as a defense means.

Interpretability. Interpretation models have been used to provide the interpretability for black-box DNNs via different techniques: back-propagation, intermediate representations, input perturbation, and meta models [3]. It is believed that that interpretability provides a sense of security in the decision-making process with human involvement. Nevertheless, recent work shows that some interpretation techniques are insensitive to DNNs or data generation processes, whereas the behaviors of interpretation models can be impacted significantly by the transformation without effect on DNNs [7].

Another recent work [17] shows the possibility of attacking IDLSes. Specifically, it proposes a new attacking class to deceive DNNs and their coupled interpretation models simultaneously, presenting that the enhanced interpretability provides a limited sense of security. In this work, we show the optimized version of the attack presented in the recent work [17] to deceive target DNNs and mislead their coupled interpretation models.

3   Fundamental Concepts

In this section, we introduce concepts and key terms used in the paper. We note that this paper is mainly focused on classification tasks, such as image classification. Let $f(x) = y \in Y$ denote a classifier (*i.e.*, DNN model f) that assigns an input (x) to a class (y) from a set of predefined classes (Y).

Let $g(x; f) = m$ denote an interpreter (g) that generates an attribution map (m) that reflects the importance of features in the input sample (x) based on the output of the classifier (f), (*i.e.*, the value the i-th element in m $(m[i])$ reflects the importance of the i-th element in x $(x[i])$).

In this regard, we note that there are two main methods to achieve interpretation of a model: ❶ **Post-hoc interpretation**: The interpretation can be achieved by regulating the complexity of DNN models or by applying methods after training. This method requires creating another model to support explanations for the current model[3]. ❷ **Intrinsic interpretation**: Intrinsic interpretability can be achieved by building self-explanatory DNN models which directly integrate interpretability into their architectures [3].

Our attacks are mainly based on the first interpretation category, where an interpreter (g) extracts information (*i.e.*, attribution map m) about how a DNN model f classifies the input x.

A benign input (x) is manipulated to generate adversarial sample $(\hat{x})$ using one of the well-known attacks (PGD [9], STADV [15]) to drive the model to misclassify the input $\hat{x}$ to a target class y_t such that $f(\hat{x}) = y_t \neq f(x)$. These manipulations, *e.g.*, adversarial perturbations, are usually constrained to a norm ball $\mathcal{B}_\varepsilon(x) = \{\|\hat{x} - x\|_\infty \lesssim \varepsilon\}$ to ensure its success and evasiveness. For example, PGD, a first-order adversarial attack, applies a sequence of project gradient descent on the loss function:

$$\hat{x}^{(i+1)} = \prod_{\mathcal{B}_\varepsilon(x)} \left(\hat{x}^{(i)} - \alpha.\ sign(\nabla_{\hat{x}} \ell_{prd}(f(\hat{x}^{(i)}), y_t)) \right) \tag{1}$$

Here, $\prod$ is a projection operator, $\mathcal{B}_\varepsilon$ is a norm ball restrained by a pre-fixed ε, α is a learning rate, x is the benign sample, $\hat{x}^{(i)}$ is the $\hat{x}$ at the iteration i, ℓ_{prd} is a loss function that indicates the difference between the model prediction $f(\hat{x})$ and y_t.

4   AdvEdge Attack

IDLSes provide a level of security in the decision-making process with human involvement. This has been a belief until a new class of attacks is presented [17]. In their work, Zhang *et al.* [17] proposed ADV^2 that bridges the gap by deceiving target DNNs and their coupled interpreters simultaneously. Our work presents new optimized versions of the attack, namely: AdvEdge and AdvEdge$^+$. This section gives detailed information on the proposed attacks and their usage against four types of interpretation models.

4.1   Attack Definition

The main purpose of the attack is to deceive the target DNNs f and their interpreters g. To be precise, an adversarial input $\hat{x}$ is generated by adding noise to the benign input x to satisfy the following conditions:

1. The adversarial input $\hat{x}$ is misclassified to y_t by f: $f(\hat{x}) = y_t$;
2. $\hat{x}$ prompts the coupled interpreter g to produce the target attribution map m_t: $g(\hat{x}; f) = m_t$;
3. $\hat{x}$ and the benign x samples are indistinguishable.

The attack finds a small perturbation for the benign input in a way that the results of the prediction and the interpretation are desirable. We can describe the attack by the following optimization framework:

$$\min_{\hat{x}} : \Delta(\hat{x}, x) \ s.t. \begin{cases} f(\hat{x}) = y_t \\ g(\hat{x}; f) = m_t \end{cases} \tag{2}$$

As there is high non-linearity in $f(\hat{x}) = y_t$ and $g(\hat{x}; f) = m_t$ for DNNs, Eq. (2) can be rewritten as the following to be more suitable for optimization:

$$\min_{\hat{x}} : \ell_{prd}(f(\hat{x}), y_t) + \lambda. \ \ell_{int}(g(\hat{x}; f), m_t) \ s.t. \ \Delta(\hat{x}, x) \le \varepsilon \tag{3}$$

Where ℓ_{prd} is the classification loss as in Eq. (1), ℓ_{int} is the interpretation loss to measure the difference between the adversarial map $g(\hat{x}; f)$ and the target map m_t. To balance the two factors (ℓ_{prd} and ℓ_{int}), the hyper-parameter λ is used. We build the Eq. (3) based on the PGD adversarial framework to compare the performance with the existing attack while other frameworks can also be utilized. Other settings are defined as follows: $\ell_{prd}(f(\hat{x}), y_t) = - \log(f_{y_t}(\hat{x}))$, $\Delta(\hat{x}, x) = \|\hat{x} - x\|_\infty$, and $\ell_{int}(g(\hat{x}; f), m_t) = \|g(\hat{x}; f) - m_t\|_2^2$. Overall, the attack finds the adversarial input $\hat{x}$ using a sequence of gradient descent updates:

$$\hat{x}^{(i+1)} = \prod_{\mathcal{B}_\varepsilon(x)} \left(\hat{x}^{(i)} - N_w \ \alpha. \ sign(\nabla_{\hat{x}} \ell_{adv}(\hat{x}^{(i)})) \right) \tag{4}$$

Here, N_w is the noise function that controls the amount and the position of noise to be added with respect to the benign input's edge weights w. ℓ_{adv} represents the equation of overall loss Eq. (3).

AdvEdge. Notice that we apply the N_w term in Eq. (4) to optimize the location and magnitude of the added perturbation. In the first attack, AdvEdge, we further restrict the added perturbation to the edges of the image that intersect with the attribution map generated by the interpreter. This means that considering the overall loss (classifier and interpreter loss), we identify the important areas of the input and then generate noise for the edges in those areas by considering the edge weights in the image obtained using the Sobel filter.

Let $\mathcal{E} : e \to \mathbb{R}^{h \times w}$ that indicates a pixel-wise edge weights matrix for an image with height h and width w, using common edge detector (e.g., Sobel filters in this settings). We apply the edge weights to the sign of the gradient update as: $\mathcal{E}(x) \otimes \alpha.\ sign(\nabla_{\hat{x}}\ell_{adv}(\hat{x}^{(i)}))$, where $\otimes$ denotes the Hadamard product. This increases the noise on the edges while decreases the noise in smooth regions of the image. Considering $N_w = \mathcal{E}(x)$, Eq. (3) can be expressed as:

$$\hat{x}^{(i+1)} = \prod_{\mathcal{B}_\varepsilon(x)} \left(\hat{x}^{(i)} - \mathcal{E}(x)\, \alpha.\ sign(\nabla_{\hat{x}}\ell_{adv}(\hat{x}^{(i)})) \right) \tag{5}$$

AdvEdge$^+$ Similar to AdvEdge, this approach also incorporates the edge weights of the input to optimize the perturbation. In this attack, we only apply the noise to the edges rather than weighting the noise in the specified areas. This is done by binarizing the edge matrix $\mathcal{E}_\delta : e \rightarrow [0,1]^{h \times w}$, . Then, obtain the Hadamard product as: $\mathcal{E}_\delta(x) \otimes \alpha.\ sign(\nabla_{\hat{x}}\ell_{adv}(\hat{x}^{(i)}))$, where the hyperparameter δ controls the threshold to binarize the edge weights. This technique allows noise values to be complete on the edges only. To improve the effectiveness of the attack, considering the restriction on the perturbation location, the threshold δ is set to 0.1. In the following subsection, we discuss the details about the attack in Eq. (4) against the representatives of four types of interpretation models, namely: *back-propagation-guided interpretation, representation-guided interpretation, model-guided interpretation,* and *perturbation-guided interpretation.*

4.2 Interpretation Models

Back-Propagation-Guided Interpretation. Back-propagation-guided interpretation models calculate the gradient of the prediction of a DNN model with reference to the given input. By doing this, the importance of each feature can be derived. Based on the definition of this class interpretation, larger values in the input features indicate higher relevance to the model prediction. In this work, as the example of this class, we consider the gradient saliency (Grad) [13]. Finding the optimal $\hat{x}$ for Grad-based IDLSes is inefficient via a sequence of gradient descent updates (as in applying Eq. (4)), since DNNs with ReLU activation functions cause the computation result of the Hessian matrix to be all-zero. The issue can be solved by calculating the smoothed value of the gradient of ReLU.

Representation-Guided Interpretation. In this type of interpreters, feature maps from intermediate layers of DNN models are extracted to produce attribution maps. We consider **C**lass **A**ctivation **M**ap (CAM) [18] as the representative of this class. The importance of the input regions can be identified by projecting back the weights of the output layer on the convolutional feature maps. Similar to the work of [17], we build g by extracting and concatenating attribution maps from f up to the last convolutional layer and a fully connected layer. We attack the interpreter by searching for $\hat{x}$ using gradient descent updates as in Eq. (4). The attack Eq. (4) can be applied to other interpreters of this class (*e.g.,* Grad-CAM [12]).

Model-Guided Interpretation. This type of interpreters trains a masking model to directly predict the attribution map in a single forward pass by masking salient positions of any input. For this type, we consider the **R**eal-**T**ime Image **S**aliency (RTS) [1]. Directly attacking RTS has been shown to be ineffective to find the desired adversarial inputs [17]. This is because the interpreter relies on both the masking model and the encoder $(enc(.))$. To overcome the issue, we add an extra loss term $\ell_{enc}(enc(\hat{x}), enc(y_t))$ to the Eq. (3) to calculate the difference of the encoder's result with the adversarial input $\hat{x}$ and the target class y_t. Then, we use the sequence of gradient descent updates as defined in Eq. (4) to find the optimal adversarial input $\hat{x}$.

Perturbation-Guided Interpretation. The perturbation-guided interpreters aim to find the attribution maps by adding minimum noise to the input and examining the shift in the model's output. For this work, we consider MASK [2] as the representative of the class. As the interpreter g is constructed as optimization procedure, we cannot directly optimize the Eq. (3) with the Eq. (4). For this issue, bi-level optimization framework [17] can be implemented. The loss function is reformulated as: $\ell_{adv}(x, m) \triangleq \ell_{prd}(f(x), y_t) + \lambda.\ \ell_{int}(m, m_t)$ by adding m as a new variable.

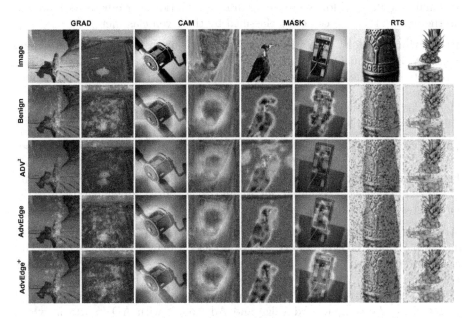

Fig. 2. Attribution maps of benign and adversarial (ADV^2, AdvEdge and AdvEdge$^+$) inputs with respect to Grad, CAM, MASK, and RTS on ResNet.

5 Experimental Setting

In this section, we explain the implementation of AdvEdge and the optimization steps to increase the effectiveness of the attacks against target interpreters (Grad, CAM, MASK, RTS). We build our two approaches (AdvEdge and AdvEdge$^+$)

based on the PGD attack that utilizes the local first order information about the network (Eq. (5)). For the parameters, we set $\alpha = 1./255$, and $\varepsilon = 0.031$ similar to previous studies [17]. To measure the proportion of the perturbation, ℓ_∞ is applied. To increase the efficiency of the attack, we apply a technique that adds noise to the edges of the images with a constant number of iterations (#iterations $= 300$). The process aims to search for perturbation points on the edges of the regions that satisfy both the classifier and interpreter.

Optimization for Both Approaches. In our attack, zero gradients of the prediction loss prevents searching the desired result with correct interpretation (Grad). To overcome the issue, the label smoothing technique with cross-entropy is proposed. In the technique, prediction loss is sampled using uniform distribution $\mathbb{U}(1 - \rho, 1)$ and during the attacking process, the value of ρ is decreased moderately. Considering $y_c = \frac{1-y_t}{|Y|-1}$, we calculate $\ell_{prd}(f(x), y_t) = -\sum_{c \in Y} y_c \log f_c(x)$.

Dataset. For our experiment, we use ImageNetV2 Top-Images [11] dataset. ImageNetV2 is a new test set collected based on the ImageNet benchmark and was mainly published for inference accuracy evaluation. For our test set, we use all the images that are correctly classified by the given classifier f.

Prediction Models. Two state-of-the-art DNNs are used for the experiments, **ResNet-50** [4] and **DenseNet-169** [6], which show 22.85% and 22.08% top-1 error rate on ImageNet dataset, respectively. The two DNNs are with different capacities (i.e., 50 and 169 layers, respectively) and architectures (i.e., residual blocks and dense blocks, respectively). Using these DNNs helps measuring the effectiveness of our attacks.

Interpretation Models. We utilize the following interpreters as the representative of Back-Propagation-Guided, Representation-Guided, Model-Guided, and Perturbation-Guided Interpretation classes: Grad [13], CAM [18], RTS [1], and MASK [2] respectively. We used the original open-source implementations of the interpreters in our experiments.

AdvEdge Attack. For the attack, we implement our attack (defined in Eq. (4) on the basis on PGD framework. Other attacks frameworks (e.g., STADV [15] can also be applied for the attack Eq. (4). For our case, we assume that our both approaches are based on the targeted attack, in which the attack forces the DNNs to misclassify the perturbed input $\hat{x}$ to a specific and randomly-assigned target class. We compare AdvEdge and AdvEdge$^+$ with ADV^2 attack, which is considered as a new class of attacks to generate adversarial inputs for the target DNNs and their coupled interpreters. For a fair comparison, we adopt the same hyperparameters (learning rate, number of iterations, step size, etc..) and experimental settings as in ADV^2 [17].

6   Attack Evaluation

In this section, we conduct experiments to evaluate the effectiveness of AdvEdge and AdvEdge$^+$. We compare our results to ADV^2 in [17]. For this comparison, we use the original implementation of ADV^2 provided by the authors. For the evaluation, we answer the following questions: ❶ Are the proposed AdvEdge and AdvEdge$^+$ effective to attack DNNs? ❷ Are AdvEdge and AdvEdge$^+$ effective to mislead interpreters? ❸ Do the proposed attacks strengthen the attacks against interpretable deep learning models?

Evaluation Metrics. We apply different evaluation metrics to measure the effectiveness of attacks against the baseline classifiers and the interpreters. Firstly, we evaluate the attack based on deceiving the target DNNs using the following metrics:

- **Misclassification confidence:** In this metric, we observe the confidence of predicting the targeted class, which is the probability assigned by the corresponding DNN to the class y_t.

Secondly, we evaluate the attacks based on deceiving the interpreter. This is done by evaluating the attribution maps of adversarial samples. We note that this task is challenging due to the lack of standard metrics to assess the attribution maps generated by the interpreters. Therefore, we apply the following metrics to evaluate the interpretability:

- $\mathcal{L}_p$ **Measure:** We use the $\mathcal{L}_1$ distance between benign and adversarial maps to observe the difference. To obtain the results, all values are normalized to $[0, 1]$.
- **IoU Test** (Intersection-over-Union): This is another quantitative measure to find the similarity of attribution maps. This measurement is widely used to compare the prediction with ground truth.

Finally, to measure the amount of noise added to generate the adversarial input, the following metric is used:

- **Structural Similarity (SSIM)**: Added noise is measured by computing the mean structural similarity index [14] between benign and adversarial inputs. SSIM is a method to predict the image quality based on its distortion-free image as reference.
  To obtain the non-similarity rate (*i.e.*, distance or noise rate), we subtract the SSIM value from 1 (*i.e.*, noise_rate = 1 − SSIM).

6.1   Attack Effectiveness Against DNNs

We first assess the effectiveness of AdvEdge and AdvEdge$^+$ as well as compare the results to the existing method (ADV^2 [17]) in terms of deceiving the target DNNs. We achieved 100% attack success rate of ADV^2, AdvEdge, and AdvEdge$^+$ against different classifiers and interpreters on 10,000 images.

Table 1. Misclassification confidence of ADV^2, AdvEdge and AdvEdge$^+$ against different classifiers and interpreters testing on 10,000 images.

	ResNet				DenseNet			
	Grad	CAM	MASK	RTS	Grad	CAM	MASK	RTS
ADV^2	92.19%	**56.52%**	53.54%	**69.83%**	87.88%	52.88%	58.05%	57.24%
AdvEdge	**93.51%**	55.53%	**59.70%**	68.99%	**88.04%**	53.79%	63.01%	57.26%
AdvEdge$^+$	92.49%	55.40%	53.48%	69.55%	86.94%	**53.54%**	**62.82%**	**57.28%**

Additionally, Table 1 presents the misclassification confidence results of the three methods against different classifiers and interpreters on 10,000 images. It should be said that due to the differences in the dataset and models, the results of ADV^2 are not consistent with the results achieved in [17]. Even though our main idea is to add a small amount of perturbation to the specific regions of images, the performance is slightly better than ADV^2 in terms of Grad (ResNet), CAM (DenseNet) and RTS (DenseNet). In other cases, the results of the models' confidence are comparable.

6.2 Attack Effectiveness Against Interpreters

This part evaluates the effectiveness of AdvEdge and AdvEdge$^+$ to generate similar interpretations to the benign inputs. We compare the interpretations of adversarial and benign inputs. We start with a qualitative comparison to check whether attribution maps generated by AdvEdge and AdvEdge$^+$ are indistinguishable from benign inputs. By observing all the cases, AdvEdge and AdvEdge$^+$ produced interpretations that are perceptually indistinguishable from their corresponding benign inputs. As for comparing with ADV^2, all methods generated attribution maps similar to the benign inputs. Figure 2 shows a set of sample inputs together with their attribution maps in terms of Grad, CAM, RTS, and MASK. As displayed in the figure, the results of our approaches provided high similarity with their benign interpretation.

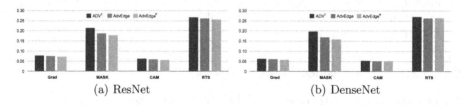

(a) ResNet   (b) DenseNet

Fig. 3. Average $\mathcal{L}_1$ distance of attribution maps generated by ADV^2, AdvEdge and AdvEdge$^+$ from those of corresponding benign samples on ResNet and DenseNet.

In addition to qualitative comparison, we use $\mathcal{L}_p$ to measure the similarity of produced attribution maps quantitatively. Figures 3(a) and 3(b) summarize the results of $\mathcal{L}_1$ measurement. As shown in the figures, our attacks generate adversarial samples with attribution maps closer to those generated for the benign samples compared to ADV^2. The results are similar across different interpreters on both target DNNs. We note that the effectiveness of our attack (against interpreters) varies depending on the interpreters. Generally, the results of Table 1, Figs. 3(a) and 3(b) show the effectiveness of our attack in generating adversarial inputs with highly similar interpretations to their corresponding benign samples.

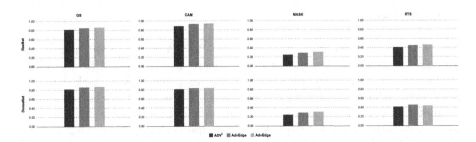

Fig. 4. IoU scores of attribution maps generated by ADV^2, AdvEdge and AdvEdge$^+$ using the four interpreters on ResNet and DenseNet. Our attacks achieve higher IoU scores in comparison with ADV^2.

Another quantitative measure to compare the similarity of attribution maps is the IoU score. As the attribution map values are floating numbers, we binarized the attribution maps to calculate the IoU. Figure 4 displays the IoU scores of attribution maps generated by ADV^2, AdvEdge and AdvEdge$^+$ using four interpreter models on ResNet and DenseNet. As shown in the figure, AdvEdge and AdvEdge$^+$ performed better than ADV^2. We note that AdvEdge$^+$ achieved significantly better results than other methods, while AdvEdge achieved a higher score on DenseNet with RTS interpreter.

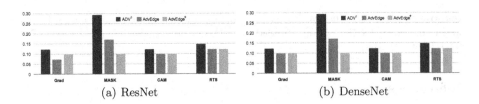

(a) ResNet   (b) DenseNet

Fig. 5. Noise rate of adversarial inputs generated by ADV^2, AdvEdge and AdvEdge$^+$ on ResNet and DenseNet.

6.3 Adversarial Perturbation Rate

To measure the amount of noise added to the adversarial image by the attacks, we utilize SSIM to measure the pixel-level perturbation. Using SSIM, we infer the parts of images that are not similar to the original images, known as noise. Figures 5(a) and 5(b) display the results of comparing the noise amount generated by ADV^2, AdvEdge, and AdvEdge$^+$. The figures show that the amount of noise added using AdvEdge and AdvEdge$^+$ is significantly lower than the amount added by ADV^2. This difference is more noticeable when using the MASK interpreter. AdvEdge$^+$ adds the least amount of noise to deceive the target DNN and the MASK interpreter.

7   Conclusion

This work presents two approaches (AdvEdge and AdvEdge$^+$) to enhance the adversarial attacks on interpretable deep learning systems (IDLSes). These approaches exploit the edge information to optimize the ADV^2 attack that generates adversarial inputs to mislead the target DNNs and their corresponding interpreter models simultaneously. We demonstrated the validity and effectiveness of AdvEdge and AdvEdge$^+$ through empirical evaluation using a large dataset on two different DNNs architectures (*i.e.,* ResNet and DenseNet). We show our results against four representatives of different types of interpretation models (*i.e.,* Grad, CAM, MASK, and RTS). The results show that AdvEdge and AdvEdge$^+$ effectively generate adversarial samples that can deceive the deep neural networks and the interpretation models.

Future Work. Besides utilizing the PGD framework, this work motivates exploring other attack frameworks, such as DeepFool, STADV [8]. Another future direction is to evaluate the effectiveness of potential countermeasures to defend against AdvEdge, such as refining the DNN and interpretation models (*e.g.,* via defensive distillation [10]) or applying an adversarial sample detector (*e.g.,* feature squeezing [16] or through utilizing the interpretation transferability property in an ensemble of interpreters [17]). We also consider to test whether our approach is applicable and tractable on various sample space (*e.g.,* numerical, text, etc.).

Acknowledgments. This work was supported by the National Research Foundation of Korea(NRF) grant funded by the Korea government(MSIT) (No. 2021R1A2C1011198).

References

1. Dabkowski, P., Gal, Y.: Real time image saliency for black box classifiers. arXiv preprint arXiv:1705.07857 (2017)
2. Fong, R.C., Vedaldi, A.: Interpretable explanations of black boxes by meaningful perturbation. In: Proceedings of the IEEE International Conference on Computer Vision, pp. 3429–3437 (2017)
3. Guidotti, R., Monreale, A., Ruggieri, S., Turini, F., Giannotti, F., Pedreschi, D.: A survey of methods for explaining black box models. ACM Comput. Surv. **51**(5), 1–42 (2018)

4. He, K., Zhang, X., Ren, S., Sun, J.: Deep residual learning for image recognition. In: Proceedings of the IEEE Conference on Computer Vision and Pattern Recognition, pp. 770–778 (2016)
5. He, Y., Meng, G., Chen, K., Hu, X., He, J.: Towards security threats of deep learning systems: a survey. IEEE Trans. Softw. Eng. 1 (2020)
6. Huang, G., Liu, Z., Van Der Maaten, L., Weinberger, K.Q.: Densely connected convolutional networks. In: Proceedings of the IEEE Conference on Computer Vision and Pattern Recognition, pp. 4700–4708 (2017)
7. Kindermans, P.-J., et al.: The (Un)reliability of saliency methods. In: Samek, W., Montavon, G., Vedaldi, A., Hansen, L.K., Müller, K.-R. (eds.) Explainable AI: Interpreting, Explaining and Visualizing Deep Learning. LNCS (LNAI), vol. 11700, pp. 267–280. Springer, Cham (2019). https://doi.org/10.1007/978-3-030-28954-6_14
8. Laidlaw, C., Feizi, S.: Functional adversarial attacks. In: NeurIPS (2019)
9. Madry, A., Makelov, A., Schmidt, L., Tsipras, D., Vladu, A.: Towards deep learning models resistant to adversarial attacks. arXiv preprint arXiv:1706.06083 (2017)
10. Papernot, N., McDaniel, P., Wu, X., Jha, S., Swami, A.: Distillation as a defense to adversarial perturbations against deep neural networks. In: 2016 IEEE Symposium on Security and Privacy (SP), pp. 582–597. IEEE (2016)
11. Recht, B., Roelofs, R., Schmidt, L., Shankar, V.: Do ImageNet classifiers generalize to ImageNet? In: International Conference on Machine Learning, pp. 5389–5400. PMLR (2019)
12. Selvaraju, R.R., Cogswell, M., Das, A., Vedantam, R., Parikh, D., Batra, D.: Grad-CAM: visual explanations from deep networks via gradient-based localization. In: Proceedings of the IEEE International Conference on Computer Vision, pp. 618–626 (2017)
13. Simonyan, K., Vedaldi, A., Zisserman, A.: Deep inside convolutional networks: visualising image classification models and saliency maps (2014)
14. Wang, Z., Bovik, A.C., Sheikh, H.R., Simoncelli, E.P.: Image quality assessment: from error visibility to structural similarity. IEEE Trans. Image Process. **13**(4), 600–612 (2004)
15. Xiao, C., Zhu, J.Y., Li, B., He, W., Liu, M., Song, D.: Spatially transformed adversarial examples. arXiv preprint arXiv:1801.02612 (2018)
16. Xu, W., Evans, D., Qi, Y.: Feature squeezing: detecting adversarial examples in deep neural networks. arXiv preprint arXiv:1704.01155 (2017)
17. Zhang, X., Wang, N., Shen, H., Ji, S., Luo, X., Wang, T.: Interpretable deep learning under fire. In: 29th {USENIX} Security Symposium ({USENIX} Security 20) (2020)
18. Zhou, B., Khosla, A., Lapedriza, A., Oliva, A., Torralba, A.: Learning deep features for discriminative localization. In: Proceedings of the IEEE Conference on Computer Vision and Pattern Recognition, pp. 2921–2929 (2016)

Incorporating Transformer Models for Sentiment Analysis and News Classification in Khmer

Md Rifatul Islam Rifat[1]([✉])[ID] and Abdullah Al Imran[2][ID]

[1] Rajshahi University of Engineering and Technology, Rajshahi, Bangladesh
[2] American International University-Bangladesh, Dhaka, Bangladesh

Abstract. In recent years, natural language modeling has achieved a major breakthrough with its sophisticated theoretical and technical advancements. Leveraging the power of deep learning, transformer models have created a disrupting impact in the domain of natural language processing. However, the benefits of such advancements are still inscribed between few highly resourced languages such as English, German, and French. Low-resourced language such as Khmer is still deprived of utilizing these advancements due to lack of technical support for this language. In this study, our objective is to apply the state-of-the-art language models within two empirical use cases such as Sentiment Analysis and News Classification in the Khmer language. To perform the classification tasks, we have employed FastText and BERT for extracting word embeddings and carried out three different type of experiments such as FastText, BERT feature-based, and BERT fine-tuning-based. A large text corpus including over 100,000 news articles has been used for pre-training the transformer model, BERT. The outcome of our experiment shows that in both of the use cases, a pre-trained and fine-tuned BERT model produces the outperforming results.

Keywords: Khmer · Deep learning · Sentiment analysis · News classification · Transformer models

1 Introduction

Last two decades have seen a growing trend towards a field named Natural Language Processing (NLP) that is concerned with the interactions between computers and human (natural) languages. Numerous applications of NLP have already been implemented in our real-life such as Sentiment Analysis, Question Answering, Chatbots, Machine Translation, Speech Recognition and so on.

In this study, we have aimed to work with the Khmer language, the official language of Cambodia. According to Wikipedia, Khmer is the second most widely spoken Austroasiatic language in the world with approximately 16 million speakers. Unlike English, French, and German, over the past two decades, few studies have been conducted for the Khmer language related to NLP. That

© Springer Nature Switzerland AG 2021
D. Mohaisen and R. Jin (Eds.): CSoNet 2021, LNCS 13116, pp. 106–117, 2021.
https://doi.org/10.1007/978-3-030-91434-9_10

is why, there are few NLP resources available for the Khmer language that have been built based on the traditional techniques.

Considering the inadequacy of Khmer NLP resources, in this work, we have provided a pre-trained BERT model for the Khmer language along with two downstream benchmarks for Sentiment Analysis and News Category Classification. BERT [1], which stands for Bidirectional Encoder Representations from Transformers, developed by Google, is the first fine-tuning based representation model. Through this architecture, the models can learn the context of a sentence from both of the direction (left and right). This model outperforms many task-specific architectures as well as achieves state-of-the-art performance on a large suite of sentence-level and token-level tasks.

Apart from English, Google also provided a multilingual BERT model that supports for 104 languages of different countries but unfortunately Khmer has not been included in the list of supported languages. This might be due to the unique grammatical architecture of the Khmer. Unlike in Latin languages such as English or French, Khmer has more complex structure to word form and permits different orders of the character components that lead to the same visual representation. So far, there is no available BERT pre-trained model, which has been trained exclusively using the Khmer language, based on which the Cambodian people can develop NLP models for different applications. That is why, we aim to provide a pre-trained BERT model which will definitely add a new dimension to Khmer language and its computational advancement.

We have pre-trained our BERT model with a large corpus composed of Khmer online newspapers and also performed some downstream tasks such as Sentiment Analysis and News Classification by fine-tuning our pre-trained BERT model. Both of these two applications, Sentiment Analysis and News Classification, have a significant impact on business. More specifically, for a consumer-centric industry, which focuses on making a positive experience by maximizing the quality of services or products, the Sentiment Analysis model can help to figure out the customers' demand based on the product or service reviews. In contrast, every day a huge number of employees of the news industry spend their time arranging the news based on the category of the news. By the utilization of our News Classification model, the news industry of Cambodia will be able to save their time as well as the manpower.

In order to validate the performance of our BERT model, we have applied 4 different evaluation metrics namely accuracy, precision, recall, and f1-score. In addition to this, we have included a comparative analysis with a context-free word embeddings named FastText. For both of the applications, we have conducted the experiment through three distinct approaches such as FastText, BERT Feature-based, and BERT Fine-tuning. Significantly, the BERT Fine-tuning approach outperforms the other two approaches by accuracy and f1-score.

2   Background Study

In recent years, there has been an increasing amount of literature on language modeling. But, there is a relatively small body of literature that is concerned with the Khmer language. Among these studies, the authors proposed some basic resources of Khmer language such as character recognition, word segmentation, and POS tagging. This section briefly discusses some of the latest and relevant background studies.

The authors in [2] proposed techniques for recognizing the character and text from Khmer ancient palm leaf documents. For isolated character recognition, they applied different types of neural network architectures such as CNN, LSTM-RNN and found the outperforming result with the combination of both convolutional and recurrent architectures. On the other hand, they used both one-dimensional and two-dimensional RNN to recognize the word/text image patches of variable length and found that two-dimensional RNN performs better than one-dimensional.

For building lemmatizers as well as extracting relation between words, POS tagging is one of the essential tasks. Considering the sentence structure and word classes' ambiguties of Khmer, in 2007, the eminent authors [3] modified the applying rule algorithms and proposed a supervised transformation-based POS tagger. Moreover, to handle the unknown words, they proposed a hybrid approach which combines rule-based and tigram approach. On the other hand, based on the Conditional Radom Fields (CRFs), in 2017, the authors [4] proposed a new approach to Khmer POS tagging by incorporating 5 groups of features such as contextual, morphological, word-shape, named-entity, and lexical features.

Like Khmer, for each of the languages, word segmentation is an integral part of all of the language modeling related tasks. Unlike English, however, there are no spaces in the writing system of the Khmer language to separate the words that make it more complex. The authors in [5–7] have worked on the word segmentation of the Khmer language. In paper [5], the authors applied Maximum Matching Algorithm and a Khmer manual corpus to make word boundaries in each sentence. Then intended to solving the unknown words, they created 21 grammar rules based on the principle of Khmer grammar books. On the other hand, to reduce the frequency of dictionary lookup and Khmer text manipulation tweaks, the paper [6] presented a study on Bi-directional Maximal Matching (BiMM). They implemented their study for Khmer word segmentation by focusing on both Plaintext and Microsoft Word document. In 2015, Chea and co-workers [7] presented a word segmenter for the Khmer language based on a supervised CRF segmentation method. Their proposed segmenter outperformed the baseline in terms of precision (=0.986), recall (0.983), and f-score (=0.985) by a wide margin. Surprisingly, they obtained substantial increases in the BLEU score of up to 7.7 points, relative to a maximum matching baseline, in their evaluation in a statistical machine translation system.

For many aspects of machine translation, the parallel corpus is essential. But, the existing parallel corpus contains some problems such as difficulty in obtaining, narrow fields, small quantity, and poor timeliness. To overcome this insufficiency of bilingual parallel corpus of low resource languages, in a recent paper [8], published in 2020, the prominent authors proposed a parallel fragment extraction method based on the Dirichlet process. On the basis of the empirical results, the authors concluded that the method based on the Dirichlet process is better than that based on the LDA model.

From the above discussion, it is obvious that there are few studies related to the Khmer language where the researchers applied traditional techniques on different NLP tasks. However, there does not exists enough NLP resources for the Khmer language that has been built based on state-of-the-art techniques. This is where this study will play a vital role in mitigating the gap by providing a state of the art resource for the Khmer language.

3   Data Description

For the three different experiments, such as creating the pre-trained BERT model for Khmer language, sentiment analysis, and news classification, we have collected three different corpora. The entire data collection process has been illustrated in the following Fig. 1.

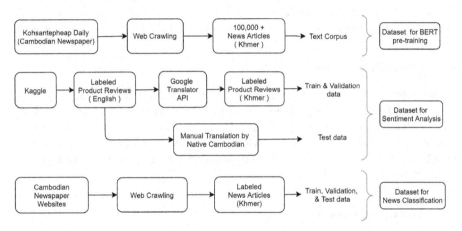

Fig. 1. Data collection pipeline.

3.1   Dataset for BERT Pre-training

In NLP, the concept behind pre-training refers to train a model with a large corpus for learning the underlying knowledge from the data. Then, the pre-trained model is used to initialize the model parameters of various downstream tasks. For creating the pre-trained BERT model of Khmer language, we have collected a large corpus from a popular Cambodian newspaper named Kohsantepheap Daily [9]. The daily newspaper Koh Santepheap was founded in 1967 by Chou Thany. The corpus contains more than 100,000 news articles that have been scraped from the official website.

3.2   Dataset for BERT Fine-Tuning

For two different downstream tasks, we have prepared 2 different datasets for fine-tuning each of the tasks.

1. **Dataset for Sentiment Analysis:** The Sentiment Analysis dataset contains 10710 instances including 5367 for positive, and 5343 for negative sentiment. For model testing purpose, we have collected additional 400 samples validated by a native Cambodian.
2. **Dataset for News Classification:** We have collected the News Classification data through scraping several Cambodian newspaper websites. The dataset includes 7418 instances with 8 different news categories as class labels. For training purpose we have used 70%(=5192), for validation 21%(=1558), and for testing 9%(=668) instances from the whole dataset.

4   Data Preprocessing

One of the crucial phases of solving a real-world problem using NLP is to pre-process the raw dataset properly. Usually, the NLP pipeline includes a bunch of tasks such as removing the punctuation, tokenization, removing the stop words, and so on. In this study, we have arranged this phase according to the form of our datasets that includes some basic preprocessing followed by word segmentation.

4.1   Basic Preprocessing

To keep only the letters and digits of the Khmer and English languages, we have applied some regular expressions on raw data using the Unicode. For basic Khmer characters, the Unicode block is U+1780 – U+17FF. After that, we have removed the punctuations of the Khmer language. The most commonly used punctuations used in the Khmer language are ។, ៕, ៖, ៗ, ៘៕, ៚, ៙ and their Unicode range is U+17D4 – U+17DA.

4.2   Word Segmentation

Word segmentation is an essential task in every application of Natural Language Processing. In the case of Khmer language, word segmentation is not a trivial task because, unlike English, spaces are not used here to separate words. Moreover, multiple Khmer words can be joined together to build up a new Khmer word (compound word) that conveys different meaning. Some examples of Khmer compound word are given in Table 1.

Table 1. Khmer Compound Words

Examples	Explanation
កឬផ្សេង	ក (or) + ឬផ្សេង (but) = កឬផ្សេង (but)
ផងដែរ	ផង (also) + ដែរ (also) = ផងដែរ (as well)
ទទួលខុសត្រូវ	ទទួល (receive) + ខុស (wrong) + ត្រូវ (right) = ទទួលខុសត្រូវ (be responsible for)

Another complexity in the writing system of the Khmer language is, a single sentence can be tokenized in several ways based on its meaning in the context [7]. Table 2 shows the distinct segmentation of an identical sentence.

Table 2. Two distinct segmentation of an identical sentence.

Khmer	English
ខ្ញុំ ចង់ឱ្យ < អ្នកស្ដាប់ > យល់ ពី បញ្ហា នេះ	I want listener to understand this problem
ខ្ញុំ ចង់ឱ្យ < អ្នក > < ស្ដាប់ > យល់ ពី បញ្ហា នេះ	I want you to listen in order to understand this problem

All these complications address many challenges in the tokenization of Khmer text. In this study, we have applied a conditional random fields (CRFs) based approach from [7] to separate the words in Khmer sentences. The authors developed a large manually-segmented corpus and also provided a set of word segmentation strategies usually used by humans.

5   Methodology

In this section, we have described the workflow of our experiment in a proper sequence. Our entire experiment can be divided into two steps: one is using the FasText and the other one using BERT. Again, BERT is composed of three distinct parts such as pre-training, fine-tuning, and feature-based. Each of these steps has been described extensively in the subsections below.

5.1   FastText

In order to make the baseline, firstly, we have conducted our experiment using the FastText word embeddings. The FastText library of the Khmer language contains 242,732 vocabs and against each of the words, it provides a vector of length 300. On the basis of these embeddings, in this phase, we have designed two DNN models for sentiment analysis and news classification.

In the architecture of a DNN model, the number of neurons in its input layer depends on the length of the feature embeddings. As the length of the feature vectors is 300, we have specified 300 neurons in the input layer for both of the models. Determining the number of hidden layers and the number of neurons in the hidden layers is a crucial task as it defines the complexity and efficiency of the network and has an enormous impact on the learning process of the model. In both the sentiment analysis and news classification model architectures, we have specified the identical number of hidden layers as well as the number of neurons in hidden layers that are shown in Table 3.

Table 3. Parameters of DNN Architectures.

Sentiment analysis			News classification		
Layers	Number of neurons	Activation function	Layers	Number of neurons	Activation function
Input Layer	300		Input Layer	300	
Dense 1	1024	tanh	Dense 1	1024	tanh
Dense 2	512	tanh	Dense 2	512	tanh
Dense 3	256	tanh	Dense 3	256	tanh
Dense 4	128	tanh	Dense 4	128	tanh
Output Layer	2	softmax	Output Layer	8	softmax

On the other hand, in the case of the classification model, the number of neurons in the output layer depends on the unique class labels. According to this terminology, we have specified the number of neurons in the output layer 2 and 8 for the sentiment analysis and news classification, respectively.

As our extracted embeddings contain both positive and negative values, thus we have applied the *tanh* as the activation function [11] of the hidden layers. The mathematical function ot *tanh* can be expressed as,

$$tanh(x) = \frac{e^x - e^{-x}}{e^x + e^{-x}} \tag{1}$$

The output of this function ranges from −1 to 1. To minimize the losses we have applied the optimization algorithm named Adam with a learning rate of 0.001.

5.2   BERT

This phase starts with pre-training the BERT model with Khmer followed by fine-tuning and feature-based approaches for the downstream tasks. The extensive discussion of this subsections are as follows:

Table 4. Configurations of BERT model

Parameters	Values
attention_probs_dropout_prob	0.1
hidden_act	gelu
hidden_dropout_prob	0.1
hidden_size	768
initializer_range	0.02
intermediate_size	307
max_position_embeddings	512
num_attention_heads	12
num_hidden_layers	12
type_vocab_size	2
vocab_size	167896

Pre-training Model Architecture. In this phase, we have built a pre-trained model for the Khmer language using BERT. BERT's model architecture is a multi-layer bidirectional transformer encoder based on [10]. In the configurations that we have applied in our experiment, the number of hidden layers (i.e.Transformer blocks) is 12, the size of each of the hidden layer is 768, and the self-attention heads are 12. Table 4 shows the configurations of our BERT model.

Fine-Tuning Model Architecture. Fine-tuning is a supervised learning process where the weights of the pre-trained model are used as the initial weights for a new model which is being trained on a similar task. This process not only speeds up the training but also creates a state-of-the-art model for a wide range of NLP tasks.

In this study, we have fine-tuned our pre-trained BERT model for two different applications such as Sentiment Analysis and News Classification. Firstly, we have initialized the fine-tuned model with the same pre-trained model parameters for both of the downstream tasks, such as Sentiment Analysis and News Classification. Then, we have fine-tuned all of the parameters end-to-end using the corresponding task-specific labeled data. Eventually, we have incorporated an additional output layer according to the target classes. Thus, the fine-tuned models for both of the downstream tasks are different, even though they have been initialized with the same pre-trained parameters. Figure 2 represents the architecture of the models.

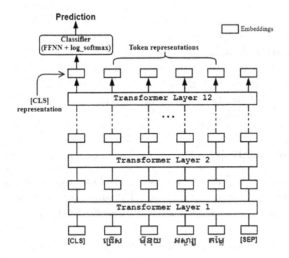

Fig. 2. Architecture of the BERT Fine-tuning Model

Feature-Based Approach. In this phase, we have extracted the feature embeddings from our pre-trained BERT model. Then, by applying these extracted features, we have designed two Deep Neural Network (DNN) models: one is for sentiment analysis and the other one is for news classification.

As the length of the BERT feature embeddings is 768, we have specified 768 neurons in the input layer for both of the models. However, we have applied 4 hidden layers for the sentiment analysis model while 5 hidden layers for the news classification model. For both of the BERT Feature-based models, the number of neurons in each of the hidden layers are shown in Table 5. According to aforementioned terminology in FastText, we have specified the number of neurons in the output layer 2 and 8 for the sentiment analysis and news classification, respectively. Like FastText, we have applied the tanh as the activation function and the Adam as the optimization algorithm in our featured-based model architectures of BERT.

Table 5. Parameters of DNN Architectures.

Sentiment analysis			News classification		
Layers	Number of neurons	Activation function	Layers	Number of neurons	Activation function
Input Layer	768		Input Layer	768	
Dense 1	1024	tanh	Dense 1	2048	tanh
Dense 2	512	tanh	Dense 2	1024	tanh
Dense 3	256	tanh	Dense 3	512	tanh
Dense 4	256	tanh	Dense 4	256	tanh
Output Layer	2	softmax	Dense 5	128	tanh
			Output Layer	8	softmax

6   Results and Analysis

In this section, we have evaluated the performance of our models, both sentiment analysis and news classification, using the classification metrics namely accuracy, precision, recall, and f1-score.

6.1   Sentiment Analysis

After the training and validation, we have tested our models with a test dataset that contains 400 instances. The obtained results for the three models of the sentiment analysis have been presented in Table 6 and Table 7.

Table 6. Accuracy of sentiment analysis models.

Accuracy			
	FastText	BERT (Feature-based)	BERT (Fine-tuning)
Training	0.79	0.73	0.83
Validation	0.78	0.71	0.83
Test	0.77	0.70	**0.81**

Table 6 shows the training, validation, and testing accuracies for the three models: FastText, BERT Feature-based, and BERT Fine-tuning of sentiment analysis. It is apparent from this table that, for each of the models, our obtained validation accuracy is very close to the training accuracy and slightly lower which indicates that each of the models has learned the underlying patterns very well from the data without overfitting. Sometimes, solely accuracy is not a good measure for the evaluation of a classification model. For this reason, we have also analyzed other metrics such as precision, recall, and f1-score during the testing phase of the sentiment analysis models that presents in Table 7.

Table 7. Precision, recall, and f1-score of sentiment analysis models.

	FastText			BERT feature-based			BERT fine-tuning		
	precision	recall	f1-score	precision	recall	f1-score	precision	recall	f1-score
Negative	0.86	0.67	0.75	0.77	0.61	0.68	0.86	0.77	0.81
Positive	0.71	0.88	0.78	0.65	0.80	0.72	0.77	0.86	0.81
macro avg	0.78	0.77	0.77	0.71	0.70	0.70	0.81	0.81	0.81
weighted avg	0.79	0.77	0.77	0.71	0.70	0.70	0.81	0.81	0.81

From both of the Table 6 and Table 7, it can be observed that the BERT Fine-tuning model outperforming the other two by the accuracy and f1-score. Another important finding is that, in the case of sentiment analysis, surprisingly, the FastText is performing better than the BERT Feature-based model.

6.2 News Classification

Similar to sentiment analysis, after the training and validation of our news classification models, we have tested our models with a dataset that contains 668 samples. The training, validation, and testing accuracies for each of the news classification models are presented in Table 8.

Table 8. Accuracy of News Classification models.

Accuracy			
	FastText	BERT (Feature-based)	BERT (Fine-tuning)
Training	0.83	0.85	0.89
Validation	0.82	0.84	0.85
Test	0.83	0.82	**0.85**

From Table 8, it can be observed that the validation accuracy is slightly lower than the training accuracy for each of the model's output. This lower differences in training and validation accuracies indicate that each of the models has learned the underlying patterns very well from the news data without overfitting. We have obtained a good accuracy for each of the news classification models while BERT Fine-tuning model outperforms the others. Unlike sentiment analysis, BERT Feature-based model also achieved higher training and validation accuracies compared to FastText. The other metrics such as precision, recall, and f1-score have also been analyzed during the testing phase of news classification models are shown in Table 9.

Table 9. Precision, recall, and f1-score of News Classification models.

Class Labels	FastText			BERT Feature-based			BERT Fine-tuning		
	precision	recall	f1-score	precision	recall	f1-score	precision	recall	f1-score
arts-and-culture	0.90	0.93	0.92	0.93	0.91	0.92	0.95	0.91	0.93
business	0.80	0.66	0.72	0.70	0.73	0.72	0.86	0.83	0.84
health	0.80	0.90	0.85	0.77	0.88	0.82	0.74	0.92	0.82
international	0.75	0.83	0.79	0.80	0.83	0.82	0.86	0.83	0.85
national	0.86	0.88	0.87	0.84	0.87	0.85	0.90	0.89	0.89
research	0.73	0.59	0.65	0.68	0.59	0.63	0.67	0.69	0.68
service	0.86	0.89	0.87	0.90	0.78	0.83	0.89	0.76	0.82
sports-news	0.97	0.94	0.96	0.90	0.96	0.93	0.96	0.96	0.96
macro avg	0.83	0.83	0.83	0.82	0.82	0.81	0.86	0.85	0.85
weighted avg	0.83	0.83	0.83	0.82	0.82	0.82	0.85	0.85	0.85

From Table 9, it is apparent that, like sentiment analysis, BERT Fine-tuning outperforms the other two models of news classification by precision, recall, and f1-score. Interestingly, for both of the applications, the test scores of the FastText models are slightly higher than the BERT feature-based models.

7   Conclusion

The purpose of this study was to employ the state-of-the-art natural language processing techniques for Khmer language, one of the most low-resourced languages currently available. In this study, we have defined two widely used application scopes such as news category classification and sentiment analysis. Three different type of experiments such as FastText (feature-based), BERT (feature-based), and BERT (fine-tuning-based) have been conducted for both of the aforementioned downstream tasks. The experimental results show that in terms of Sentiment Analysis, BERT fine-tuning based approach outperformed the other approaches with a test accuracy of 81%. Similarly, in terms of News Classification, again BERT fine-tuning based approach stood out as the best performer with a test accuracy of 85%. In future, we would like to investigate other state-of-the-art variants of BERT such as RoBERT, DistilBERT, XLM-RoBERTa and the new giant GPT-3 for Khmer language.

References

1. Devlin, J., Chang, M. W., Lee, K., Toutanova, K.: BERT: pre-training of deep bidirectional transformers for language understanding. CoRR, abs/1810.04805 (2018). http://arxiv.org/abs/1810.04805
2. Valy, D., Verleysen, M., Chhun, S., Burie, J.C.: Character and text recognition of Khmer historical palm leaf manuscripts. In: 16th International Conference on Frontiers in Handwriting Recognition (ICFHR), pp. 13–18. IEEE (2018). https://doi.org/10.1109/ICFHR-2018.2018.00012
3. Sangvat, S., Pluempitiwiriyawej, C.: Khmer POS tagging using conditional random fields. In: Hasida, K., Pa, W.P. (eds.) PACLING 2017. CCIS, vol. 781, pp. 169–178. Springer, Singapore (2018). https://doi.org/10.1007/978-981-10-8438-6_14
4. Nou, C., Kameyama, W.: Khmer POS tagger: a transformation-based approach with hybrid unknown word handling. In: International Conference on Semantic Computing (ICSC), pp. 482–492. IEEE (2007). https://doi.org/10.1109/ICSC.2007.104
5. Long, P., Boonjing, V.: Longest matching and rule-based techniques for Khmer word segmentation. In: 10th International Conference on Knowledge and Smart Technology (KST), pp. 80–83. IEEE (2018). https://doi.org/10.1109/KST.2018.8426109
6. Bi, N., Taing, N.: Khmer word segmentation based on bi-directional maximal matching for plaintext and microsoft word document. In: Signal and Information Processing Association Annual Summit and Conference (APSIPA), Asia-Pacific, pp. 1–9. IEEE (2014). https://doi.org/10.1109/APSIPA.2014.7041822
7. Chea, V., Thu, Y. K., Ding, C., Utiyama, M., Finch, A., Sumita, E.: Khmer word segmentation using conditional random fields. In: Khmer Natural Language Processing, pp. 62–69 (2015)
8. Ning, S., Yan, X., Nuo, Y., Zhou, F., Xie, Q., Zhang, J.P.: Chinese-Khmer parallel fragments extraction from comparable corpus based on Dirichlet process. Procedia Comput. Sci. **166**, 213–221 (2020)
9. Koh Santepheap Daily. https://kohsantepheapdaily.com.kh/. Accessed 28 Aug 2020
10. Vaswani, A., et al.: Attention is all you need. In: Advances in Neural Information Processing Systems, pp. 5998–6008. Curran Associates, Inc. (2017). http://papers.nips.cc/paper/7181-attention-is-all-you-need.pdf
11. Nwankpa, C., Ijomah, W., Gachagan, A., Marshall, S.: Activation functions: comparison of trends in practice and research for deep learning. CoRR, abs/1811.03378 (2018). http://arxiv.org/abs/1811.03378

Deep Bangla Authorship Attribution Using Transformer Models

Abdullah Al Imran[1]([✉]) [ID] and Md Nur Amin[2] [ID]

[1] American International University-Bangladesh, Dhaka, Bangladesh
[2] University Jean Monnet, Saint Etienne, France

Abstract. Authorship attribution is one of the renowned problems in the domain of Natural Language Processing (NLP). Leveraging the state-of-the-art (SOTA) techniques of NLP such as transformer models, this problem domain has achieved a considerable advancement. However, this progress is unfortunately only bound to the well-resourced languages like English, French, and German. Under-resourced language like Bangla is yet to leverage such SOTA techniques to make a breakthrough in this domain. In this study, we address this research gap and aim to contribute to the Bangla authorship attribution problem by building highly accurate models using several SOTA variants of transformer models like mBERT, bnBERT, bnElectra, and bnRoBERTa. Using the pre-trained weights of these models we have performed fine-tuning and tackled the task of authorship attribution of 16 prominent Bangla writers. Outcomes show that our bnBERT model can classify the authors with superior accuracy of 98% and also outperform all the existing models available in the literature.

Keywords: Bangla · Authorship attribution · Text classification · Natural Language Processing · Transformer models · BERT

1 Introduction

Because of the dramatic rise in the usage of the internet and its easy access through smart devices, a large amount of textual content is being written or posted to the web in digital form at a rapid rate. The growing prevalence of text digitization makes the detection of authorship extremely challenging. As a consequence, the automated identification or attribution of authorship has garnered considerable research importance over the last few years.

The fundamental notion underlying authorship attribution is discriminating writing of various authors by measuring specific textual characteristics that is followed consistently throughout author's work. A text of unknown author is attributed to one of a specified set of possible authors for whom text samples are provided in the general authorship attribution. Authorship identification has a wide range of applications, including plagiarism detection, forensic linguistics,

© Springer Nature Switzerland AG 2021
D. Mohaisen and R. Jin (Eds.): CSoNet 2021, LNCS 13116, pp. 118–128, 2021.
https://doi.org/10.1007/978-3-030-91434-9_11

military intelligence, civil law, and many more. As a result, authorship attribution has gained a considerable attention to the Natural Language Processing (NLP) researchers.

Every author possesses a distinguished manner of writing, separating one author from the other. Traditionally, stylometry has been used to detect distinguishable features of an author's work. However, the recent innovations in the domain of NLP is empowering us to tackle the task of finding distinguishable writing patterns more accurately and efficiently from a large text corpus. One such state-of-the-art (SOTA) NLP technique is called transformer model. A transformer model is an attention-based stacked encoder–decoder architecture that is pre-trained at scale over a large text corpus. These models are designed to have sophisticated internal attention mechanisms that enable these models to dynamically select subsets of the input to focus.

The current research trend shows that transformer models are being applied for various NLP downstream tasks including the authorship attribution problem. However, such research practices can be easily found for well-resourced languages like English, French, German, and Chinese. On the other hand, the low-resourced language like Bangla is far away from leveraging the SOTA models such as transformers in different natural language tasks.

In this study, we are aiming fill up this research gap through tackling the authorship attribution problem in Bangla by leveraging several variants of the SOTA transformer model. To achieve the purpose, we have chosen four different variants of the Bidirectional Encoder Representations from Transformers (BERT) model namely BERT Multilingual (mBERT), BERT Bangla Base (bnBERT), Electra Bangla Base (bnElectra), and RoBERTa Bangla Base (bnRoBERTa). All of these are pre-trained models that has been trained over a large text corpus. We have utilized the pre-trained weights of these models and implement the fine-tuning methods with a classification layer to perform authorship attribution as a downstream task. We have used a portion of the benchmark dataset named BAAD16 (Bangla Authorship Attribution Dataset) [9] for the fine-tuning purpose and a portion for the evaluation purpose. For the task of classifying 16 different authors, all of the fine-tuned models have achieved an accuracy over 87%. Among all models, the performance of bnBERT (=98% accuracy) found to be superior even to all available models in the literature.

2   Related Works

Being a low-resourced language, Bangla lacks a significant amount of study in the field of Natural Language Processing, particularly in the classic problem of authorship attribution. Statistical and classical machine learning approaches using traditional feature extraction algorithms are employed in the majority of the studies among the few that were available. Some of them have been found employing neural networks and deep learning-based approaches in their studies. In this section, we'll briefly discuss the relevant studies and the progress of natural language processing for the authorship attribution problem in the Bangla language.

One common and classic way to classify authorship is by using stylometry; a linguistic approach that takes into account an author's unique literary styles that can not be repressed by conscious mind. M. Tahmid et al. [1] employed traditional statistical analysis to identify the most effective stylometric features such as most frequent words, most frequent bigrams, and word lengths to create a modified stylometric feature set that has significant differences among authors writing patterns. Using these features, they proposed a voting system that can detect the original author of an unknown document with an average 90.67% success rate. The training and evaluation of their system were performed over two corresponding datasets of 700 and 300 writings of six Bangladeshi columnists of the current time. Likewise, Md. Ashikul et al. [2] also employed a similar approach where they performed statistical analysis on the experiment of authorship attribution problem using multilayer perceptron (MLP) model. For this statistical analysis and experiment they built a corpus containing articles of five Bangla writers from which a total number of 1381 articles were used for training, 242 for validation and 350 for testing purposes. Their proposed MLP model achieved above 85% accuracy rate for each of the writers.

Characters are the smallest unit of text from which stylometric signals can be extracted to find the author of a text. Taking this into account, Khatun et al. [3] investigated character-level and word-level CNN approaches to identify writing patterns and distinguish between authors. They have applied both of these approaches on the authorship classification problem with author classes from 6–14. They found that the character-level CNN approach is more efficient than the word-level CNN approach in terms of the model training time and memory efficiency. Although it comes at the cost of compromising accuracy of 2–5% than the best performing word-level models. They also mentioned that using the pre-trained word embedding models the performance can be improved up to 10%. The best accuracy (=98%) they achieved is for classifying 6 authors using fastText and skip-gram. However, in classifying 14 authors, they achieved 81.2% accuracy with fastText and skip-gram.

Another common practice in tackling authorship attribution problem is the use of pre-trained word embeddings for feature extraction. Hemayet et al. [4], used pre-trained word embedding models (word2vec, Glove and fasText) in order to extract features from texts considering the context and co-occurrence of the words. Later the extracted features were fed into their proposed convolutional and recurrent neural network based classifiers. Finally, they performed a comparative analysis and showed the model performance on each of these word embedding methods and during a classification task on blog articles of 6 authors. The fastText word embedding coupled with convolutional architecture achieved the best accuracy of 92.9%. Again, in another paper, Hemayet et al. [5] performed a comparative study among fastText's hierarchical classifier, Naive Bayes, and SVM along with n-gram based feature sets. They showed that fasText's hierarchical classifier along with uni-gram features achieved the highest accuracy of 82.4% over SVM and Naive Bayes.

In [6], Ahmed et al. proposed a profile-based technique that uses supervised learning methods to identify the lyricist of Bangla songs composed by two renowned poets and novelists, Kazi Nazrul Islam and Rabindranath Tagore. Working on song lyrics made it a challenging task as the selection of words by the novelists relies on factors such as rhythms and completeness. Out of the supervised models (Naïve Bayes, Simple Logistic Regression, Decision Tree, Support Vector Machine, and Multilayer Perceptron) used in this problem, the Simple Logistic Regression model achieved the highest accuracy of 86.29%. As every writer has his own way of writing, thus, the probability to use some adjacent word is more likely by a particular author. Similarly, such traditional approaches were also performed by D. M. et al. [7] where they proposed a hybrid approach of n-gram amalgamating with Naive Bayes for authorship attribution. In this approach, bigram and unigram counts for adjacent and single words are calculated by combining the n-Gram algorithm with Naïve Bayes. This approach outperforms the sole Naive Bayes model by a margin of 9% accuracy. Again, Sumnoon Ibn et al. [8] also used n-gram and parts of speech as features and proposed a Multilayer Perceptron (MLP) based neural architecture for classifying 23 authors from a text corpus consisting of 12,142 writings. Their MLP model produced an accuracy of 94%.

From the above discussion, it is apparent that similar types of approaches and models have been studied over and over again. The most used traditional approaches are n-gram based feature extraction with classical machine learning algorithms and multilayer perceptrons. On the other hand, few studies investigated the some advanced but most common pre-trained word embedding models such as word2vec, fastText, and Glove. It is apparent that the authorship attribution problem in Bangla language still lacks studies concerned with the state-of-the-art transformer models from the recent advancements of natural language processing research. In this study, we have addressed this lacking and showed how the latest transformer models can be used to outperform all the previous approaches in the authorship attribution problem in Bangla. Hence, we are aiming to contribute towards making Bangla from a low-resourced to a well-resourced language in NLP research.

3   Data Description

In this study, we have used the BAAD16 (Bangla Authorship Attribution Dataset) [9] dataset. BAAD16 includes Bangla text samples from 16 prominent Bangla authors containing a total of 13.4+ million words. The dataset is equally partitioned with each document having the same length of 750 words. The dataset is partitioned into two parts training data and test data. The training set includes 14374 and the testing set includes 3592 data samples. From the training samples we have drawn a sample of 2875 instances through stratified sampling as a validation set. So, finally 11499 instances remain in the training dataset. The class distribution of training, validation and test datasets are given in the below Table 1.

Table 1. Class distribution of the datasets

No	Class Labels	Train	Train %	Validation	Validation %	Test	Test %
1	humayun_ahmed	2889	25.12%	723	25.15%	906	25.22%
2	shunil_gongopaddhay	1256	10.92%	314	10.92%	393	10.94%
3	shomresh	901	7.84%	225	7.83%	282	7.85%
4	*shorotchandra*	*841*	*7.31%*	*210*	7.30%	261	7.27%
5	robindronath	806	7.01%	201	6.99%	252	7.02%
6	m_zafar_iqbal	704	6.12%	176	6.12%	220	6.12%
7	shirshendu	670	5.83%	168	5.84%	210	5.85%
8	toslima_nasrin	*596*	5.18%	149	5.18%	186	5.18%
9	shordindu	569	4.95%	142	4.94%	177	4.93%
10	shottojit_roy	544	4.73%	136	4.73%	169	4.70%
11	tarashonkor	496	4.31%	124	4.31%	155	4.32%
12	bongkim	360	3.13%	90	3.13%	112	3.12%
13	nihar_ronjon_gupta	305	2.65%	76	2.64%	95	2.64%
14	manik_bandhopaddhay	301	2.62%	75	2.61%	93	2.59%
15	nazrul	143	1.24%	36	1.25%	44	1.22%
16	zahir_rayhan	118	1.03%	30	1.04%	37	1.03%

4   Methodology

In this section, we will briefly discuss the complete workflow of the experiment that has been performed in this study. Firstly, we have performed some basic data preprocessing steps over our datasets. Secondly, we have chosen the pre-trained transformer models which will be fine-tuned later on our datasets for the downstream classification task. Thirdly, we have evaluated the performance of all the fine-tuned models using a comprehensive list of evaluation metrics. Finally, we have analysed and compared the performance of the models and identified the best classification model for our authorship attribution problem. A detailed flow-diagram of the entire workflow is presented in the Fig. 1 below.

4.1   Data Preprocessing

In a natural language processing task, textual data always needs to be preprocessed before moving into modelling. In this study, we have performed a few basic preprocessing operations on our datasets. We have used regular expressions to remove all the noises from the texts such as special characters, and punctuations. We have built a comprehensive list of stopwords of Bangla language and used the list to remove all the stopwords from the texts.

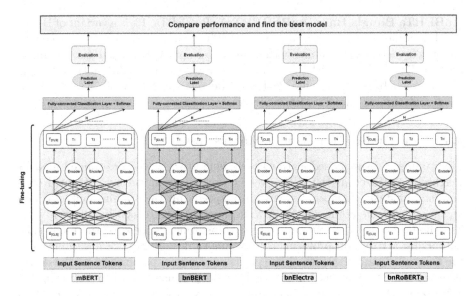

Fig. 1. Workflow used in this study

4.2 Pre-trained Transformer Models

Since Bangla is a resource constrained language, every variant of transformer models has not been pre-trained for Bangla language. However, we have found few transformer models that have pre-trained weights for Bangla language. Among all these models we have chosen four variants for our fine-tuning purpose. The pretrained models that have been chosen for fine-tuning are described below.

BERT Multilingual (mBERT). mBERT [10] is a transformers model pre-trained on a large corpus of multilingual data in a self-supervised fashion. In particular, it has been pre-trained over Wikipedia content with a shared vocabulary across 104 different languages. In total it has 12 layers, output dimension of 768, 12 multi-headed attentions, and 110M parameters.

BERT Bangla Base (bnBERT). bnBERT [11] is a pre-trained language model of Bangla language using Mask Language Modeling. It has been trained over two large Bangla corpus namely OSCAR, and Bangla Wikipedia dump dataset. In total bnBERT has 12 layers, output dimension of 768, 12 multi-headed attentions, and 110M parameters.

Electra Bangla Base (bnElectra). bnElectra [12] is a pre-trained model of Bangla language that has employed the method of self-supervised language representation learning. It has been trained over two large Bangla corpus namely OSCAR, and Bangla Wikipedia dump dataset. In total bnElectra has 12 layers, and an output dimension of 256.

RoBERTa Bangla Base (bnRoBERTa). RoBERTa [13] is a variant of BERT which modifies key hyperparameters of BERT, removing the next-sentence pre-training objective and training with much larger mini-batches and learning rates. bnRoBERTa [14] is a RoBERTa model which was pre-trained on a large Bangla text corpus. bnRoBERTa has 12-layers, output dimension of 768, 12 multi-headed attentions, and 125M parameters.

4.3   Fine-Tuning Transformer Models

Fine-tuning is a supervised learning technique in which the pre-trained model's weights are utilized as the initial weights for a new model being trained on a downstream task. This process not only speeds up the training but also creates a state-of-the-art model for a wide range of NLP tasks. In this study, we have fine-tuned all the four pre-trained transformer models for the multi-class authorship attribution problem. During fine-tuning, firstly, we have initialized the fine-tuned model with the same pre-trained model parameters for the downstream task. Then, we have fine-tuned all the parameters end-to-end using the corresponding task-specific labeled data. Eventually, we have incorporated an additional output layer according to the target classes. Usually, the BERT model generates two different outputs concurrently: one is "sequence output" (token representation) and the other one is "pooled output" ([CLS] representation). As the final hidden state corresponding to [CLS] token is used as the aggregate sequence representation for classification tasks, in our study, we have fed the [CLS] representation into the output layer.

Classifier Architecture. Apart from the output layer, the same architecture has been used for fine-tuning, which is a distinctive feature of transformer models. Besides, the self-attention mechanism in the transformer makes the fine-tuning process more straightforward than standard language models. That is why, in this study, we have fine-tuned our pre-trained models with just one additional output layer. For, mBERT, bnBERT, and RoBERTa we have used a fully connected linear layer with 768 input features and 16 output features. For bnElectra, we have used a fully connected linear layer with 256 input features and 16 output features.

Training Parameters. We have used the same training parameters for each of the models. The models were fine-tuned with 4 epochs and a batch size of 8. We have also used 500 warmup steps, and a weight decay of 0.01. The total optimization steps were 5752.

4.4   Evaluation Metrics

Since we are performing a multi-class classification task, we have recorded and analyzed the performance in terms of overall performance and class-level performance. For overall performance we have used Accuracy, F1 score with macro average, F1 score with micro average, and Cohen's Kappa as evaluation metrics. And for class-level performance, we have used AUC score and F1 score as metrics.

5 Result and Analysis

After the training and validation, we have tested our models on the test dataset that contains 3592 instances. The obtained results are presented from two perspectives: overall performance and class-level performance. The overall performance has been shown for both testing and validation datasets. The results of overall performance are presented in the below Table 2.

Table 2. Overall model performance

Model	Result type	Accuracy	F1 (Macro)	F1 (Micro)	Kappa
mBERT	Validation	0.922	0.877	0.922	0.912
	Test	0.919	0.876	0.919	0.909
bnBERT	Validation	0.983	0.977	0.983	0.980
	Test	**0.980**	**0.971**	**0.980**	**0.978**
bnElectra	Validation	0.915	0.781	0.915	0.904
	Test	0.912	0.778	0.912	0.901
bnRoBERTa	Validation	0.877	0.833	0.877	0.862
	Test	0.878	*0.831*	0.878	0.862

From the overall performance table we can see that all the models have a similar result both in test and validation in terms of all metrics. bnBERT has the highest accuracy (=0.98) where mBERT and bnElectra have a similar (=0.92) result both in test and validation. For F1 (macro), bnBERT again has the highest performance of about 0.97 and mBERT being the closest of about 0.87 in test and validation. bnBERT keeps up the highest performance (=0.97) in F1 (Micro) where mBERT and bnElectra produced similar results in test and validation of about 0.92. Likewise, bnBERT has the highest Kappa value (=0.98) where mBERT and bnElectra yielded somewhat similar results both in test and validation. bnRoBERTa has the worst performance in all metrics of all models. Overall, bnBERT yields outperforming results of approximately 0.98 for all performance metrics both in test and validation set.

The following Table 3 presents the class-wise performance of all the models. To measure individual class performance we have used AUC and F1 scores.

Breaking down the results of class level performance we can see that, the best model for overall performance, bnBERT has the highest result in AUC and F1 metrics for all classes. Focusing on the individual classes, 'zahir_rayhan' achieved the best classification result in AUC (=1.0) and F1 (=0.97) by bnBERT where mBERT was very close to that of bnBERT. At the same time, bnElectra had

Table 3. Class-wise model performance

Class	mBERT		bnBERT		bnElectra		bnRoBERTa	
	AUC	F1	AUC	F1	AUC	F1	AUC	F1
humayun_ahmed	0.984	0.976	0.995	0.994	0.985	0.981	0.962	0.951
shunil_gongopaddhay	0.973	0.957	0.993	0.991	0.974	0.964	0.958	0.916
shomresh	0.975	0.933	0.993	0.989	0.990	0.959	0.940	0.903
shorotchandra	0.977	0.930	0.993	0.979	0.991	0.957	0.973	0.922
robindronath	0.932	0.852	0.977	0.962	0.934	0.864	0.883	0.782
m_zafar_iqbal	0.981	0.945	0.997	0.980	0.967	0.952	0.956	0.879
shirshendu	0.945	0.906	0.992	0.983	0.945	0.906	0.939	0.858
toslima_nasrin	0.956	0.902	0.991	0.984	0.962	0.911	0.937	0.877
shordindu	0.905	0.873	0.957	0.947	0.896	0.837	0.922	0.846
shottojit_roy	0.934	0.905	0.996	0.977	0.997	0.949	0.889	0.849
tarashonkor	0.932	0.863	0.990	0.974	0.898	0.791	0.907	0.841
bongkim	0.896	0.836	0.985	0.952	0.894	0.798	0.797	0.694
nihar_ronjon_gupta	0.914	0.836	0.963	0.957	0.929	0.841	0.881	0.772
manik_bandhopaddhay	0.922	0.836	0.994	0.974	0.960	0.737	0.871	0.711
nazrul	0.703	0.507	0.965	0.921	0.500	0.000	0.896	0.745
zahir_rayhan	1.000	0.961	1.000	0.974	0.500	0.000	0.957	0.756

the worst performance both in AUC (=0.50) and F1 (=0.00) for 'zahir_rayhan'. It produced the same result for the author class 'nazrul' which is also the worst performing class out of all classes in all metrics by all models. From the above result analysis, it is apparent that regardless of overall or class wise performance, bnBERT has the outperforming result out of all models.

Table 4. Performance comparison with existing literature

	No. author class	No. of test documents	Accuracy
Hossain et al. [1]	6	300	90.67%
Islam et al. [2]	5	592	85.00%
Khatun et al. [3]	14	6566	81.20%
Chowdhury et al. [4]	6	300	92.90%
Chowdhury et al. [5]	3	150	82.40%
Al Marouf et al. [6]	2	148	82.69%
Anisuzzaman et al. [7]	3	300	95.00%
Ahmad et al. [8]	23	3043	94.12%
bnBERT	**16**	**3592**	**98.00%**

Along with the comparison between the models implemented in this study, we have also compared the performance of our best model, bnBERT, with the existing models in the literature. The Table 4 summarizes the comparative results.

Again, it is evident that our best performing model, bnBERT outperformed all the other models proposed by above mentioned authors in the literature.

6   Conclusion and Future Work

In this study, we performed the task of authorship attribution on famous Bengali authors using several state-of-the-art transformer models as well as comprehensive and comparative analysis of the performance of these models. From the analysis of the result, we found that all models have performed over 87% in terms of classification accuracy, and F1 (Micro) score. As per our comparative analysis, the bnBERT model has outperformed all the models with a test performance of 98% accuracy, 0.971 F1 (Macro), 0.980 F1 (Micro), and 0.978 Kappa score. Furthermore, the bnBERT model outperforms all the other models proposed in the existing studies and the closest model by a margin of 3% accuracy. In future, we would like to develop the model further using a multimodal modeling approach.

References

1. Hossain, M.T., et al.: A stylometric analysis on Bengali literature for authorship attribution. In: 2017 20th International Conference of Computer and Information Technology (ICCIT). IEEE (2017)
2. Islam, Md.A., et al.: Authorship attribution on Bengali literature using stylometric features and neural network. In: 2018 4th International Conference on Electrical Engineering and Information and Communication Technology (iCEEiCT). IEEE (2018)
3. Khatun, A., Rahman, A., Islam, Md.S., Marium-E-Jannat: Authorship attribution in Bangla literature using character-level CNN. In: 2019 22nd International Conference on Computer and Information Technology (ICCIT), Dhaka, Bangladesh, pp. 1–5. IEEE (2019). https://doi.org/10.1109/ICCIT48885.2019.9038560
4. Chowdhury, H.A., Imon, Md.A.H., Islam, Md.S.: A comparative analysis of word embedding representations in authorship attribution of Bengali literature. In: 2018 21st International Conference of Computer and Information Technology (ICCIT). IEEE (2018)
5. Chowdhury, H.A., Imon, Md.A.H., Islam, Md.S.: Authorship attribution in Bengali literature using fastText's hierarchical classifier. In: 2018 4th International Conference on Electrical Engineering and Information and Communication Technology (iCEEiCT). IEEE (2018)
6. Al Marouf, A., Hossian, R.: Lyricist identification using stylometric features utilizing BanglaMusicStylo dataset. In: 2019 International Conference on Bangla Speech and Language Processing (ICBSLP). IEEE (2019)
7. Anisuzzaman, D.M., Salam, A.: Authorship attribution for Bengali language using the fusion of N-gram and Naïve Bayes algorithms. Int. J. Inf. Technol. Comput. Sci. (IJITCS) 10(10), 11–21 (2018)

8. Ibn Ahmad, S., Alam, L., Hoque, M.M.: An empirical framework to identify authorship from Bengali literary works. In: Bhuiyan, T., Rahman, M.M., Ali, M.A. (eds.) ICONCS 2020. LNICST, vol. 325, pp. 465–476. Springer, Cham (2020). https://doi.org/10.1007/978-3-030-52856-0_37

9. Khatun, A.: BAAD16: Bangla authorship attribution dataset (2020). https://doi.org/10.17632/6D9JRKGTVV.4. https://data.mendeley.com/datasets/6d9jrkgtvv/4

10. Devlin, J., Chang, M.-W., Lee, K., Toutanova, K.: BERT: pre-training of deep bidirectional transformers for language understanding. arXiv:1810.04805 [cs] (2019)

11. Sarker, S.: BanglaBERT: Bengali mask language model for Bengali language understanding. https://github.com/sagorbrur/bangla-bert. Accessed 24 July 2021

12. Clark, K., Luong, M.-T., Le, Q.V., Manning, C.D.: ELECTRA: pre-training text encoders as discriminators rather than generators. arXiv:2003.10555 [cs] (2020)

13. Liu, Y., et al.: RoBERTa: a robustly optimized BERT pretraining approach. arXiv:1907.11692 [cs] (2019)

14. Saifullaah, K.: Roberta Bengali Base. https://huggingface.co/khalidsaifullaah/roberta-bengali-base. Accessed 24 July 2021

A Deep Learning Based Traffic Sign Detection for Intelligent Transportation Systems

Bao-Long Le[1,2], Gia-Huy Lam[1,2], Xuan-Vinh Nguyen[1,2], The-Manh Nguyen[1,2], Quoc-Loc Duong[1,2], Quang Dieu Tran[3], Trong-Hop Do[1,2], and Nhu-Ngoc Dao[4(✉)]

[1] University of Information Technology, Ho Chi Minh City, Vietnam
{19521782,18520832,18521655,18521084,18521006}@gm.uit.edu.vn,
hopdt@uit.edu.vn
[2] Vietnam National University, Ho Chi Minh City, Vietnam
[3] Ho Chi Minh National Academy of Politics, Hanoi, Vietnam
dieutq@hcma.vn
[4] Department of Computer Science and Engineering, Sejong University,
Seoul, South Korea
nndao@sejong.ac.kr

Abstract. Automatic detection and classification of traffic signs [1] bring convenience and caution to drivers on the road. It provides drivers with accuracy and timeliness in compliance as well as notifications in the route they are on. In the field of computer vision, the problem of detecting and classifying traffic signs has attracted great attention from research communities, because of the consequences it can bring if any mistake is made. In this problem, we have built a highly realistic data set with many challenges for Vietnam's traffic. In addition, we also solve the problem of automatic detection [2] and classification of traffic signs on the dataset that we have built using YOLOv4 and YOLOv5 algorithms with fine-tuned parameters. The results obtained in this paper are that the accuracy in detecting and classifying signs is quite high and the error is very low compared to outside traffic in Vietnam. The article is expected to benefit the development of practical applications and bring certain contributions in further developing this issue.

Keywords: Traffic sign detection · Smart car · Intelligent transportation system

1 Introduction

In recent years, self-driving car have become a hot topic that receive lots of attention from both academic and industry. One of the key components in a self-driving car is the computer vision module used for obtaining various type of traffic data from environment. Traffic sign is among crucial data for self-driving

© Springer Nature Switzerland AG 2021
D. Mohaisen and R. Jin (Eds.): CSoNet 2021, LNCS 13116, pp. 129–137, 2021.
https://doi.org/10.1007/978-3-030-91434-9_12

Fig. 1. System design.

car to operate properly. For example, based on the instruction on traffic signs, the vehicle know if it can turn left or right, or if it must reduce the speed. Therefore, a traffic sign detection system is a must for any self-driving car system [3].

Detecting a single traffic sign is not a difficult problem as most traffic signs has simple patterns with features easy to extract. Detecting and differentiating many traffic signs [4], however, is a challenging problem as many traffic signs have similar patterns. Beside accuracy, processing time is another factor to concern. For such an application like self-driving car, any mistake or delay in detecting and classifying a traffic sign might lead to serious consequences. The problem is exacerbated in developing countries with modest traffic infrastructure where traffic signs are usually blocked by many obstacles.

Thanks to the development of advance object detection algorithms, traffic sign detection has become a much approachable compare to it was just less than 10 years ago. Among possible approaches for traffic sign detection, deep learning based algorithms are likely to have the best performance in terms of accuracy and processing time. A tremendous number of experiments has shown that deep learning based techniques like You Only Look Once (YOLO) [5], Single Shot Detection (SSD) [6] perform very well in manky object detection tasks. However, compared to normal object detection tasks, traffic sign detection [7] is different in that the number of object class, which is the number of types of traffic signs, is much larger. The larger number of classes, the higher possibility of misclassifying the detected object [8].

This paper focuses on building a traffic sign detection application to detect popular traffic signs in Vietnam. This application receives a traffic video as input. It then locates the regions of the traffic signs in the videos and recognizes these signs. To train the traffic sign detection model, a large dataset consisting of 16770 images of 54 types of traffic signs has been built. The performance of the proposed application has been tested and evaluated in various metrics. Based on the experiment results, an analysis of detection errors in the application has also been provided.

2   Proposed System Architecture

System Design: The design of the proposed system is described in Fig. 1. The input of the system are traffic video frames. A transfer learning model based on YOLOv4 is used for detecting the traffic signs in each video frames to obtain

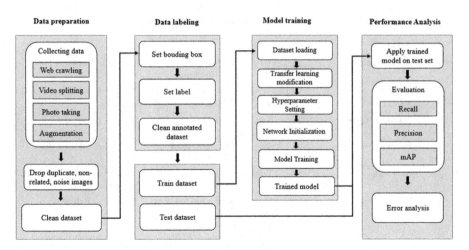

Fig. 2. Experimental procedure.

the labels of these signs. Then, the contents of these labels is shown to users through the web based interface of the system.

Transfer Learning Model Based on Yolov4: YOLOv4 [9] has many special enhancements that increase the accuracy and speed of its brother YOLOv3 [10] on the same COCO dataset and on the V100 GPU. The structure of v4 is divided into four parts: Backbone, Neck, Dense prediction, Sparse Prediction.

The backbone network for object recognition is usually pre-trained through the ImageNet classification problem. Pre-train means that the weights of the network have been adjusted to identify relevant features in an image, although they will be fine-tuned in the new task of object detection. The author considers using the backbone: CSPResNext50, CSPDarknet53, EfficientNet-B3.

Neck is responsible for mixing and matching feature maps learned through feature extraction (backbone) and identification process (YOLOv4 called Dense prediction).

YOLOv4 allows customization of Neck structures such as: FPN, PAN, NAS-FPN, BiFPN, ASFF, SFAM, SSP.

3   Experiment

The procedure of the experiment in this paper is described in Fig. 2. The experiment includes four steps: data preparation, data labeling, model training, and performance analysis.

Table 1. Correlation table between *class_id* and *label*.

Class	Label	Class	Label	Class	Label
0	No thoroughfare	18	No proceeding unspecified direction	36	Road/lane reserved for cars
1	No entry	19	Road junction	37	One way
2	No cars	20	Roundabout ahead	38	Parking
3	No motorcycles	21	Road junction with priority	39	U-turn permitted
4	No motor vehicles	22	Give way to main road	40	Direction allowed per lane
5	No trucks	23	Traffic signals ahead	41	Road/lane reserved for specified vehicles
6	No bus or trucks	24	Railway level crossing ahead that has-automatic gates or other barriers	42	Direction indicating
7	No motorcycles, tricycles	25	Railway level crossing ahead that does not have any automatic gate or other barriers	43	Alternative route where a turn is prohibited
8	No pedestrians	26	Pedestrian crossing ahead	44	Star/end of a built-up area
9	Gross vehicle weight limit	27	Children	45	Pedestrian crossing
10	Height limit	28	Construction	46	Overpass/underpass route for pedestrian
11	Stop	29	Slow	47	Hospital
12	No left/right turn	30	Direction	48	Traffic police station
13	No U-turn	31	Keep pass side	49	Gas station
14	No overtaking	32	Roundabout	50	Bus stop/station
15	No overtaking bytrucks	33	Walking path	51	Disabled parking only
16	Maximum speed limit	34	Minimum speed limit	52	Overpass route
17	No stopping/parking or waiting	35	Overpass route	53	Others

3.1   Datasets

In this paper, the dataset of traffic signs was collected in two ways: image collection from Google search page and video recording. Most of the data is collected by video recording because of its closeness to reality, the variety of contexts as well as the noise that the images available on Google rarely bring. The video recording is divided into two directions, once is the actual battle (out to the street to shoot traffic signs), the other is based on the image projected from the satellite on Google Maps and then back to the screen. For the first direction is collected images will more than for the second direction. However, The second direction is used to supplement data for signs that are difficult to encounter in real life because it is not possible to correctly locate the remaining signs. If this second direction still does not meet the quantity, the sample signs will be stitched

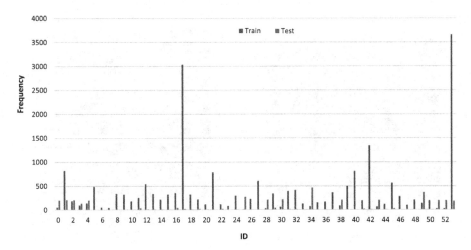

Fig. 3. Number of photos per set.

into the actual context to create a realistic image and ensure the quantity for the signs.

The collected signs are common signs that can be encountered in life with the label names based on the traffic manual, a total of about 54 labels of which 53 are single signs, one category contains images deemed complex or absent from the selected number of signs as shown in Table 1.

This label was added for the later developed problem. After labeling the images and videos, there are 16770 images in total, of which 13439 are for the training set and 3331 for the test set. Figure 3 illustrates the statistical chart of the number of each assigned label.

3.2 Data Preprocessing

Each image has many different features. Therefore, to be used in the model, the image data has to go through several preprocessing steps. Below are the preliminary preprocessing steps on the image dataset:

- Read the image, then convert the color channels of all images to RGB format to create consistency in the number of color channels for all images to match the model input.
- Resize the photo to the appropriate size - height: 416 pixels and width: 416 pixels. So all images have been converted to size $416 * 416 * 3$.

After preprocessing, we use yolov4 to train labeled images from the dataset with the following parameters: Yolov4 using the model yolov4 Pre-trained. The parameters used are: batch $= 64$, subdivisions $= 16$, max_batches $= 108000$, steps $= 86400,97200$, filters $= 177$, classes $= 54$, width $= 416$, height $= 416$.

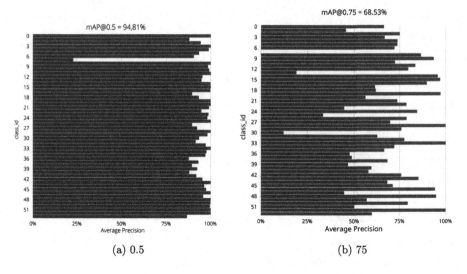

Fig. 4. mAP.

3.3   Evaluation Methods

Performance metrics of object detection problem include:

- **IoU (Intersection over union)** is the ratio between measuring the degree of intersection between two contours (usually the predicted contour and the actual contour) to determine if two frames are overlapping. This ratio is calculated based on the area of intersection of 2 contours with the total area of intersection and non-intersection between them.
- **Precision** measure how accurate is the model's prediction i.e. percentage of model's prediction is correct.
- **Recall** measure how well the model finds all positive patterns.

From the precision and recall defined above, we can also evaluate the model based on changing a threshold and observing the values of Precision and Recall. The concept of Area Under the Curve (AUC) is similarly defined. With Precision-Recall Curve, the AUC has another name, **Average precision (AP)**. Suppose there are N thresholds for precision and recall, with each threshold for a pair of precision values, recall is $R_n, n = 1, 2, \ldots, N$. Precision-Recall curve is drawn by drawing each point with coordinates (P_n) on the coordinate axis and connecting them together. AP is defined by:

$$AP = \sum_{n=0}^{N} [R_n - R_{n-1}] * P_n \tag{1}$$

In multiple-classes object detection, mAP is the average of AP calculated for all classes.

SIGN DETECTION

Fig. 5. SIGN detection demo.

3.4 Results

During the long training period (specifically, training around 4000 rounds/day, with approximately 27 days for training to complete), there were many models saved at rounds 10000, 20000, ... 10000; along with the models saved from the calculation of mAP at each small round. And we compared the obtained models. In the end, the best model is the one with mAP@0.5 = 94.81% and mAP@0.75 = 68.53%.

Derived from Figs. 4a, 4b and Table 1, it can be seen that the overall model evaluation results for the dataset are very good with mAP@0.5 up to 94.81% and mAP@0.75 = 68.53%, only a few cases are not high, like class_id = 7 is a sign that prohibits motorcycles and tricycles with very low accuracy AP = 22.85% at rating mAP@0.5 and AP = 0 at rating mAP@0.75.

3.5 SIGN Detection Application

To build this application, we use python language with the main library Flask, that capable of creating an interface that can be accessed by the website. After detecting the signs and determining their classes, the parts containing the signs in the video frames will be shown in the right panels to show the signs and their contents as shown in Fig. 5. Next, transmit to the web the cropped image with the category of the image after the model predicts it. To transmit information on the device interface so that the driver can observe. This application builds for users a list of 20 consecutive signs that help drivers have more information about signs and the next section.

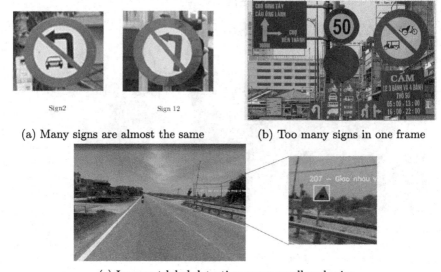

(a) Many signs are almost the same   (b) Too many signs in one frame

(c) Incorrect label detection cause small scale sign

Fig. 6. Typical error of detecting and labeling.

3.6 Error Analysis

Some pairs of figures have a lot of detail look similar to each other, which is explained in Fig. 6a. This issue easily causes confusion for detection model. Frames with a large number of objects (signs) or objects with overlap as shown in Fig. 6b also make the model difficult to detect and recognize. Another issue is about detecting small objects in large scenes. In figures c, because the object is so small in proportion to the frame, it is mistaken as sign #21 instead of sign #25 (Figs. 6c). There are a few road signs that are not ordinarily utilized, so collecting data will be a troublesome issue for us as class imbalance also affects the predictive performance of the model. Some of the reasons for the difference in accuracy of identifying signs is due to the unevenness in the complexity of them, or some signs have a highly correlated appearance with others, or the rate involved in data set construction.

4   Concluding Remarks

In the future, we plan to improve and expand the dataset by recording more videos of different routes to make it closer to real life. We also develop other methods to improve the quality of identifying QR codes by combining more frames. In this article, we used the method Yolov4. The method gave a high result of 94.18% for mAP @0.5, but it gave a quite low result of 68.53% for mAP@75 due to some cases of unable to identify such as prohibit motorcycle signs or prohibit three-wheeled vehicle sign. The accuracy of identifying those mentioned signs were extremely low: AP $= 22.85\%$ for mAP@0.5, and AP $= 0$ for mAP@0.75. This happened because the amounts of different signs in the dataset are quite uneven. In the future, we plan to improve and expand the dataset for those types of signs that have a small number of images. In addition to identifying the signs, we would develop the problem to also provide instructions or warnings based on the collected images from the dash camera. We hope to contribute this dataset to the community in order to motivate the research of identifying traffic signs, improve the efficiency of identifying with better methods.

Acknowledgment. This work was supported by the National Research Foundation of Korea (NRF) grant funded by the Korea government (MSIT) (No. 2021R1G1A1008105).

References

1. Shneier, M.: Road sign detection and recognition. In: Unmanned Systems Technology VIII, vol. 6230, p. 623016. International Society for Optics and Photonics (2006)
2. Wali, S.B., Hannan, M.A., Hussain, A., Samad, S.A.: An automatic traffic sign detection and recognition system based on colour segmentation, shape matching, and SVM. Math. Probl. Eng. **2015**, 1–11 (2015)
3. Zaki, P.S., William, M.M., Soliman, B.K., Alexsan, K.G., Khalil, K., El-Moursy, M.: Traffic signs detection and recognition system using deep learning. arXiv preprint arXiv:2003.03256 (2020)
4. Tabernik, D., Skočaj, D.: Deep learning for large-scale traffic-sign detection and recognition. IEEE Trans. Intell. Transp. Syst. **21**(4), 1427–1440 (2019)
5. Huang, R., Pedoeem, J., Chen, C.: YOLO-LITE: a real-time object detection algorithm optimized for non-GPU computers. In: 2018 IEEE International Conference on Big Data (Big Data), pp. 2503–2510. IEEE (2018)
6. Zuo, Z., Yu, K., Zhou, Q., Wang, X., Li, T.: Traffic signs detection based on faster R-CNN. In: 2017 IEEE 37th International Conference on Distributed Computing Systems Workshops (ICDCSW), pp. 286–288. IEEE (2017)
7. Zhang, F., Zeng, Y.: D-FCOS: traffic signs detection and recognition based on semantic segmentation. In: 2020 IEEE International Conference on Power, Intelligent Computing and Systems (ICPICS), pp. 287–292. IEEE (2020)
8. Yu, J., Liu, H., Zhang, H.: Research on detection and recognition algorithm of road traffic signs. In: 2019 Chinese Control and Decision Conference (CCDC), pp. 1996–2001. IEEE (2019)
9. Bochkovskiy, A., Wang, C.-Y., Liao, H.-Y.M.: YOLOv4: optimal speed and accuracy of object detection. arXiv preprint arXiv:2004.10934 (2020)
10. Zhang, B., Wang, G., Wang, H., Xu, C., Li, Y., Xu, L.: Detecting small Chinese traffic signs via improved YOLOv3 method. Math. Probl. Eng. **2021**, 1–10 (2021)

Detecting Hate Speech Contents Using Embedding Models

Phuc H. Duong[1(✉)], Cuong C. Chung[1], Loc T. Vo[1], Hien T. Nguyen[2],
and Dat Ngo[3]

[1] Artificial Intelligence Laboratory, Faculty of Information Technology,
Ton Duc Thang University, Ho Chi Minh City, Vietnam
duonghuuphuc@tdtu.edu.vn, {518H0603,518H0102}@student.tdtu.edu.vn
[2] Department of Economic Mathematics, Banking University of Ho Chi Minh City,
Ho Chi Minh City, Vietnam
hiennt.mis@buh.edu.vn
[3] NewAI Research, Ho Chi Minh City, Vietnam
dat.ngo@newai.vn

Abstract. The rise of hate speech contents on social network platforms
has recently become a topic of interest. There have been a lot of studies to
develop systems that can automatically detect hate speech contents. In
this paper, we propose a knowledge-rich solution to hate speech detection
by incorporating hate speech embeddings to generate a more accurate
representation of the given text. To obtain the hate speech embeddings,
we construct a hate speech dictionary in a semi-supervised fashion. We
conduct experiments on two popular datasets, which show that the combination of word embeddings and hate speech embeddings can produce
promising results when compared with the methods that employ large-
scale pre-trained language models.

Keywords: Natural language processing · Hate speech detection ·
Deep learning

1 Introduction

Social network platforms are places where people can express their freedom of
speech as regulated within the framework of specific laws and community standards. However, the rapid growth of social network platforms has facilitated
the propagation of hate speech contents. For example, the attacks on Asian-
Americans have escalated concerns that racist contents on social network platforms would lead to real-world violence[1]. We thus can see that a real-world event
could escalate rapidly into an online hate, and vice versa. As defined in Oxford
dictionary[2], hate speech is a sort of content that attacks or threatens a particular group of people, including disability, ethnicity, gender identity, national

---

[1] https://www.usatoday.com/story/tech/2021/03/24/asian-american-hate-crimes-co
vid-harassment-atlanta-google-facebook-youtube/6973659002/
[2] https://www.oxfordlearnersdictionaries.com/definition/english/hate-speech

© Springer Nature Switzerland AG 2021
D. Mohaisen and R. Jin (Eds.): CSoNet 2021, LNCS 13116, pp. 138–146, 2021.
https://doi.org/10.1007/978-3-030-91434-9_13

origin, race, religious affiliation, sex. Facebook[3], which is a popular social network platform, has developed its community standards[4] to outline what activity is and is not allowed on its platform. Despite its well-defined policies, the content review process should be done by both people and technology[5]. From a practical viewpoint, there is a problem of ambiguity caused by a variety of hate speech labels (e.g., abuse, offensive language, hate speech, cyberbullying, harassment), since those labels have quite similar meanings. Another problem is the lack of contextual information, since a social media post is often a short text, noise, and not self-contained. Hate speech content also depends on temporal and historical contexts.

The process of automatically determining hate speech contents is not often straightforward. In [1–4], the authors exploit a lot of linguistic features to represent a given text before feeding it into machine learning algorithms, e.g., lemmatization, n-gram, bag-of-word, part-of-speech tagging, sentiment lexicon. By employing hand-crafted rules/features to capture both semantic information and syntactic structures from a given text, these methods can learn characteristics of what hate speech contents would be. However, these sort of methods, also known as feature engineering, are difficult, time-consuming and domain-specific, since they depend on expert knowledge.

With the ability of learning representations of features automatically, deep learning models have been widely adopted in recent years to overcome the limitations of feature-engineering approaches [4–6]. However, some popular neural network models such as CNN, RNN, LSTM are quite difficult to acquire explicit knowledge about hate speech contents from the training data. Therefore, incorporating background knowledge into deep learning models is considered as an important add-on [3,7,8]. In this paper, we propose a knowledge-rich solution to hate speech detection with a specific focus on leveraging hate speech embeddings in the deep learning models. In general, we can summarize the contributions of this paper as follows:

- We present the procedure to construct a hate speech dictionary to obtain hate speech embeddings.
- The word embeddings and hate speech embeddings are concatenated and fed into neural network models, i.e., multilayer perceptron, BiLSTM, CNN to train the classification models.
- We evaluate the proposed model on two benchmark datasets, i.e., HASOC-2019 [9], HSOF-3 [2].

The rest of this paper is organized as follows: Sect. 2 briefly reviews the relevant literature methods in hate speech detection; Sect. 3 describes our proposed model; Sect. 4 presents the datasets and describes the experimental setup and results; and we draw a conclusion in Sect. 5.

---

[3] https://about.fb.com/

[4] https://www.facebook.com/communitystandards/

[5] https://about.fb.com/news/2020/08/how-we-review-content/

2   Related Work

A considerable number of studies have been conducted for detection of hate speech contents. From our viewpoint, we categorize the previous studies into two sorts of approaches, i.e., (1) feature engineering and (2) deep learning. Feature engineering requires expert knowledge to extract features from raw data, and then select an optimal set of ones. In [10], the authors proposed a list of 11 criteria to conclude whether a given tweet is hate speech or not, e.g., a tweet is considered as hate speech if it contains a sexist or racial slur. Instead of defining these hand-carfted rules as in [10], methods proposed in [1,2,4] employ NLP techniques to extract features from text, then feeding them into machine learning algorithms. We observe that features like bag-of-word, n-gram, lemmatization, part-of-speech tagging, grammatical structure, TF-IDF are often employed as semantic and syntactic features. Besides that, some studies [3,8] also exploits external knowledge sources, e.g., LIWC dictionary, Hatebase, emotion lexicon corpus to expand the feature space.

Feature engineering is considered costly, time-consuming, domain-specific, and depends on expert knowledge. On the other hand, deep learning models are capable of learning representations of features automatically. The obtained representations can then be fed into downstream models such as CNN [5,11, 12], RNN/LSTM [5,12] to train classification models. Explicitly incorporating external knowledge into deep learning is also considered as an important add-on for many NLP tasks. As mentioned in [7], neural network models are hard to acquire explicit knowledge from training data, since it is not explicitly annotated. In this paper, we propose a knowledge-rich solution to the task of hate speech detection, in particular, we leverage the representation of hate speech concepts as an additional input to the model.

3   Proposed Method

In Fig. 1, we present the overall architecture of proposed model. Given a sequence of words, first, each word is represented as a d-dimensional word vector by concatenating word embeddings and hate speech embeddings. The word vectors are then fed into a neural network model, followed by the softmax function to normalize the output to a probability distribution over predicted labels.

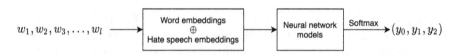

Fig. 1. The overall architecture of proposed model.

Input text	w_1	w_2	w_3	w_4	...	w_l
Word embeddings	E_1^{WE}	E_2^{WE}	E_3^{WE}	E_4^{WE}	...	E_l^{WE}
	$\oplus$	$\oplus$	$\oplus$	$\oplus$	$\oplus$	$\oplus$
Hate speech embeddings	E_0^{HSE}	E_1^{HSE}	E_1^{HSE}	E_0^{HSE}	...	E_0^{HSE}
Input representation	I_1	I_2	I_3	I_4	...	I_l

Fig. 2. The input representation is a concatenation of word embeddings and hate speech embeddings.

3.1 Input Representation

The input representation is a concatenation of word embedding and hate speech embedding vectors, as presented in Fig. 2. Both of the layers transform a word into a vector representation of d-dimensional. For word embeddings, we employ the word2vec[6] model to generate a 300-dimensional word vector for each given word. The hate speech embedding layer is represented by a boolean scalar value to indicate whether a given word is a hate speech term or not. Technically, if a word is a hate speech term, the model will set and assign $scalar = 1$ to each corresponding position, and vice versa. To obtain the hate speech embeddings, we first need to construct the hate speech dictionary as the following procedure:

- We extract instances labeled as hate speech or offensive language from popular hate speech datasets to construct a hate speech dictionary, denoted by $Dict_{HS}$. A dictionary containing neutral instances is also constructed, denoted by $Dict_{Neutral}$.
- To eliminate irrelevant words in the $Dict_{HS}$, we assume that a word appearing in both the $Dict_{HS}$ and $Dict_{Neutral}$ is not a hate speech or offensive word. To determine these words, we measure the similarity for any pair of words by employing the Wu-Palmer and Path-Similarity measurements which are implemented in the NLTK[7] library. A word will be removed from the $Dict_{HS}$ if its similarity scores are larger than or equal to 0.5.
- Finally, the $Dict_{HS}$ is delivered to annotators, and with the approval rating at 70%, the hate speech dictionary contains 766 terms.

In summary, given a text of length l, denoted by $w_{1:l}$, first, the word embedding vector is initialized with the word2vec model. Next, the model looks up each w_i in the hate speech dictionary to generate a hate speech embedding vector corresponding to each word position. The dimension of the word embedding vector is 300 and hate speech embedding vector is 1. Finally, the input representation is a concatenation of the two embedding vectors, denoted by $I \in \mathbb{R}^{l \times 301}$.

---

[6] https://code.google.com/archive/p/word2vec/
[7] https://www.nltk.org/

3.2   Neural Network Models

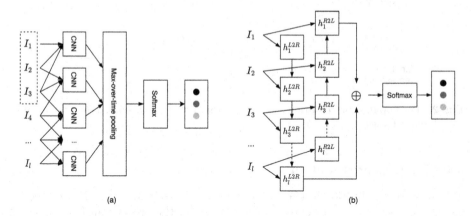

(a)   (b)

Fig. 3. (a) is the architecture of CNN model with $h = 3$; and (b) is the architecture of BiLSTM model, where the $\oplus$ symbol denotes the concatenation operator.

Multilayer Perceptron Model. Given an input representation, we first employ the bag-of-word model to generate a vector representation by performing an element-wise summation over the embedding vectors, as presented in Eq. 1 where $x_i \in I$. The vector is then fed into a multilayer perceptron model consisting of two hidden layers with dropout and ReLU activation function.

$$s = \sum_{i=1}^{|x|} x_i \tag{1}$$

CNN Model. As shown Fig. 3(a), we set up the filter of the CNN model as an n-gram feature. Given the input representation, $I \in \mathbb{R}^{l \times d}$, we define a filter $W_f \in \mathbb{R}^{h \times d}$, where $d = 301$ and h acts as an n-gram feature, i.e., $h = 1, 2, 3$. The convolution operation of a filter on h consecutive word vectors outputs a scalar. Thus, m different filters output a vector $c_i \in \mathbb{R}^m$, as presented in Eq. 2. Given a text of length l, we obtain $[l - h + 1]$ vectors after performing the convolution operations over the text. These vectors are then fed into a max-over-time pooling layer to compute the vector $s \in \mathbb{R}^m$, followed by the softmax function.

$$c_i = \text{ReLU}\left(W_f \cdot I_{i:i+h-1} + b_f\right) \tag{2}$$

BiLSTM Model. The vanilla LSTM model processes a sequence of words from one direction, we thus stack the two LSTM models on top of each other to enable the model to look at a text from both directions, as shown in Fig. 3(b). Specifically, for the ith word, we first simultaneously compute the hidden state vectors from the left-to-right ($h_i^{L2R} \in \mathbb{R}^k$) and right-to-left ($h_i^{R2L} \in \mathbb{R}^k$) directions, where k is the dimension of hidden layer. The final output of the BiLSTM model is the concatenation of the last hidden state vectors of each direction, denoted by $s \in \mathbb{R}^{2k}$. We then feed the final hidden state vector into the softmax function to squash the value into a probabilistic range, as described in Eq. 3.

$$y = \text{softmax}\left(W_s \cdot s + b_s\right) \tag{3}$$

4 Experiments

4.1 Datasets

We evaluate the proposed model on the HASOC-2019 [9] and HSOF-3 [2] datasets. The instances from both datasets are extracted from social network platforms, and manually labeled by humans. The HASOC-2019 dataset has 5,853 training instances and 1,154 test instances; each instance is labeled as hate speech or not. The HSOF-3 dataset has 24,802 instances and three labels, i.e., hate speech, offensive language, and neither. We also combine the HASOC-2019 and HSOF-3 datasets to construct a new one, named HS2-2021. To combine the two datasets, we first conflate the offensive language and hate speech labels of the HSOF-3 dataset. Then, we merge the new HSOF-3 and HASOC-2019 datasets together to form the single HS2-2021 dataset. As a result, the HS2-2021 dataset has 23,169 instances labeled as hate speech, and the rest have 8,619 instances. For the HSOF-3 and HS2-2021 datasets, we evaluate the model by using 5-fold cross validation and holding out 10% of the sample for evaluation.

4.2 Experimental Results

For input representation, we consider two cases: (i) using only word embeddings (WE), and (ii) concatenating word embeddings and hate speech embeddings (WE + HSE). For neural network models, we evaluate three sorts of models: (a) multilayer perceptron (MLP), (b) CNN, and (c) BiLSTM. We compare the performance of our proposed model with BERTweet [13], Wang et al. [14], Kovács et al. [6] and Davidson et al. [2], as reported in Table 3. We report the hyperparameters of neural network models in Table 1, and the experimental results in Table 2, including accuracy and macro F1-score metrics.

The experimental results show that the CNN model has achieved the best performance in both two cases (2 and 5) for all datasets, in terms of macro F1 measure. We figure out that the proposed model performs better when we combine both word embeddings and hate speech embeddings. For HASOC-2019, the method in [6] outperforms the other ones, since its model consists of 2-CNN-and-3-LSTM layers with the RoBERTa language model [15], and this model is by far larger than our proposed one. For HSOF-3, the authors in [2] manually consider each linguistic feature, it thus can help the model to be well fit on the dataset. For HS2-2021, the combination of HASOC-2019 and HSOF-3 datasets

Table 1. The hyperparameters of neural network models.

CNN	Learning rate	2e−5
	Number of filters	300
	Dropout	0.6
	Activation function	ReLU
	Optimizer	Adam
BiLSTM	Hidden dimension	64
	Learning rate	2e−5
	Dropout	0.6
	Optimizer	Adam
MLP	Number of hidden layers	2
	Hidden layer dimension	100
	Learning rate	2e−5
	Dropout	0.6
	Activation function	ReLU
	Optimizer	Adam

Table 2. Experimental results for the HASOC-2019, HSOF-3 and HS2-2021 datasets, where (1, 2, 3) only consider word embeddings, (4, 5, 6) combine word embeddings and hate speech embeddings.

#	Models	HASOC-2019		HSOF-3		HS2-2021	
		Acc	F1	Acc	F1	Acc	F1
1	WE + MLP	0.7804	0.7172	0.9076	0.8243	0.8459	0.831
2	WE + CNN	0.7923	**0.7425**	0.9308	**0.8798**	0.8948	**0.8873**
3	WE + BiLSTM	0.797	0.7365	0.9199	0.854	0.8588	0.8415
4	[WE + HSE] + MLP	0.7909	0.7309	0.9303	0.8784	0.8903	0.8814
5	[WE + HSE] + CNN	0.8098	**0.752**	0.9309	**0.8824**	0.8951	**0.8878**
6	[WE + HSE] + BiLSTM	0.8025	0.7366	0.9287	0.8758	0.8588	0.8415

Table 3. A comparison between our best model and other models from the literature.

Models	HASOC-2019		HSOF-3		HS2-2021	
	Acc	F1	Acc	F1	Acc	F1
BERTweet + Softmax	0.8326	0.7839	0.9758	0.9567	0.8956	0.9175
Wang et al. [14]	–	0.7882	–	–	–	–
Kovács et al. [6]	–	0.7945	–	–	–	–
Davidson et al. [2]	–	–	–	0.9	–	–
Our model						
[WE + HSE] + CNN	0.8098	0.752	0.9309	0.8824	0.8951	0.8878

significantly increase the macro F1 scores. In summary, we have shown that the combination of word embeddings and hate speech embeddings can produce promising results in comparison with the methods that employ large-scale pretrained language models [6,13].

5 Conclusion

In this paper, we have proposed a hate speech detection model that combines word embeddings and hate speech embeddings. By incorporating the hate speech embeddings, we can generate a more accurate representation of the given text, in particular, we use a boolean scalar value to indicate whether a given word is a hate speech term or not. The experimental results have shown that our proposed model achieves promising results even though it is quite simple when compared to the others that employ large-scale pre-trained language models. In future work, we plan to expand the hate speech dictionary with more fine-grained labels and assign weights to hate speech terms. The resources of this study are publicly available on our GitHub repository[8].

References

1. Warner, W., Hirschberg, J.: Detecting hate speech on the world wide web. In: Proceedings of the Second Workshop on Language in Social Media, pp. 19–26 (2012)
2. Davidson, T., Warmsley, D., Macy, M.W., Weber, I.: Automated hate speech detection and the problem of offensive language. In: ICWSM, pp. 512–515. AAAI Press (2017)
3. Gao, L., Huang, R.: Detecting online hate speech using context aware models. In: Mitkov, R., Angelova, G. (eds): Proceedings of the International Conference Recent Advances in Natural Language Processing, RANLP 2017, Varna, Bulgaria, 2–8 September 2017, pp. 260–266. INCOMA Ltd (2017)
4. de Gibert, O., Pérez, N., Pablos, A.G., Cuadros, M.: Hate speech dataset from a white supremacy forum. In: Fiser, D., Huang, R., Prabhakaran, V., Voigt, R., Waseem, Z., Wernimont, J., (eds.) Proceedings of the 2nd Workshop on Abusive Language Online, ALW@EMNLP 2018, Brussels, Belgium, 31 October 2018, pp. 11–20. Association for Computational Linguistics (2018)
5. Qian, J., Bethke, A., Liu, Y., Belding, E.M., Wang, W.Y.: A benchmark dataset for learning to intervene in online hate speech. In: Inui, K., Jiang, J., Ng, V., Wan, X. (eds.) Proceedings of the 2019 Conference on Empirical Methods in Natural Language Processing and the 9th International Joint Conference on Natural Language Processing, EMNLP-IJCNLP 2019, Hong Kong, China, 3–7 November 2019, pp. 4754–4763. Association for Computational Linguistics (2019)
6. Kovács, G., Alonso, P., Saini, R.: Challenges of hate speech detection in social media. SN Comput. Sci. $\mathbf{2}$(2), 1–15 (2021)

---

[8] https://github.com/duonghuuphuc/hate-speech-detection

7. Ma, Y., Peng, H., Cambria, E.: Targeted aspect-based sentiment analysis via embedding commonsense knowledge into an attentive LSTM. In: McIlraith, S.A., Weinberger, K.Q. (eds.) Proceedings of the Thirty-Second AAAI Conference on Artificial Intelligence, (AAAI-18), the 30th innovative Applications of Artificial Intelligence (IAAI-18), and the 8th AAAI Symposium on Educational Advances in Artificial Intelligence (EAAI-18), New Orleans, Louisiana, USA, 2–7 February 2018, pp. 5876–5883. AAAI Press (2018)
8. Founta, A.M., et al.: Large scale crowdsourcing and characterization of twitter abusive behavior. In: ICWSM, pp. 491–500. AAAI Press (2018)
9. Mandl, T., et al.: Overview of the HASOC track at FIRE 2019: hate speech and offensive content identification in Indo-European languages. In: Majumder, P., Mitra, M., Gangopadhyay, S., Mehta, P., eds.: FIRE 2019: Forum for Information Retrieval Evaluation, Kolkata, India, December 2019, pp. 14–17. ACM (2019)
10. Waseem, Z., Hovy, D.: Hateful symbols or hateful people? Predictive features for hate speech detection on Twitter. In: SRW@HLT-NAACL, The Association for Computational Linguistics, pp. 88–93 (2016)
11. Gambäck, B., Sikdar, U.K.: Using convolutional neural networks to classify hate-speech. In: Waseem, Z., Chung, W.H.K., Hovy, D., Tetreault, J.R. (eds.) Proceedings of the First Workshop on Abusive Language Online, ALW@ACL 2017, Vancouver, BC, Canada, 4 August 2017, pp. 85–90. Association for Computational Linguistics (2017)
12. Zampieri, M., Malmasi, S., Nakov, P., Rosenthal, S., Farra, N., Kumar, R.: Predicting the type and target of offensive posts in social media. In: Burstein, J., Doran, C., Solorio, T. (eds.) Proceedings of the 2019 Conference of the North American Chapter of the Association for Computational Linguistics: Human Language Technologies, NAACL-HLT 2019, Minneapolis, MN, USA, 2–7 June 2019, Volume 1 (Long and Short Papers), pp. 1415–1420. Association for Computational Linguistics (2019)
13. Nguyen, D.Q., Vu, T., Nguyen, A.T.: BERTweet: a pre-trained language model for English Tweets. In: Proceedings of the 2020 Conference on Empirical Methods in Natural Language Processing: System Demonstrations, pp. 9–14 (2020)
14. Wang, B., Ding, Y., Liu, S., Zhou, X.: YNU_Wb at HASOC 2019: ordered neurons LSTM with attention for identifying hate speech and offensive language. In: Mehta, P., Rosso, P., Majumder, P., Mitra, M. (eds.) Working Notes of FIRE 2019 - Forum for Information Retrieval Evaluation, Kolkata, India, 12–15 December 2019, Volume 2517 of CEUR Workshop Proceedings, pp. 191–198. CEUR-WS.org (2019)
15. Liu, Y., et al.: RoBERTa: a robustly optimized BERT pretraining approach. CoRR abs/1907.11692 (2019)

MIC Model for Cervical Cancer Risk Factors Deep Association Analysis

Tiehua Zhou[1] , Yingxuan Tang[1] , Ling Gong[2], Hua Xie[3], Minglei Shan[1] ,
and Ling Wang[1(✉)]

[1] Department of Computer Science and Technology, School of Computer Science,
Northeast Electric Power University, Jilin, China
{thzhou,smile2867ling}@neepu.edu.cn
[2] Department of Nursing, Beihua University, Jilin City, Jilin, China
gongling@beihua.edu.cn
[3] Jilin Central General Hospital, Jilin City, China

Abstract. Prevention of cervical cancer (CC) is challenging due to unobvious early symptoms and the complexity of influencing factors. There are many risk factors for CC including the direct risk factors and indirect risk factors that may be caused by other diseases or reasons. In this paper, we proposed a MIC (Multiple Indicators Correlation) model to resolve the problem of analyzing risk factors by establishing the indicators structure. Based on the close relationship of indicators shown in the literature, the strength of relationship among indicators is calculated through the method of correlation analysis, and then strong association rules of CC were dug by computing the relationship of indicators combination and disease from electronic medical records (EMR). Experiment shows that though calculating the strength of multi-indicator joint influence, MIC model solves the problem of unstructured and many missing values for the data of EMR, it has better accuracy in prediction of CC.

Keywords: Cervical cancer risk factors · Association rules · Deep association analysis · Data mining · NLP

1 Introduction

Risk factors are important reference indicators for disease prevention and early screening [1]. Among the many risk factors, some are directly related to cervical cancer (CC), such as high-risk HPV infection [2], and some may be complications caused by other diseases or reasons. Therefore, it is necessary to analyze the main inducible factors of cervical cancer to provide a reference for cervical cancer prevention [3].

This work was supported by the National Natural Science Foundation of China (No. 62102076), and by the Education Department Foundation of Jilin Province, China (No. JJKH20210066KJ).

© Springer Nature Switzerland AG 2021
D. Mohaisen and R. Jin (Eds.): CSoNet 2021, LNCS 13116, pp. 147–155, 2021.
https://doi.org/10.1007/978-3-030-91434-9_14

Statistical analysis methods such as Case-control studies [4], cohort studies [5], and cross-sectional studies [6] are used to assess the correlation between risk factors and diseases by questionnaires. However, each risk factor has different effects on disease. The importance of risk factors was analyzed by quantifying risk factors, which is conducive to targeted prevention. Methods of quantifying risk factors have also made some progress in recent years. Yang, W. et al. analyzed the weight of risk factors for CC by the random forest and Pearson correlation coefficient [7]. Sunil Kumar et al. proposed a framework to identify strongly predictive attributes for depressed patients, The framework used the Pearson product-moment correlation coefficient to study the relationship between the target and independent attributes [8]. Alwidian, J. et al. proposed the WCBA algorithm, the weight of each attribute was assigned from the domain knowledge [9]. Huang, W. et al. analyzed the direct and indirect associations between dietary magnesium intake and breast cancer risk by multivariate logistic regression model [10]. Piotr, P. et al. Extracted risk levels from the context of risk factors in medical texts and assigned weights [11].

An important issue is that the interaction of risk factors affects the occurrence of diseases. Therefore, correlation analysis of multiple risk factors has become a challenge. The polynomial logistic regression model [12] and the COX model [10] found that several risk factors cause cervical diseases together. Masoudi Sobhanzadeh, Y. et al. used If-Then association rules to extract the relationship between drugs and diseases, drugs and drugs and applied discrete algorithms (Trader) to find synthetic lists for controlling hypertension [13]. Zhou, L. et al. proposed a neural network model, which selected the influence of multiple indicators that was calculated by Pearson. The accumulation of low correlation was considered to make risk prediction with a better accuracy [14]. Therefore, the correlation analysis of indicators and the mining combination of key risk factors are very necessary for the research of disease risk factors.

In this paper, we proposed a model named MIC (Multiple Indicators Correlation), which deeply analyzes direct and indirect risk factors and their influence relationships of CC by established the indicators structure. Through the analysis of electronic medical records (EMR), abnormal indicators were found and the association rules were mined. The association analysis of multiple risk factors has better find the key combination of CC.

2   MIC Model

The MIC model (Multiple Indicators Correlation) includes a correlation analysis algorithm and a strong association rule mining algorithm. The goal is that the key combination of indicators is extracted from the EMR and the experiment is verified in the prediction of cervical cancer.

2.1   Multiple Indicators Correlation Model

Symbols and their meanings used in this paper are shown in Table 1. The details of the MIC model are shown in Table 2.

Table 1. Symbols in this paper.

Notation	Description
C	The set of expression of CC
I	The set of expression of risk factors
I_A	The subset of abnormal risk factors of EMR
$f_{(C,I)}$	The number of occurrences of CC and risk factors in all paragraphs
$f_{(I,I)}$	The number of occurrences of risk factors and risk factors in all paragraphs
$R_{(C,I)}$	The weight of CC and risk factors
R_C	The set of $R_{(C,I)}$
$R_{(I,I)}$	The weight of risk factors and risk factors
R_I	The set of $R_{(I,I)}$
T_C	The threshold of $R_{(C,I)}$
T_I	The threshold of $R_{(I,I)}$
$R1_{(C,I)}$	The weight of CC and risk factors
$R2_{(C,I)}$	The synthetic weight of CC and single indicators
$R1_C$	The set of $R1_{(C,I)}$
$E_{(I,I)}$	The set of expression of CC
E_I	The weight of risk factors and abnormal risk factors
$S_{(C,I_k)}$	The synthetic weight of CC and risk factors
M_A	The weight of I_A
W	The minimum value of M_A in all I_A
G	The network graph of CC and risk factors

It is an effective method to extract the relationship between CC and indicators from medical literature. Organize the collection of expressions of CC through cervical cancer literature and data set $C = \{C_1, C_2, \ldots \ldots, C_z\}$, set of indicators is $I = \{I_1, I_2, \ldots \ldots, I_n\}$ and set of related expressions for each indicator is $I_k = \{I_{k1}, I_{k2}, \ldots \ldots, I_{km}\}$ $k \in (1, n)$, The number of occurrences of C and I in all paragraphs are counted as $f_{(C_i, I_{kj})}$ $i \in (1, z), j \in (1, m)$, and the number of occurrences of all pairwise indicators in the same paragraph is $f_{(I_{xi}, I_{yj})}$ $x \in (1, n), y \in (1, n), j \in (1, m), x \neq y$.

Definition 1. *The weight of CC and risk factors $R_{(C,I_k)}$.*

$$R_{(C, I_k)} = exp\left(\frac{\sum\limits_{i=1}^{z} \sum\limits_{j=1}^{m} f_{(C_i, I_{kj})}}{\sum\limits_{k=1}^{n} \sum\limits_{i=1}^{z} \sum\limits_{j=1}^{m} f_{(C_i, I_{kj})}} \right) \tag{1}$$

Definition 2. *The weight of risk factors and risk factors* $R_{(I_x, I_y)}$.

$$R_{(I_x, I_y)} = exp\left(\frac{\sum_{i=1}^{z}\sum_{j=1}^{m} f_{(I_{xi}, I_{yj})}}{\sum_{k=1}^{n}\sum_{i=1}^{z}\sum_{j=1}^{m} f_{(C_{xi}, l_{kj})}}\right) \tag{2}$$

According to formula (1), the set $R_C = \{R_{(C,I_1)}, R_{(C,I_2)}, \cdots \cdots, R_{(C,I_n)}\}$ is calculated. According to formula (2), the set of relational strength of indicator is calculated, which could express as $R_{I_k} = \{R_{(I_k,I_1)}, R_{(I_k,I_2)}, \cdots \cdots, R_{(I_k,I_n)}\}$.

Table 2. The details of the MIC model.

MIC Model (Multiple Indicators Correlation)
Input: C, I, $f_{(C,I)}$, $f_{(I,I)}$, D(EMR Data)
Output: Satisfactory I_A
1:**for** each c in C, i in I **do**
2:   $R_{(C,I)} \leftarrow$ WeightOfCandI($f_{(c,i)}$)
3:   $R_{(I,I)} \leftarrow$ WeightOfCandI($f_{(x,y)}$)
4:**end for**
5:T_C, $T_I \leftarrow$ threshold($R_{(C,I)}$, $R_{(I,I)}$)
6:$N \leftarrow$ the number of D
7:**for** each i in I **do**
8:   $N_i \leftarrow$ the number of i is abnormal value in D
9:   $R1_{(C,I)} \leftarrow$ Weight1OfIandI(N_i, N)
10:   Satisfactory I: V1 $\leftarrow R_{(c,i)} > T_C$
11:   Satisfactory I: V2 $\leftarrow R_{(I,i)} > T_I$
12:**end for**
13:G((V1,V2),($R_{(C,I)}$,$R_{(I,I)}$))
14:**for** any d in D **do**
15:   $I_A \leftarrow$ Extracting Exception Attributes from D
16:   **for** any A in all the subset of I_A, a i in A **do**
17:     $E_I \leftarrow$ WeightOfIand(a, G, $R_{(I,I)}$)
18   **if** $|R_{(C,a)} - R1_{(C,a)}| >$avg of $|R_{(C,I)}-R1_{(C,I)}|$ **then**
19:       $R2_{(C,I)} = (R_{(C,a)} + R1_{(C,a)}) / 2$
20:     **else**
21:       $R2_{(C,I)} = \max(R_{(C,a)}, R1_{(C,a)})$
22:     $M_a \leftarrow$ MultiIndicatorsCorrelation($R2_{(C,I)}$, E_I)
23:     $W \leftarrow$ min value of all M_a
24:     **return** a
25:     **end for**
26:**end for**

Definition 3. *Threshold T_C of R_C.*

$$T_C = \frac{1}{n}\sum_{p=1}^{n} R_{(C,\ I_k)} \tag{3}$$

Definition 4. *Threshold T_{I_K} of R_{I_K}.*

$$T_{I_k} = \frac{1}{n-1}\sum_{q=1}^{n-1} R_{(I_k,\ I_1)} \tag{4}$$

The indicators structure is established as a network $G = (V, E)$ with CC as the center point. V represents the set of CC and all indicators, E represents the relationship between CC and the indicators, and the relationship between the indicators and the indicators. Each indicator is a node in the graph. The indicators whose relational strength between CC and each indicator is greater than the threshold T_C is used as direct risk factors of cervical cancer, which is directly connected to the node CC in the network graph. The relationship of the indicators in the direct risk factors and other indicators is greater than the threshold value T_{I_k} in the network graph is connected as indirect risk factors until all indicators and relationships appear in the network, the system is established.

In addition, it is also very important to analyze the relationship between a single indicator and the disease from the data of EMR. It can calculate the relationship between a single indicator and CC from another angle. In all the data, the number of EMR is N, and the number of abnormal indicators I_k in EMR is N_k. The strength of the relationship between CC and each indicator is calculated by the formula (5), and the relational strength set between CC and I_k is obtained $R1_C = \{R1_{(C,I_1)}, R1_{(C,I_2)}, \cdots \cdots, R1_{(C,I_n)}\}$.

Definition 5. *The weight of CC and risk factors $R1_{(C,I_k)}$.*

$$R1_{(C,\ I_k)} = exp\left(\frac{N_k}{N}\right) \tag{5}$$

Due to the different conditions of each patient, the content presented by the electronic medical record is different, it is difficult to structure and analyze the data of EMR uniformly. Abnormal indicators extracted from EMR form an abnormal indicator set I_A could solve this problem well. The extraction of abnormal attributes follows some rules, authoritative medical literature was referred to define the value rules of outliers. From the other point of view, all the values of a continuous attribute in the data set were analyzed, and the value of its quartile was taken as the boundary. According to the content of the attribute, binary data was set the value of 1 or 0 as an exception attribute.

The relational strength of all indicators and abnormal indicators could reflect the interaction. The set of indicators I_S is screened as all indicators that have a strong relationship with I_k and appear in I_A. The influence level is calculated by standard deviation, denoted as $E_{(I_K,I_A)}$, $E_I = \{E_{(I_1,I_A)}, E_{(I_2,I_A)}, \cdots \cdots, E_{(I_n,I_A)}\}$, Where μ is the average value of relational strength in I_S and I_k, N_S is the number of elements in I_S.

$$E_{(I_k,\ I_s)} = \sqrt{\frac{1}{N_s} \sum_{h=1}^{N_s} \left(R_{(I_k,\ I_h)} - \mu\right)^2} \tag{6}$$

The degree of a single indicator on the prevalence of CC is calculated from R_C, $R1_C$ and E_I together.

Synthetic Weight of CC and Single Indicators Algorithm. R_C comes from medical literature and $R1_C$ comes from EMR, both of them reflect the relationship between CC and risk factors to a certain extent. If the difference between $R_{(C,I_K)}$ and $R1_{(C,I_K)}$ of an indicator is greater than the average of the difference about $R_{(C,I_K)}$ and $R1_{(C,I_K)}$ of all indicators, the largest of the two is selected, otherwise, the average of the two is selected, denoted as $R2_C = \{R2_{(C,I_1)}, R2_{(C,I_1)}, \cdots \cdots, R2_{(C,I_1)}\}$. The calculation of $R_{(C,I_K)}$ and $R1_{(C,I_K)}$ could obtain the synthetic weight of CC and single indicator.

$$\delta = \frac{1}{n} \sum_{p=1}^{n} \left| R_{(C,\ I_p)} - R1_{(C,\ I_{1p})} \right| \tag{7}$$

Definition 6. *The synthetic weight of CC and single indicators* $S_{(C,I_K)}$.

$$S_{(C,I_k)} = \left(max\left(R_{(C,I_k)}, R1_{(C,I_k)}\right) + E_{(I_k,\ I_A)} \right) 1 \left[\left| R_{(C,I_1)} - R1_{(C,I_1)} \right| < \delta \right] \tag{8}$$

$$S_{(C,I_k)} = \left(\frac{R_{(C,I_k)} + R1_{(C,I_k)}}{2} + E_{(I_k,\ I_A)} \right) 1 \left[\left| R_{(C,I_1)} - R1_{(C,I_1)} \right| < \delta \right] \tag{9}$$

Multiple Indicators Correlation Algorithm. In the data of EMR, there may be multiple abnormal indicators. To explore the key combination of abnormal indicators that will have the greatest impact on CC, We proposed a Multiple Indicators Correlation algorithm and the next example is used to illustrate the idea of the algorithm as shown in Fig. 1. In the previous step, the indicators'

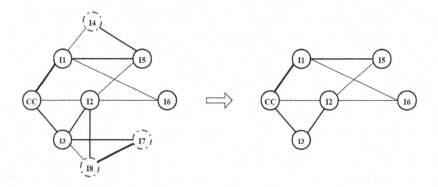

Fig. 1. Mining strong multi-node combinations.

network has been constructed. Assuming that the abnormal indicator of a data of EMR is $I_A = \{I_1, I_2, I_3\}$, and all its non-empty subsets are $I_{A1} = \{I_1\}$, $I_{A2} = \{I_2\}$, $I_{A3} = \{I_3\}$, $I_{A4} = \{I_1, I_2\}$, $I_{A5} = \{I_1, I_3\}$, $I_{A6} = \{I_2, I_3\}$, $I_{A7} = \{I_1, I_2, I_3\}$, N_A is the number of subsets. Assuming that all indicators in the network G are abnormal, according to formula (8) or (9), $S_{(C,I_k)}$ of each point is computed. We remove the nodes of these abnormal indicators in the subset from the network and observe the change after removal. The subset with the smallest change is the greatest impact on CC.

Definition 7. *Algorithm for The weight of I_A.*

$$\varphi_1 = \frac{1}{n}\sum_{k=1}^{n} S_{(C,\ I_k)} \tag{10}$$

$$\varphi_2 = \frac{1}{n}\sum_{k=1}^{n} S_{(C,\ I_k)} \tag{11}$$

$$M_{A1} = \sqrt{\frac{1}{n}\sum_{k=1}^{n}\left(S_{(C,\ I_k)} - \varphi_1\right)^2} - \sqrt{\frac{1}{n-n_A}\sum_{k=1}^{n-n_A}\left(S_{(C,\ I_k)} - \varphi_2\right)^2} \tag{12}$$

$$w = min\left(M_{A1}, M_{A2}, \ldots \ldots, M_{A7}\right) \tag{13}$$

Subset related to w is the greatest impact on the prevalence of CC. Predictions of cervical cancer were achieved by matching a strong correlation subset from EMR data.

3   The Experimental Analysis

The experimental environment was based on windows 10, GeForce GTX 1080Ti. The processing language is python. The data set of the model is divided into two parts, one is medical literature, the other is cervical cancer case data set. Medical literature comes from NEJM, Lancet, JAMA, BMJ related to cervical cancer, a total of 317 articles. The cervical cancer case data set comes from two sources, one is from the UCI cervical cancer risk factor data set [15], and the other is the cervix from TCGA-CESC obtained on The Cancer Imaging Archive (TCIA) Cancer clinical data set [16]. The data sets of EMR were divided into 80% for training and 20% for testing. The Mixed Logit model (ML) is selected as a comparative experiment [17]. In these two different data sets, the TCGA data set are all positive samples, and we use the recognition rate of the positive samples in the test set as the accuracy rate. The classification accuracy rate for predicting the prevalence of cervical cancer is shown in Table 3.

MIC model has a good performance in accuracy, and the precision of prediction is slightly higher than the Mixed Logit model. It has a good prediction effect on the accuracy rate for predicting the prevalence of cervical cancer.

Table 3. Classification accuracy of two methods.

Accuracy \ Data Method	TCGA data	risk factors data		
	Accuracy%	Accuracy%	Precision%	Recall%
ML model	85.7	82.7	77.8	81.3
MIC model	91.9	88.2	84.1	84.5

Table 4. The key sets of risk factors of CC.

The key sets of risk factors of CC
HPV
Menopause
Age
Contraceptives, menopause
Contraceptives, HPV
Age, keratinizing squamous cell carcinoma
Smoke
Age, number of sexual partners, smoking year, HPV
Number of sexual partners
Age of first sexual intercourse, HPV
Age, smoke HPV, keratinizing squamous cell carcinoma
Age, the number of pregnancy
Smoke, contraceptives

52 indicators of risk factors are used for analysis and experiments. Prioritize the frequency of the key association sets extracted by the multiple indicators association algorithm, and the results are shown in Table 4.

High-risk HPV is still the main factor of CC. Many people with cervical cancer are in menopause or perimenstrual period, the use of contraceptives will increase the risk of cervical cancer. The high number of sexual partners, the early first sexual intercourse, pregnancy, childbirth, miscarriage, and ectopic pregnancy will increase the possibility of HPV infection and then increase the likelihood of CC. Smoking is also a risk factor worth noting. People who smoke are more likely to have the disease than don't smoke.

4   Conclusion

MIC model adopted a new risk factors assessment method, it resolve the problem of difficulty in quantifying the indicators relationship and mining the key combination of factors. In the future, our goal is establish a more complete indicators structure by collecting more and more comprehensive data.

References

1. Cohen, P.A., Jhingran, A., Oaknin, A., Denny, L.: Cervical cancer. Lancet **393**(10167), 169–182 (2019)
2. Torre, L.A., Islami, F., Siegel, R.L., Ward, E.M., Jemal, A.: Global cancer in women: burden and trends. Cancer Epidemiol. Prev. Biomark. **26**(4), 444–457 (2017)
3. Sung, H., et al.: Global cancer statistics 2020: GLOBOCAN estimates of incidence and mortality worldwide for 36 cancers in 185 countries. CA Cancer J. Clin. **71**(3), 209–249 (2021)
4. Kashyap, N., Krishnan, N., Kaur, S., Ghai, S.: Risk factors of cervical cancer: a case-control study. Asia Pac. J. Oncol. Nurs. **6**, 308–314 (2019)
5. Castle, P.E., et al.: Effect of several negative rounds of human papillomavirus and cytology co-testing on safety against cervical cancer. Ann. Internal Med. **168**(1), 20–29 (2018)
6. Barchitta, M., Maugeri, A., Quattrocchi, A., Agrifoglio, O., Scalisi, A., Agodi, A.: The association of dietary patterns with high-risk human papillomavirus infection and cervical cancer: a cross-sectional study in Italy. Nutrients **10**(4), 469 (2018)
7. Yang, W., Gou, X., Xu, T., Yi, X., Jiang, M.: Cervical cancer risk prediction model and analysis of risk factors based on machine learning. In: Proceedings of the 2019 11th International Conference on Bioinformatics and Biomedical Technology, ICBBT 2019, New York, NY, USA, pp. 50–54. Association for Computing Machinery (2019)
8. Kumar, S., Chong, I.: Correlation analysis to identify the effective data in machine learning: prediction of depressive disorder and emotion states. Int. J. Environ. Res. Public Health **15**(12), 2907 (2018)
9. Alwidian, J., Hammo, B.H., Obeid, N.: WCBA: weighted classification based on association rules algorithm for breast cancer disease. Appl. Soft Comput. **62**, 536–549 (2018)
10. Huang, W.Q., et al.: Direct and indirect associations between dietary magnesium intake and breast cancer risk. Sci. Rep. **9**, 1–10 (2019)
11. Przybyła, P., Brockmeier, A.J., Ananiadou, S.: Quantifying risk factors in medical reports with a context-aware linear model. J. Am. Med. Inform. Assoc. JAMIA **26**(6), 537–546 (2019)
12. Islam, A.-U., Ripon, S.H., Bhuiyan, N.Q.: Cervical cancer risk factors: classification and mining associations. APTIKOM J. Comput. Sci. Inf. Technol. **4**(1), 8–18 (2019)
13. Masoudi-Sobhanzadeh, Y., Masoudi-Nejad, A.: Synthetic repurposing of drugs against hypertension: a datamining method based on association rules and a novel discrete algorithm. BMC Bioinform. **21**(1), 313 (2020)
14. Zhou, L., Cai, L., Jiang, L., Chen, L.: Power grid enterprise intelligent risk identification model considering multi-attribute and low correlation data. IEEE Access **7**, 111324–111331 (2019)
15. UCI machine learning repository. http://archive.ics.uci.edu/ml/index.php
16. The cancer imaging archive (TCIA) public access. https://wiki.cancerimagingarchive.net
17. Overgoor, J., Pakapol Supaniratisai, G., Ugander, J.: Scaling choice models of relational social data, New York, NY, USA, pp. 1990–1998. Association for Computing Machinery (2020)

Power Grid Cascading Failure Prediction Based on Transformer

Tianxin Zhou[ID], Xiang Li[(✉)][ID], and Haibing Lu[ID]

Santa Clara University, Santa Clara, CA 95053, USA
xli8@scu.edu

Abstract. Smart grids can be vulnerable to attacks and accidents, and any initial failures in smart grids can grow to a large blackout because of cascading failure. Because of the importance of smart grids in modern society, it is crucial to protect them against cascading failures. Simulation of cascading failures can help identify the most vulnerable transmission lines and guide prioritization in protection planning, hence, it is an effective approach to protect smart grids from cascading failures. However, due to the enormous number of ways that the smart grids may fail initially, it is infeasible to simulate cascading failures at a large scale nor identify the most vulnerable lines efficiently. In this paper, we aim at 1) developing a method to run cascading failure simulations at scale and 2) building simplified, diffusion based cascading failure models to support efficient and theoretically bounded identification of most vulnerable lines. The goals are achieved by first constructing a novel connection between cascading failures and natural languages, and then adapting the powerful transformer model in NLP to learn from cascading failure data. Our trained transformer models have good accuracy in predicting the total number of failed lines in a cascade and identifying the most vulnerable lines. We also constructed independent cascade (IC) diffusion models based on the attention matrices of the transformer models, to support efficient vulnerability analysis with performance bounds.

Keywords: Power grid · Smart grid · Cascading failure · Transformer · Independent cascade model

1 Introduction

In smart grids, the integration of cyber and physical processes on one hand enhanced the accessibility to all the functionality of the power grid, but on the other hand, it leads to potential threat to the grid from the cyber surface, since for attacks, attackers now may access the grid via internet connections; for accidents, the cyber surface opens up more possibilities. The damage level of the potential attacks and accidents can be escalated because of power grid cascading failures (PGCF) [12], where the failure of one transmission line may lead to failures of other lines and eventually large blackouts. Many real-world blackouts, for example, the Northeast America blackout and Italy blackout in

© Springer Nature Switzerland AG 2021
D. Mohaisen and R. Jin (Eds.): CSoNet 2021, LNCS 13116, pp. 156–167, 2021.
https://doi.org/10.1007/978-3-030-91434-9_15

2003, Brazil and Paraguay blackout in 2009, and India blackout in 2012 are all related to cascading failure [1,3,4,13]. Because of the catastrophic impact of cascading failures in smart grids, a key infrastructure network, it is important to understand cascading failures and perform protection actions.

To prioritize the allocation of protection resources on the transmission lines, it is crucial to understand what are the most important lines in a cascading failure. We consider two types of lines as important: 1) the most *critical* lines: the failure of those lines could cause the largest scale of cascading failure and 2) the most *vulnerable* lines: the lines that are most likely to fail by the failure of other lines. In order to identify those important lines, the approach of running simulations of cascading failures is studied. One widely used model for simulation is the OPA model, which was first introduced in [9,10,14,28], and many of its variations are studied later [21–23,26]. Other cascading failure simulation models include the hidden failure model [11] and the cascading failure model [6]. One essential component of all the models is the calculation of the power flow equation [5], which is needed for each round in cascading failure. The existing simulation models face two challenges: 1) Since the number of possible failed line combinations is huge ($\binom{N}{k}$ for an $N - k$ analysis), it is infeasible to do cascading failure simulations at scale. 2) there exists no efficient way to identify the most critical/vulnerable lines with theoretical performance guarantee, as the cascading failure models are too complicated.

To deal with the first challenge, Machine Learning (ML) models are considered in literature [15,25,29]. The existing models can predict the severity of a cascading failure given the initial failures, however, it is hard to extract information like the actual lines failed in a cascade, which is important for analysis. We will consider more powerful models that can predict the whole cascading failure process instead of the severity of cascading failure. The reason why it is possible is a novel connection between cascading failures and natural languages: both the lines failed in a cascading failure and the words in a sentence are sequences of elements, which makes it possible to adapt the sequence-to-sequence models in NLP and use them on cascading failure prediction tasks. Among the sequence-to-sequence deep learning models, the transformer based models [30] are the state-of-the-art. Comparing to the traditional recurrent neural network, the transformer sacrifices the focus on the order of the elements in the sequence but gained stronger ability to learn the correlations between elements. This disadvantage may compromise the performance on the pure NLP problems but it does not affect the performance for the PGCF problem because the order of failed lines in each set in a cascading failure stage has very little effect on calculating or predicting the set of failing lines for the next stage. To the best of our knowledge, there exists no research on using sequence-to-sequence models for the PGCF problem.

The second challenge can be addressed with an intrinsic feature of transformer models: the attention mechanism. The correlations of elements represented by the attention matrix indicates the percentage that the elements "attend" to each other. In a transformer model trained for PGCF, it means how likely a line will fail after the failure of another line. This possibility representation can be applied to the independent cascade (IC) diffusion model [17]. After converting the attention matrix to a probability matrix, it is possible to simulate PGCF with an IC model, which greatly simplifies the process and provides further performance boost to cascading failure simulation.

To verify the effectiveness of the transformer and IC models in cascading failure simulation, we trained transformer models for three power grid networks, including two IEEE test cases and the SciGrid network. The cascading failure samples are generated using the model from [6]. The IC models are then derived from the trained transformer models. Both models are capable of doing cascading failure prediction tasks, the f1 score can go as high as 0.77 for the transformer model in SciGrid. In terms of efficiency, the transformer model can generate cascading failures up to 56 times faster than the classical power flow based models, while the IC model can be several orders of magnitude faster.

Our contributions are summarized as follows.

- We propose a new approach of simulating PGCF with the transformer model, based on a novel connection between cascading failures and natural languages.
- We utilize the parameters from the transformer model to build an IC model to greatly simplify the simulation of the PGCF process and support vulnerability analysis with theoretical performance guarantee.
- We trained the transformer model and constructed the IC model in multiple widely used power network data sets, including IEEE test cases and SciGrid. Experiment results on PGCF simulation tasks show that the transformer and IC models have good accuracy and greatly boost efficiency, when comparing to the power flow based cascading failure models.

Organization. The rest of the paper is organized as follows. Section 2 reviews the related works. Section 3 explains the cascading failure model, the transformer model and the IC model. Section 4 provides the evaluation and comparison between the three models. We conclude the paper in Sect. 5.

2 Literature Review

The analysis of the vulnerability of power grid has been a focus of studies to improve the security of smart grid. Many of the studies are based on the deterministic models [6,8–11,14,22,23,26,28], and references therein. Other studies are based on stochastic models [16,20,24,27,31,32]. Furthermore, there are limited number of studies utilize ML techniques to analyze the PGCF [15,25,29]. All of those models have their own advantages and limitations.

The foundation of the deterministic models were the power flow equation from [5]. The model in [11] provides the fundamental template for the cascading failure which is extended in [6–8] with vulnerability analysis and control implication modules. The OPA model [9,10,14,28] enriches the template with the complex factors that dynamically changing the configuration of the grid. The variants of the OPA model make the efforts with different point of view. The improved OPA model [22] makes the improvement with the concept of the hidden failure. The OPA model with slow process [26] add on the factors of tree contacts and temperature variation to the original OPA model. The AC OPA model [21,23] changes the DC OPF calculation to the AC OPF calculation. The deterministic models can reveal details of PGCF, however, they may experience performance issues due to extensively resolving the power flow equations.

The stochastic factors are introduced to simplify the calculation with Markov Chain or probability density function [16,20,24,27,31,32]. In [15,25,29], multiple ML techniques are used to make statistical analysis from a more general perspective. Both stochastic and ML models lack the ability to describe the status of individual components in a cascading failure.

The transformer model [30] has been proven to be the foundation for the state-of-art Deep Learning (DL) techniques for natural language processing (NLP), especially for the sequence-to-sequence problem. In [19], the information diffusion problem for the social network was addressed by a transformer based model. However, since the mechanism of information diffusion and PGCF are very different, the model is not applicable in PGCF simulation.

DL techniques have been widely used to solve different power grid tasks [18]. The BiLSTM with Attention, for example, is used to analyse the stability of the power grid [33]. However, the model only predicts a binary results of whether if the grid is stable or not. To the best of our knowledge, no study has applied the transformer based model to simulate the PGCF process.

PGCF simulation may also be addressed using the diffusion models [2,17], in which the state of nodes in a network can be impacted by the state of the neighboring nodes in a stochastic manner. However, it is pointed out in [6] that cascading failure may propagate non-locally, hence, a diffusion model based on the smart grid topology cannot be directly applied to simulate PGCF and some transformation is needed.

3   Models

In this section, we first describe the cascading failure model, which is used to generate the data set for the training and testing for the transformer model. Then, we introduce the text generation task in NLP and show how it is related to PGCF simulation, and describe the transformer model. In the end, we discuss an approach to construct an IC model with the attention matrix from the transformer model.

3.1   Cascading Failure Model

To generate cascading failure samples for model training and testing, a simplified cascading failure model from [6] is used in this paper. The power grid can be described as a graph G with as set of nodes N, which can be further divided into two groups: the supply nodes $S \subseteq N$ and the demand nodes $D \subseteq N$. For node $i \in S$, s_i represents the active power generated at i, d_i represents the demand power of $i \in D$ and θ_i represents the phase angle of i. $\delta^+(i)(\delta^-(i))$ represents the set out(in)-neighbors of node i. We use a tuple (i, j) to represent a transmission line between nodes i and j, with f_{ij} indicating the power flow, x_{ij} as the reactance, and u_{ij} as the capacity.

The cascading failure model has the following steps:

1. A set of lines is randomly selected to fail as the initial state.
2. The power flow of the grid is calculated by the Eq. 1 and 2.

$$\sum_{j \in \delta^+(i)} f_{ij} - \sum_{j \in \delta^-(i)} f_{ji} = \begin{cases} s_i, & i \in S \\ -d_i, & i \in D \\ 0, & \text{otherwise} \end{cases} \tag{1}$$

$$\theta_i - \theta_j - x_{ij} f_{ij} = 0, \forall (i, j) \tag{2}$$

3. The lines with power flow higher than the capacity ($f_{ij} < u_{ij}$) are set to failed.
4. If no lines failed in step 3, the cascade ends and all the failed lines are recorded as the final state. Otherwise, repeat steps 2 and 3.

3.2 Transformer Model

Text Generation vs. PGCF. The text generation task is one of the most classical NLP problems which is normally solved by a sequence-to-sequence model. The model is "asked" with a sequence of words as input then "answer" with another sequence of words as the output. This is the same as a simplified PGCF process which has a sequence of initial failed lines and a sequence of final failed lines. Since almost all the state-of-art sequence-to-sequence models for NLP problems are based on the transformer model, it could be a great fit for the PGCF analysis (Fig. 1).

Attention Mechanism. The detailed structure of the transformer model can be found in [30]. The most powerful feature of this model is the attention mechanism (Eq. 3) which calculates the correlation between all elements in the sequence [30].

$$Attention(Q, K, V) = softmax(\frac{QK^T}{\sqrt{d_k}})V \tag{3}$$

The matrix multiplication of QK^T represents the relationship between *Query* matrix Q and *Key* matrix K. d_k is the dimension of the matrix K. $\sqrt{d_k}$ is used for scaling, which does not have direct impact to the result, but may improve model training efficiency [30]. By taking the softmax of the matrix multiplication result and multiply with the *Value* matrix V, we obtain the level of the attention between each pair of elements in the *Query* and *Key*. The complexity for Eq. 3 is $O(n^2)$ which is a great improvement to the power flow based models. However, with multi-layer structure and the recurrent calculation to simulate the PGCF, the transformer model is still computational expensive.

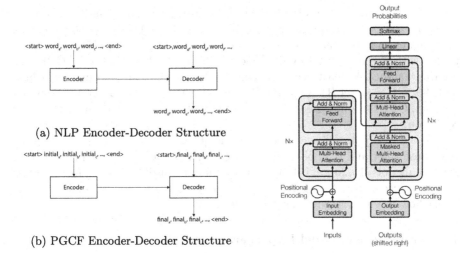

Fig. 1. The general structure of Encoder-Decoder Model for NLP and PGCF problem. "⟨start⟩" and "⟨end⟩" are the tokens to indicate the start and the end of the sequence.

Fig. 2. Transformer structure [30]

3.3 Independent Cascade Model

Since the transformer model is still "heavy" for prediction, with the attention matrix extracted from the trained transformer model, it is possible to construct an IC model that greatly simplifies the PGCF simulation. If we converted the set of all transmission lines into a complete directed graph $G(N, E)$. For edge (i, j) from node i to j, its weight w_{ij} determines how likely node j will fail after node i's failure. The weight can be seen as the attention paid by i to j. If i attends j significantly, it is more possible that j will be failed by the failure of i. The attention mechanism of the transformer has exactly same purpose.

To summarize, the IC Model simulates the cascading failure with the following steps: (1) assign scaled $Attention_{ij}^{\theta}$ to w_{ij}; (2) randomly fail a set of lines $R_m, m = 0$; (3) fail set of lines $R_{m+1} = \{j | w_{ij} > P(ij), \forall i \in R_m, (i, j) \in E \backslash R_m\}$, where $P(ij)$ is uniformly randomly sampled in $[0, 1]$ independently for each (i,j); (4) terminate if $R_{m+1} = \emptyset$, else increment m and repeat step (3).

Because the calculation for the state of each node is just one comparison, the complexity is only $O(n)$ which is another great improvement than the transformer model. Also, due to the simplicity, many optimization problems defined on the IC model can have theoretically bounded solutions (e.g. [17]), which makes the IC model valuable in cascading failure vulnerability analysis for future studies.

4   Experiments

To validate the performance of the proposed approaches, we train the transformer model and construct the IC model on three widely used synthetic power grids, and generate cascading failure samples using the model in [6]. The stats of the networks and samples are summarized in Table 1. We use 80% samples for training and 10% each for testing and validation. A virtual Google compute engine with 4 vcpus plus 15 GB memory and one NVIDIA Tesla T4 GPU was used in training. When testing the computational efficiency, we use a machine with 80 CPUs (Intel(R) Xeon(R) CPU E5-4650v2 @ 2.40 GHz) and 566 GB memory, GPUs are not used to ensure all models are evaluated under the same condition.

We use the power flow based model in [6] as a baseline to compare with the transformer and IC models. The reason for not comparing with the existing ML/DL models [15,25,29] is that they are fundamentally different, for example, they may use power flow features to train the parameters, or combining the power flow calculation with the ML techniques.

4.1   Transformer Model Hyperparameter

The structure of the transformer model is shown in Fig. 2. Considering the "vocabulary" size (total number of lines in our cases) is a lot less than the common NLP problems, and because the improvements are limited with heavier model according to the results of our experiments, we chose to only have 2 encoder layers and 2 decoder layers. For the embedding and attention matrix, the dimensionality is set to 128, the same as the inner feed-forward layer.

4.2   Total Number of Failed Lines Prediction

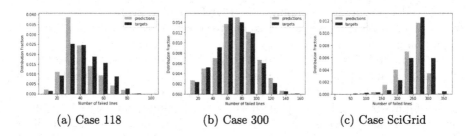

(a) Case 118   (b) Case 300   (c) Case SciGrid

Fig. 3. Total number of failed lines predictions - transformer model

The distribution of the total number of failed lines is shown in Fig. 3. It is obvious to see that the more lines in the grid, the larger scale of PGCF may occur. For case 300, it appears our prediction is more consist with the targets comparing to other cases. However, it is also closer to a normal distribution for both the predictions and targets. That could mean the vulnerability is more normally distributed throughout the grid. Especially comparing the results with case SciGrid, there are reasons to believe some of the lines may always trigger more lines to fail.

Table 1. Dataset description

Case	Lines	Total samples
IEEE-118	173	1,000,000
IEEE-300	283	100,000
SciGrid	852	191,479

Table 2. SSD of line failure frequency

Case	Predict SSD	Target SSD
IEEE-118	0.227	0.187
IEEE-300	0.196	0.174
SciGrid	0.335	0.335

4.3 Line Failure Frequency

In Table 2, the scaled standard deviation (SSD) of the failure frequency (f) for each line in three cases is calculated by

$$SSD = SD(f)/S$$

where $SD(.)$ is the function for standard deviation and S is the size of test set.

Since SSD for case 300 is the lowest, the failed frequency for each line does not deviate much which is consist with the result we obtained from Sect. 4.2 that the vulnerability is more normally distributed for case 300. We could also expect the prediction of actual failed lines can be more difficult for case 300 and more accurate and reliable for the case SciGrid.

In Table 3, 8/10 predicted most *vulnerable* lines are the same as the target set for case 118 and case 300, and 6/10 predictions are correct for case SciGrid. But, the general error distribution (Fig. 4) indicates that the error rate for most of the predictions are within $[0, 0.1]$, especially for case SciGrid. If the above expectation was correct, this distribution could mean the transformer model performs well for the most *vulnerable* lines prediction for complex power grids.

Table 3. 10 most *vulnerable* lines

Case	Prediction	Target
IEEE-118	73, 65, 30, 31, 32, 129, 67, 141, 142, 144	73, 65, 67, 30, 31, 32, 144, 143, 129, 158
IEEE-300	202, 230, 20, 164, 123, 153, 93, 5, 279, 143	202, 164, 230, 93, 20, 123, 153, 76, 143, 99
SciGrid	71, 26, 65, 8, 3, 252, 141, 253, 38, 86	26, 71, 8, 65, 38, 86, 25, 30, 13, 126

4.4 Line Failure Magnitude

To predict the most *critical* lines, we use the concept "magnitude" as defined by Eq. 4. For each cascading failure sample, the contribution of one initial failed line can be considered as the total number of failed lines divided by the number of initial failed lines. For each transmission line, its magnitude can be the average contribution out of all cascading failure samples that line had contributed to.

$$magnitude(line_i) = \frac{\sum_{i \in initial_j} \frac{num\_of(cascade_j)}{num\_of(initial_j)}}{frequency_i} \quad (4)$$

The transformer model performs even better for the most *critical* lines prediction (Fig. 5). The higher error rate for the case 118 implies the transformer model may perform worse for simpler grids.

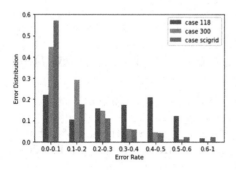

Fig. 4. Line failure frequency   **Fig. 5.** Line failure magnitude

4.5 F1 Score

The precision, recall, and f1 scores for three cases are listed in Table 4. It is obvious to see that the transformer model performs better with the SciGrid case (f1: 0.77) which is still consistent with the observation in previous experiments. The reason that case 300 did not perform better than case 118 could also be the normally distributed vulnerability.

Table 4. F1 score - transformer

Cases	Precision	Recall	F1
IEEE-118	0.46	0.67	0.55
IEEE-300	0.41	0.72	0.52
SciGrid	0.70	0.87	0.77

Table 5. Time consumption (sec/sample)

Cases	Power Flow	Transformer	IC
IEEE-118	5.35	2.91	0.017
IEEE-300	9.93	4.75	0.021
SciGrid	103.17	1.82	0.067

4.6 IC Model Simulation

The IC model simplifies PGCF simulation at the cost of lower accuracy. Hence, the prediction of the total number of failed lines (Fig. 6) could be worse comparing to the result from the transformer model (Fig. 3). The higher distribution for the smaller scale cascading failure prediction implies the conversion between the attention and the weights needs to be more sophisticated that the potential large scale cascading failures won't be missed. The implication is also supported by the peak distribution for case SciGrid because the oversimplified conversion may encourage the cascading failure with the deactivation of the *vulnerable* lines.

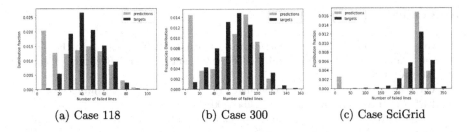

(a) Case 118   (b) Case 300   (c) Case SciGrid

Fig. 6. Total number of failed lines predictions - IC model.

4.7   Computational Efficiency

Complexity Analysis. From the Eqs. 1 and 2, the linearized power flow model has time complexity $O(n^3)$ for the worst case scenario. From the Eq. 3, we know the complexity for the attention calculation is n^2 (Sect. 3.2). And, the complexity for the IC model is $O(n)$ as explained in Sect. 3.3.

It is obvious that the transformer model will perform much faster when n is larger. However, when n is smaller, the difference won't be that significant because the other factors in the transformer model may contribute more to the computational complexity. For example, when n is close to the dimensionality of the embedding matrix d, the complexity can be close to $O(n^3)$.

Computing Time. In Table 5, we can see when the power grid gets more complex (852 lines vs. 173 lines and 283 lines), the power flow model takes significantly longer time. The IC model is the fastest as expected. These results are consistent with the discussion above. Besides, because the computing speed for transformer model will be affected by the dimensionality of the feature matrix, there is no exponential difference between different cases.

5   Conclusion

In this paper, we studied the problem of predicting cascading failures with transformer models and further construct an IC model as a simplified cascading model, which can be used for both prediction and theoretical analysis. By considering line failures in cascading failure as a sequence, we trained transformers on cascading failure data, and then built IC models using attention matrices in the transformers. Comparison with the power-flow based cascading failure model in three widely used power grid test cases showed that the transformer and IC models have acceptable accuracy and can greatly improve simulation efficiency. Also, it is possible to use the trained models to support identification of the most *critical* and *vulnerable* lines in cascading failure, which can contribute to protection planning.

References

1. Abraham, S., et al.: Final Report on the August 14, 2003 Blackout in the United States and Canada: Causes and Recommendations, Natural Resources Canada, Ottawa (2004)
2. Asavathiratham, C., Roy, S., Lesieutre, B., Verghese, G.: The influence model. IEEE Control Syst. Mag. **21**(6), 52–64 (2001)
3. Bacher, D.R., Näf, U.: Report on the blackout in Italy on 28 September 2003 (2003)
4. Bakshi, S.A., Velayutham, S.A., Srivastava, D.S.C., Agrawal, S.K.K.: Report of the Enquiry Committee on Grid Disturbance in Northern Region on 30th July 2012 and in Northern, Eastern & North-Estern Region on 31st July 2012 (2012)
5. Bergen, A., Vittal, V.: Power Systems Analysis. Prentice Hall, Hoboken (2000)
6. Bernstein, A., Bienstock, D., Hay, D., Uzunoglu, M., Zussman, G.: Power grid vulnerability to geographically correlated failures – analysis and control implications. In: IEEE INFOCOM 2014 - IEEE Conference on Computer Communications, pp. 2634–2642 (2014)
7. Bienstock, D.: Optimal control of cascading power grid failures. In: 2011 50th IEEE Conference on Decision and Control and European Control Conference, pp. 2166–2173 (2011)
8. Bienstock, D., Verma, A.: The $n - k$ problem in power grids: new models, formulations, and numerical experiments. SIAM J. Optim. **20**(5), 2352–2380 (2010)
9. Carreras, B.A., Lynch, V.E., Dobson, I., Newman, D.E.: Critical points and transitions in an electric power transmission model for cascading failure blackouts. Chaos: Interdiscip. J. Nonlinear Sci. **12**(4), 985–994 (2002)
10. Carreras, B.A., Newman, D.E., Dobson, I., Degala, N.S.: Validating OPA with WECC data. In: 2013 46th Hawaii International Conference on System Sciences, pp. 2197–2204 (2013)
11. Chen, J., Thorp, J.S., Dobson, I.: Cascading dynamics and mitigation assessment in power system disturbances via a hidden failure model. Int. J. Electr. Power Energy Syst. **27**(4), 318–326 (2005)
12. Chung, H.M., Li, W.T., Yuen, C., Chung, W.H., Zhang, Y., Wen, C.K.: Local cyber-physical attack for masking line outage and topology attack in smart grid. IEEE Trans. Smart Grid **10**(4), 4577–4588 (2019)
13. Decker, I.C., Agostini, M.N., e Silva, A.S., Dotta, D.: Monitoring of a large scale event in the brazilian power system by wams. In: 2010 IREP Symposium Bulk Power System Dynamics and Control - VIII (IREP), pp. 1–8 (2010)
14. Dobson, I., Carreras, B., Lynch, V., Newman, D.: An initial model for complex dynamics in electric power system blackouts. In: Proceedings of the 34th Annual Hawaii International Conference on System Sciences, pp. 710–718 (2001)
15. Gupta, S., Kambli, R., Wagh, S., Kazi, F.: Support-vector-machine-based proactive cascade prediction in smart grid using probabilistic framework. IEEE Trans. Industr. Electron. **62**(4), 2478–2486 (2015)
16. Gupta, S.R., Kazi, F.S., Wagh, S.R., Singh, N.M.: Probabilistic framework for evaluation of smart grid resilience of cascade failure. In: 2014 IEEE Innovative Smart Grid Technologies - Asia (ISGT ASIA), pp. 255–260 (2014)
17. Kempe, D., Kleinberg, J., Tardos, E.: Maximizing the spread of influence through a social network. In: Proceedings of the Ninth ACM SIGKDD International Conference on Knowledge Discovery and Data Mining, KDD '03, pp. 137–146. Association for Computing Machinery, New York (2003)

18. Liao, W., Bak-Jensen, B., Pillai, J.R., Wang, Y., Wang, Y.: A review of graph neural networks and their applications in power systems. arXiv preprint arXiv:2101.10025 (2021)
19. Liu, C., Wang, W., Jiao, P., Chen, X., Sun, Y.: Cascade modeling with multi-head self-attention. In: 2020 International Joint Conference on Neural Networks (IJCNN), pp. 1–8 (2020)
20. Ma, Z., Shen, C., Liu, F., Mei, S.: Fast screening of vulnerable transmission lines in power grids: a PageRank-based approach. IEEE Trans. Smart Grid 10(2), 1982–1991 (2019)
21. Mei, S., Ni, Y., Wang, G., Wu, S.: A study of self-organized criticality of power system under cascading failures based on AC-OPF with voltage stability margin. IEEE Trans. Power Syst. 23, 1719–1726 (2008)
22. Mei, S., He, F., Zhang, X., Wu, S., Wang, G.: An improved OPA model and blackout risk assessment. IEEE Trans. Power Syst. 24(2), 814–823 (2009)
23. Mei, S., Yadana, Weng, X., Xue, A.: Blackout model based on OPF and its self-organized criticality. In: 2006 Chinese Control Conference, pp. 1673–1678 (2006)
24. Nakarmi, U., Rahnamay-Naeini, M.: A Markov chain approach for cascade size analysis in power grids based on community structures in interaction graphs. In: 2020 International Conference on Probabilistic Methods Applied to Power Systems (PMAPS), pp. 1–6 (2020)
25. Pi, R., Cai, Y., Li, Y., Cao, Y.: Machine learning based on Bayes networks to predict the cascading failure propagation. IEEE Access 6, 44815–44823 (2018)
26. Qi, J., Mei, S., Liu, F.: Blackout model considering slow process. IEEE Trans. Power Syst. 28(3), 3274–3282 (2013)
27. Rahnamay-Naeini, M., Wang, Z., Ghani, N., Mammoli, A., Hayat, M.M.: Stochastic analysis of cascading-failure dynamics in power grids. IEEE Trans. Power Syst. 29(4), 1767–1779 (2014)
28. Ren, H., Dobson, I., Carreras, B.A.: Long-term effect of the n-1 criterion on cascading line outages in an evolving power transmission grid. IEEE Trans. Power Syst. 23(3), 1217–1225 (2008)
29. Shuvro, R.A., Das, P., Hayat, M.M., Talukder, M.: Predicting cascading failures in power grids using machine learning algorithms. In: 2019 North American Power Symposium (NAPS), pp. 1–6 (2019)
30. Vaswani, A., et al.: Attention is all you need. In: Guyon, I., et al. (eds.) Advances in Neural Information Processing Systems, vol. 30, pp. 5998–6008. Curran Associates, Inc. (2017)
31. Wang, Z., Scaglione, A., Thomas, R.J.: A Markov-transition model for cascading failures in power grids. In: 2012 45th Hawaii International Conference on System Sciences, pp. 2115–2124 (2012)
32. Zhang, X., Zhan, C., Tse, C.K.: Modeling cascading failure propagation in power systems. In: 2017 IEEE International Symposium on Circuits and Systems (ISCAS), pp. 1–4 (2017)
33. Zhang, Y., Zhang, H., Zhang, J., Li, L., Zheng, Z.: Power grid stability prediction model based on BiLSTM with attention, pp. 344–349. Association for Computing Machinery, New York (2021)

Measurements of Insight from Data

Security Breaches in the Healthcare Domain: A Spatiotemporal Analysis

Mohammed Al Kinoon[1]([✉]), Marwan Omar[1], Manar Mohaisen[2], and David Mohaisen[1]

[1] University of Central Florida, Orlando, USA
malkinoon@knights.ucf.edu
[2] Northeastern Illinois University, Chicago, USA

Abstract. Over the past several years, data breaches have grown and become more expensive in the healthcare sector. Healthcare organizations are the main target of cybercriminals due to the sensitive and valuable data, such as patient demographics, SSNs, and personal treatment records. Data breaches are costly to breached organizations and affected individuals; hospitals can suffer substantial damage after the breach, while losing customer trust. Attackers often use breached data maliciously, e.g., demanding ransom or selling patient's information on the dark web. To this end, this paper investigates data breaches incidents in the healthcare sector, including community, federal, and non-federal hospitals. Our analysis focuses on the reasoning and vulnerabilities that lead to data breaches, including the compromised information assets, geographical distribution of incidents, size of healthcare providers, the timeline discovery of incidents, and the discovery tools for external and internal incidents. We use correlation to examine the impact of several dimensions on data breaches. Among other interesting findings, our in-depth analysis and measurements revealed that the average number of data breaches in the United States is significantly higher than in the rest of the world, and the size of the health provider, accounting for factors such as the population and number of adults in a region, highly influences the level of exposure to data breaches in each state.

Keywords: Healthcare data breaches · Confidentiality · Data security

1 Introduction

Electronic health records (EHR) can be described as "a longitudinal electronic record of patient health information generated by one or more encounters in any care delivery setting. Included in this information are patient demographics, progress notes, problems, medications, vital signs, past medical history, immunizations, laboratory data, and radiology reports" [15]. The adoption of EHR improves the healthcare industry and patients alike, and the transformation of healthcare organizations from paper-based to digital has increased healthcare quality by improving patient care and participation, care coordination, diagnostics and patient outcomes, and practice efficiency. However, despite the numerous benefits of EHR, this transformation has led to numerous privacy and security issues which may arise from vulnerabilities (e.g. software vulnerabilities, insider threats, human error, etc.) increasing the possibility of cyber-attacks [11]. The alarming

© Springer Nature Switzerland AG 2021
D. Mohaisen and R. Jin (Eds.): CSoNet 2021, LNCS 13116, pp. 171–183, 2021.
https://doi.org/10.1007/978-3-030-91434-9_16

surge in healthcare data breaches has caused huge concerns in the healthcare sector due to the illegitimate and unauthorized disclosure of private healthcare data [2, 20].

Healthcare Data breaches can be classified as either internal or external, and they can occur as a result of theft of private health records, hacking, loss of sensitive patient data, and unauthorized access to patient's private information [27]. External cybersecurity incidents are typically committed by cybercriminals operating in the dark web, while internal data breaches result from something internal to an organization such as disgruntled employees, malicious insiders, employee negligence, and human error. Patient medical records and personal information are often targeted in healthcare data breaches due to their sensitivity and value. External attacks aim to steal those records and demand a ransom or sell those records for hundreds of dollars per single patient on the dark web [22].

Data breaches are devastating and can cause significant damage to healthcare organizations; all the research in this domain demonstrates that the healthcare industry is the most targeted sector due to the attractive financial return of selling sensitive patient records on the dark web [26]. Additionally, the lenient security controls deployed by healthcare organizations further complicate matters and make the healthcare domain a favorite target for hackers. The cost of recovering from such breaches varies greatly by the nature of the incident and number of compromised health records. To better understand the cost aspect, we can break down the cost of data breaches for healthcare entities into two categories: direct costs and indirect costs. Direct expenses include activating incident response teams, engaging forensic experts, outsourcing hotline support, and providing free credit monitoring subscriptions and discounts for future products and services. On the other hand, indirect costs include in-house investigations and communication, as well as the extrapolated value of customer loss resulting from turnover or diminished customer acquisition rates [10]. Given these facts, it's compelling to conduct extensive research studies into the causes, effects, and consequences of healthcare data security incidents. Perhaps more importantly, gaining insights into the different trends and the landscape, and understanding, analyzing, and measuring the statistics in data breaches is crucial for combating such incidents. This is the motivation of this paper and we wish to also motivate the research community in this space to extend the body of knowledge by conducting more studies to be able to better understand data breach and propose solutions in the fight against cybercrimes.

Contributions. To understand the landscape of healthcare data breaches against several attributing characteristics, we provide a detailed measurement-based study of the VERIS (Vocabulary for Event Recording and Incident Sharing) and the Office of Civil Rights (OCR) datasets. To understand attackers' intents and motives, we analyze the type of assets targeted during breaches over various characteristics to investigate their effect. We also analyzed data breaches considering multiple views looking at their distribution, affected entities, breached information, location of the breach, etc.

2 Data Sources

One of the challenges with analyzing cybersecurity incidents, in general, and in the healthcare sector, in particular, is that most datasets are proprietary [25]. Additionally, most breached healthcare organizations shy away from disclosing their vulnerabilities after a breach due to a variety of concerns, including public image, reputation, and patient-trust. The other challenge lies in the fact that each victim healthcare entity tends to take a different approach in analyzing and documenting a data breach [26]. This, in turn, complicates research efforts because data breach statistics are not stored in a central online repository and thus inaccessible to the broader research community. To address the above challenges and conduct our measurements and analysis of data breaches, we turn to the largest publicly available datasets of cybersecurity incidents, namely, the VERIS dataset, and the OCR dataset, which we describe below.

VERIS. We obtained a reliable data source to conduct our research, namely, the Vocabulary for Event Recording and Incident Sharing (VERIS). Veris provides a common language for reporting data breaches incidents in an organized and repeatable manner [13]. Thus, Veris plays a significant role in providing a solution to one of the most critical and persistent challenges in the security industry; lack of quality information. Veris contributes to the solution of this problem by helping organizations collect helpful incident-related details and share them anonymously and responsibly with others. Veris's primary goal is to lay a foundation to constructively and cooperatively learn from our experiences to ensure the proper measurements and managing risk [3].

Office of Civil Rights (OCR). Our second dataset is obtained from the U.S. Department of Health and Human Services Office of Civil Rights. The U.S. Department of Health and Human Services (HHS) Office for Civil Rights (OCR) enforces federal civil rights laws, conscience and religious freedom laws, the Health Insurance Portability and Accountability Act (HIPAA) Privacy, Security, Breach Notification Rules, and the Patient Safety Act and Rule, which together protect your fundamental rights of nondiscrimination, conscience, religious freedom, and health information privacy [16]. The OCR has its breach portal, where data breaches are reported. The website contains data breaches that are currently under investigation within the last 24 months by the OCR. There is also an archived dataset, where resolved data breaches and/or those older than 24 months are archived. All the data breaches reported by the OCR are in the U.S. only. Additionally, all records in the subsequent data breaches affect 500 or more individuals as minor data breaches that affect less than 500 individuals are not reported by the OCR.

3 Studied Dimensions and Variables

This study aims to examine healthcare data breaches considering different aspects of threat characterization and modeling.

- **Geographical mapping:** Section 4.1 analyzes the geographical mapping and distribution of incidents around the world. Analyzing the geographical mapping of the incidents is necessary for several purposes: (i) it provides us with an understanding of the areas most targeted by adversaries for an affinity characterization, (ii) identifying locations around the world where the number of incidents varies due to valuable medical information, particular age group, banking details, etc. We can use this analysis for correlation and prediction capabilities.
- **State-level distribution:** Section 4.2 measures the state distribution of incidents in the U.S. This analysis is necessary for (i) identifying the hot spots targeted by attackers and (ii) conducting correlation analysis between states.

- **Compromised assets**: Section 4.3 details the targeted assets by breaches such as media, server, terminal, etc. Alongside, we will categorize the assets into groups, then dive into their varieties by an individual group against the number of incidents.
- **State-level correlation:** Section 4.4 carries a correlation analysis of the number of incidents within the top ten states with characteristics such as population, Gross Domestic Product (GDP), number of adults, etc. This correlation provides us with essential insights into the reasoning and bearings for each state.
- **Healthcare provider size:** Section 4.5 analyzes the number of breaches versus the size of organizations in terms of the number of employees. We intend to discover if the number of employees influences the frequency of data breach incidents.
- **Timeline discovery:** Section 4.6 examines the response time for incidents affecting healthcare organizations. We will measure the amount of taken time until the discovery of incidents. This analysis helps us determine the organization's security level, and whether more extended discoveries cause more damage.
- **Discovery methods:** Section 4.7 aims to identify the discovery mechanisms used by healthcare entities. Then, we will measure the reported tools and their use in data breaches in our dataset. This analysis can help with determining the appropriate tools needed to be implemented in organizations
- **Adversary demography—The threat intent:** Section 4.8 measures the intention of attackers during data breaches. We intend to acknowledge whether the incidents are targeted or opportunistic.

4 Measurement Results and Discussions

4.1 The Global Distribution of Incidents

Mapping incidents is explicitly provided in our dataset. The dataset uses the ISO 3,166 country codes for each country variable [7], where the codes are generated based on the physical location of the hospital targeted by the attack. Based upon this analysis, we discovered that 1,955 incidents out of the total incidents (2,407) had taken place in the United States, representing 81% of the total incidents. The United Kingdom comes in second, with 157 incidents, representing 7%, and Canada comes in third with 152 incidents, representing only (6%). Figure 1 presents the results for the remaining highest ten countries, while the rest of the world represents (2%) comprising 58 incidents.

As a result of the geographical mapping analysis, we decided to conduct our in-depth analysis study on the United States since most incidents occurred in this country. Several reasons explain why the majority of the incidents are in the United States. First, the Health Insurance Portability and Accountability Act (HIPPA) requires healthcare entities to notify the Department of Health and Human Services (DHHS) whenever a data breach occurs. Second, covered entities must notify affected individuals following the discovery of a breach of unsecured protected health information [16]. In addition to that, covered entities must notify the Secretary of breaches of unsecured protected health information if the affected individuals are 500 or more [16]. Third, covered entities that experience a breach affecting more than 500 residents of a State or jurisdiction are, in addition to notifying the affected individuals, required to provide notice to prominent media outlets serving the State or jurisdiction [16]. Moreover, breach notification is

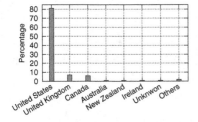

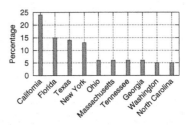

Fig. 1. Incidents by country.   **Fig. 2.** Incidents by state.

also required for vendors and third-party service providers under the Health Information Technology for Economic and Clinical Health Act (HITECH) [14]. Finally, the HIPPA Security Rule requires healthcare organizations to create a risk management plan protecting all personal health data against security incidents (Office of Civil Rights 2015), which may explain the significant number of reported incidents in the United States [1].

4.2 Number of Incidents by State

Following the global distribution of incidents, we moved into the mapping of incidents on the state level. We analyzed the number of incidents by state. As a result of this analysis, we noticed that California is the highest state with the number of incidents comprising 241 incidents, representing 24% of the overall. Florida comes in second with 147 incidents, representing 15%, and Texas with 145 incidents, representing 14%. Figure 2 shows the remaining results of this analysis.

4.3 Analyzing the Compromised Assets

This section investigates the compromised information assets in the Veris dataset. We harnessed the power of Natural Language Processing (NLP) models to help with analyzing the data gathered from breaches. Information assets fall into six main groups: media, server, terminal, network, user, and people. Each group comprises different varieties [18]. First, the network group includes access control readers such as badge and biometrics, camera or surveillance system, firewall, intrusion detection system (IDS) or intrusion prevention systems, and others. Second, the media group comprises disk media such as CDs or DVDs, flash drives or cards, hard disk drives, identity smart cards, and others. Third, the people group includes administrator, auditor, cashier, customer, former employee, guard, and others. Fourth, the server includes authentication, backup, database, Dynamic Host Configuration Protocol (DHCP), DNS, mail, and others. Fifth, the terminal group includes an automated Teller Machine (ATM), detached PIN pad or card reader, gas "pay-at-the-pump" terminal, self-service kiosk, and others. Finally, the user group includes an authentication token or device, desktop or laptop, media player or recorder, mobile phone or smartphones, and many others.

The existence of assets depends on several reasons and conditions during each incident. We will measure each asset group based on their occurrences in the incidents, and then, we get into the measurement of their varieties to look into the most targeted type of each asset group. This analysis is essential, and its primary purpose is to adequately

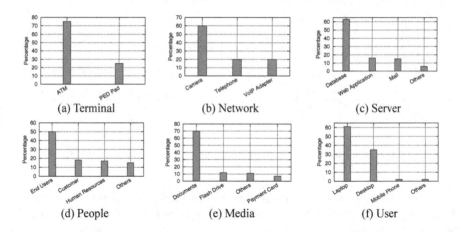

Fig. 3. Information asset groups and their varieties.

describe the incidents, assess control weaknesses and vulnerabilities, determine impact, and identify mitigation strategies.

Usually, during a data breach incident, one or more assets get compromised by hackers [9]. A compromised asset refers to any loss of confidentiality, integrity, availability during or after the incidents. In the following section, we seek to analyze and measure the asset groups and the total incidents for each group; then, we move to their different asset groups. Based on this analysis, we noticed that media assets are the clear leader comprising 564 incidents of the overall, representing 33.97%, and server comes in second, comprising 560 incidents, representing 33.73%. Table 1 shows the remaining asset categories and their number of incidents.

After measuring the number of incidents for each asset group as a whole, we moved into measuring their varieties. Based on the analysis done, we found that 61% of the incidents in the user group are through laptops, followed by the terminal group with 75% of the incidents through ATMs. In the server asset group, we found out that 63% of the incidents happened through exploiting the database. While for the people asset group, 50% of the incidents are because of the end-

Table 1. Assets varieties with the number of incidents during data breaches.

Asset Group Type	# Incidents	Percentage
Media	564	33.97%
Server	560	33.73%
User	493	29.69%
People	34	2.04%
Network	5	0.30%
Terminal	4	0.24%
Overall	1660	100 %

user. Most of the incidents that happen in the network are throughout cameras, with represent 60%. Lastly, 70% of the incidents in the media group are through documents. In Fig. 3, we present the remaining results for the other asset groups and their varieties.

4.4 State Level Correlation

This section will conduct a state-level correlation between the number of reported incidents and hospitals, staffed beds, population, and gross domestic product (GDP) for the top 10 states. GDP is the gross domestic product and is represented in billion U.S. dollars. To address the following question, we conducted a state-level analysis considering these factors related to the reported incidents in our dataset. We decided to run this analysis on the highest 10 states in terms of the number of reported incidents. We started by collecting the specified statistics for each state, including population, GDP, staffed beds, and hospitals. The relationship between two variables can be a positive relationship (1), no relationship (0), and an inverse relationship (−1). Upon this analysis, we discovered that the population and adults are highly correlated with the number of incidents (0.96). Followed by the GDP (0.95). The remaining results of the correlation are shown in Table 2.

4.5   Organizations Size

The following section investigates the size of healthcare entities and how organization's size might contribute to a data breach. Using Veris, we performed the analysis by looking into the scope of healthcare organizations at the time of the incident. We classified healthcare organizations into two main groups: small and large. A small group includes a size of up to 1,000 employees, while a large organization would be over 1,000 employees. Upon this analysis, there were a total of 1,361 incidents divided into two groups. Our analysis revealed that 57% of the incidents are in the small group, while 43% are in large groups.

Table 2. State level correlation. Numbers of incidents (I), hospitals (H), employees (E), staffed beds (B), GDP (G), population (P), and adults (A) are considered.

	I	H	E	B	G	P	A
I	1.00						
H	0.88	1.00					
E	0.92	0.91	1.00				
B	0.94	0.92	0.97	1.00			
G	0.95	0.86	0.92	0.89	1.00		
P	**0.96**	0.95	0.94	0.94	0.95	1.00	
A	**0.96**	0.88	0.94	0.96	0.89	0.90	1.00

4.6   Timeline Discovery

Timeline discovery of data breaches varies depending on the type of industry, geography, and level of security of an organization. According to a recent study conducted by the IBM security team in the healthcare sector, the average time to discover a data breach is 329 days, and 93 days are required to regain control. Unfortunately, prior work fails to provide in-depth analysis on the timeline discovery of the data breaches, including discovery tools for external and internal incidents. To fill this gap, we analyzed the timeline discovery of the reported incidents and went over the tools used for incident discovery for both internal and external discovery methods. This analysis is essential to address the lessons learned during the incidents and remediation process and provide organizations with insights and corrective actions to improve their detection and defensive capabilities. Our analysis found out that organizations fail to identify data breaches early enough, resulting in more damage. From the reported incidents, we discovered that 3% of the incidents took minutes until discovery, 9% took hours, 15% took days, 6% took weeks, 52% took months, and 15% took years. In the coming section, we will address different discovery methods and whether there is a difference between internal attacks and external attacks.

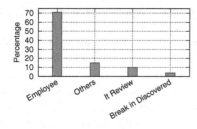
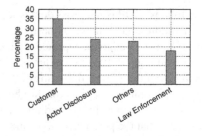

Fig. 4. Internal discovery   **Fig. 5.** External discovery

4.7   Internal and External Discovery Methods

Discovery methods fall into two main categories; internal and external. Organizations use several tools to discover an incident depending on the type of data breach. External and internal data breaches are different, and each one of them requires special discovery tools. First, healthcare organizations use numerous tools to discover incidents for internal incidents, such as Host IDS or file integrity monitoring, network IDS, and IPS alerts. In contrast, practices including law enforcement, actor disclosure, and customer notifications can help discover external incidents. Our analysis found out that most of the internal incidents are discovered by employees, representing 71% of the total incidents. In contrast, customers discover 35% of the external incidents, and actor disclosure comes in second, representing 24%. The remaining results of this analysis are shown in Figs. 4 and 5.

4.8   Targeted vs Opportunistic

To understand the nature of the data breach incidents and whether they are intentional or non-intentional, we conducted a measurement analysis to investigate the number of targeted incidents and opportunistic ones. This classification is uniquely relevant to deliberate and malicious actions. There are two main categories: targeted and opportunistic. First, opportunistic incidents occur when the victim exhibits a weakness that the actor has the knowledge to exploit. Second, targeted incidents happen when the adversary chooses the victim as a target, and then the actor will investigate possible vulnerabilities to exploit. Using our exclusively given records in our dataset, we found that more than half of healthcare data breaches are opportunistic, representing 80%, while, on the other hand, 20% are targeted.

5   Analysis of the OCR Dataset

Type of Breach. We analyzed the causes of healthcare data breaches based on the reported incidents and observed that most incidents occur due to hacking or IT-related disclosure comprising 1,069 incidents, representing 31% of the overall incidents. Unauthorized access and disclosure came in second, holding 934 incidents overall, representing 27%. Finally, the theft category came in the third place, comprising 909 incidents, accounting for 26% of the total incidents.

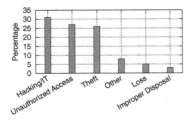

Fig. 6. Type of breach.

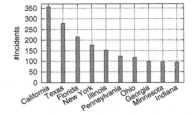

Fig. 7. Distribution by state.

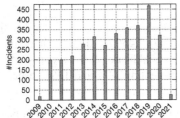

Fig. 8. Incidents by year.

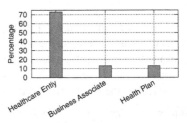

Fig. 9. Covered entities.

State Distribution. The following section addresses the distribution of the incidents for the U.S states. Using the OCR data, we measured the incidents for each state; this analysis is essential for trends and comparison. Following this analysis, we have observed that states with large population, high Gross Domestic Product (GDP), and large adult population are more targeted than others, as shown in Sect. 4.4. California was the most affected, totalling 357 incidents, followed by Texas with 279 incidents, while Florida was the third largest with 215 incidents (Figs. 6 and fig:UsspsStateshhs).

Distribution of Incidents by Year. Using the ORC dataset, and over the period between 2009 and the time of conducting this study in 2021, we measured the reported incidents in the dataset affecting 500 or more victims and reported to the HHS OCR. Following this analysis, we notice that the number of incidents surged over time, indicating a lack in implementing stringent security controls by organizations in the healthcare industry. As shown in Fig. 8, there is a massive increase in the number of incidents in 2019, as it was the year with the highest number of breaches in the whole dataset.

Covered Entity. We analyzed the distribution of incidents by organization type. According to the OCR dataset, there are three main targeted entities. First, healthcare entities that provide health care services and engages in professional review activity through a formal peer review process for the purpose of furthering quality health care, a committee of that entity, a professional society, a committee or agent thereof, including those at the national, state, or local level, physicians, dentists, or other health care practitioners that engage in professional review activity through a formal peer review process to further quality health care [17]. Second, a business associate, which is a person or entity that performs certain functions or activities that involve the use or disclosure of protected health information on behalf of or provides services to a covered entity [24]. Third, health plan, which constitutes individual or group health plans that provide or

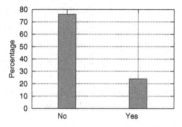

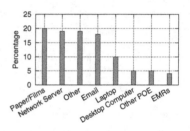

Fig. 10. Business associate.   **Fig. 11.** Information breached.

pay the cost of medical care [5]. Following this analysis, we observed that healthcare entities are most targeted during the incidents, having 2,450 incidents which represents 73% of the total incidents, business associate and healthcare plan came in second and third comprising 451 and 439 incidents, and representing 14% and 13%, respectively. Figure 9 depicts the results of this analysis.

Business Associates. We further analyzed the existence of incidents when a business associate is present or not. According to HIPPA, any covered entities and business associates enter into a contract to ensure the safety of protected healthcare information. A business associate may use or disclose protected health information only as permitted or required by its business associate contract or as required by law [23]. Our analysis revealed that 2,532 incidents had no business associates included, representing 76%, while only 819 incidents had a business associate, representing 24% of the incidents as shown in Fig. 10.

Location of Breached Information. When a data breach occurs, private and confidential patient information gets disclosed due to either unauthorized access or human error. Healthcare system keeps record of valuable information and medical records, containing sensitive personally identifiable information (PII) such as address history, financial information, social security numbers, and patient medical treatment records. This sensitive information is often targeted by hackers due to its outstanding value. Hackers can easily use that data to set up a line of credit or take out a loan under patients' names. Unfortunately, healthcare organizations often lack the stringent security measures (e.g., encryption, robust anti-virus software, multi-factor authentication, etc.) required to secure medical records. To this end, we analyzed the most targeted information to gain insight into the type of medical and personal data prioritized by hackers in healthcare data breaches. We observed that paper/films are the most breached information comprising 662 of the overall incidents, representing 20%. Closely, the network server came in second, comprising 643 incidents, accounting for 19%. The other category came in third, comprising 641 incidents, representing 19% as well. The remaining attributes and results of this analysis are presented in Fig. 11.

6   Related Work

In the past few years, numerous studies have analyzed data breaches in the healthcare sector. Choi *et al.* [4] estimate the relationship between data breaches and hospital advertising expenditures. They concluded that teaching hospitals were associated with significantly higher advertising expenditures two years after the breach. Another study [12] investigated the privacy-protected data collection and access in IoT-based healthcare applications and proposed a new framework called PrivacyProtector to preserve the privacy of patients' data. Another study [6] found that the healthcare industry was being targeted for two main reasons: being a rich source of valuable data and its weak defenses. Other study [28] suggested a framework to examine the accuracy of automatic privacy auditing tools. Siddartha *et al.* [21] suggested that current healthcare security techniques miss data analysis improvements, e.g., data format-preserving, data size preserving, and other factors. Most related to our work, the 2021 Data Breach Investigations Report [8] summarized the findings and determined that external actors are behind 61% of data breaches while 39% of data breaches involved internal actors. According to the same report, personal information is the most compromised, comprising 66% of data breaches. In contrast to our work, authors of [19] conducted a comprehensive analysis of HIPPA data breach reports. They found that the main disclosure types of protected healthcare information were hacking incidents, unauthorized access (internal), theft or loss, and improper disposal of unnecessary data. The authors used the Simple Moving Average (SMA) and Simple Exponential Smoothing (SES) time series methods. They applied them to the data to determine the trend of healthcare data breaches and their cost to the healthcare industry. Our comprehensive study comprises but is not limited to analyzing compromised assets, internal and external discovery methods, discovery timeline of data breaches, distribution of the incident globally and in the united states, and breached information. In addition, we used correlation as a mathematical tool to determine healthcare data breaches and quantify the effects of different factors like GDP, population, number of hospitals, and their sizes in terms of the staffed beds on data breaches.

7 Conclusion

Our study revealed that the number of adults and the state population highly influence the exposure to data breach incidents, with California, Florida, and Texas being the lead targets. We show that the media group was the most breached asset, followed by the Server and User group. Interestingly, we found that the majority of incidents occur in small size organization – 57%. In contrast, 43% of the incidents occur in large organizations, suggesting that large healthcare organizations tend to have better security systems. Our timeline discovery revealed that most of the incidents, approximately 52%, were discovered within months, while 15% of the incidents took years to be discovered. Employees discovered the majority of the incidents for internal incidents. Based on a long-term dataset analysis, most of the incidents, 80%, tend to be opportunistic, while 20% are targeted. In the future, it would be interesting to conduct research harnessing the power of machine learning to enable information sharing on data breaches.

Acknowledgement. This work was supported by NRF-2016K1A1A2912757.

References

1. Adebayo, A.O.: A foundation for breach data analysis. J. Inf. Eng. Appl. **2**(4), 17–23 (2012)
2. Alkinoon, M., Choi, S.J., Mohaisen, D.: Measuring healthcare data breaches. In: Kim, H. (ed.) WISA 2021. LNCS, vol. 13009, pp. 265–277. Springer, Cham (2021). https://doi.org/10.1007/978-3-030-89432-0_22
3. Chernyshev, M., Zeadally, S., Baig, Z.: Healthcare data breaches: implications for digital forensic readiness. J. Med. Syst. **43**(1), 1–12 (2019)
4. Choi, S.J., Johnson, M.E.: Understanding the relationship between data breaches and hospital advertising expenditures. Am. J. Manag. Care **25**(5), e14–e20 (2019)
5. Employers Council: What is the definition of a health plan under HIPAA? (2015). https://bit.ly/3Aherpb
6. Coventry, L., Branley, D.: Cybersecurity in healthcare: a narrative review of trends, threats and ways forward. PubMed, April 2018. https://doi.org/10.1016/j.maturitas.2018.04.008
7. Developers: International organization for standardization: 3,166 country codes (2021). https://bit.ly/3Eoem5J
8. Verizon Enterprise: Verizon data breach investigations report (2021). https://vz.to/3AvCNfn
9. Gwebu, K., Barrows, C.W.: Data breaches in hospitality: is the industry different? J. Hosp. Tour. Technol. (2020)
10. (HC3), Health Sector Cybersecurity Coordination Center: A cost analysis of healthcare sector data breaches (2019). https://bit.ly/3hHpJMj
11. Kamoun, F., Nicho, M.: Human and organizational factors of healthcare data breaches: the swiss cheese model of data breach causation and prevention. Int. J. Healthc. Inf. Syst. Inform. (IJHISI) **9**(1), 42–60 (2014)
12. Luo, E., Bhuiyan, M.Z.A., Wang, G., Rahman, M.A., Wu, J., Atiquzzaman, M.: PrivacyProtector: privacy-protected patient data collection in IoT-based healthcare systems. IEEE Commun. Mag. **56**(2), 163–168 (2018)
13. Makridis, C., Dean, B.: Measuring the economic effects of data breaches on firm outcomes. J. Econ. Soc. Meas. **43**(1–2), 59–83 (2018)
14. McLeod, A., Dolezel, D.: Cyber-analytics: modeling factors associated with healthcare data breaches. Decis. Support Syst. **108**, 57–68 (2018). https://doi.org/10.1016/j.dss.2018.02.007
15. Menachemi, N., Collum, T.H.: Benefits and drawbacks of electronic health record systems (2011). https://bit.ly/3EscQ2k
16. Office for Civil Rights: Breach notification rule (2013). https://bit.ly/3jCpHXI
17. Rank, N.P.D.: NPDB guide book (2021). https://bit.ly/2XwgTdw
18. Sarabi, A., Naghizadeh, P., Liu, Y., Liu, M.: Risky business: fine-grained data breach prediction using business profiles. J. Cybersecur. **2**(1), 15–28 (2016)
19. Seh, A.H., et al.: Healthcare data breaches: insights and implications. Healthcare **8**, 133 (2020). https://doi.org/10.3390/healthcare8020133
20. Seh, A.H., et al.: Healthcare data breaches: insights and implications. In: Healthcare. vol. 8, p. 133. Multidisciplinary Digital Publishing Institute (2020)
21. Siddartha, B.K., Ravikumar, G.K.: Analysis of masking techniques to find out security and other efficiency issues in healthcare domain. In: Third International conference on I-SMAC, pp. 660–666 (2019). https://doi.org/10.1109/I-SMAC47947.2019.9032431
22. Smith, T.: Examining data privacy breaches in healthcare. Ph.D. thesis, Walden U. (2016)
23. U.S. HHS: Business associate contracts (2013). https://bit.ly/3ChsJH9
24. U.S. HHS: Business associates (2019). https://bit.ly/3tM4PQV
25. Walker-Roberts, S., Hammoudeh, M., Aldabbas, O., Aydin, M., Dehghantanha, A.: Threats on the horizon: understanding security threats in the era of cyber-physical systems. J. Supercomput. **76**(4), 2643–2664 (2020). https://doi.org/10.1007/s11227-019-03028-9

26. Walker-Roberts, S., Hammoudeh, M., Dehghantanha, A.: A systematic review of the availability and efficacy of countermeasures to internal threats in healthcare critical infrastructure. IEEE Access **6**, 25167–25177 (2018)
27. Wikina, S.B.: What caused the breach? An examination of use of information technology and health data breaches. Perspect. Health Inf. Manag. **11**(Fall), 1–16 (2014)
28. Yesmin, T., Carter, M.W.: Evaluation framework for automatic privacy auditing tools for hospital data breach detections: a case study. Int. J. Med. Inform. **138**, 104123 (2020)

Social and Motivational Factors for the Spread of Physical Activities in a Health Social Network

NhatHai Phan[1(✉)], David Kil[2], Brigitte Piniewski[3], and Dejing Dou[4]

[1] New Jersey Institute of Technology, Newark, USA
phan@njit.edu
[2] HealthMantic Inc., Los Altos, USA
david.kil@healthmantic.com
[3] PeaceHealth Laboratories, Eugene, USA
BPiniewski@peacehealthlabs.org
[4] University of Oregon, Eugene, USA
dou@cs.uoregon.edu

Abstract. Identifying the effects of social and motivational factors is critical to understanding how healthy behaviors, i.e., physical activities, spread in digital therapeutics programs. We evaluated a comprehensive interconnected social network of 254 overweight and obese individuals across 335 days. Daily physical activities, social activities, biomarkers, and biometric measures were available for all subjects. We improved proportional hazards models to characterize the impact of self-motivation, influence, and susceptibility in the spread of physical activities. After 6 months, the YesiWell users increased leisure walking minutes by 164% on average compared with 47% among the control participants ($P < 0.05$). The YesiWell users also lost more weight than the controls (5.2 pounds vs. 1.5 pounds) ($P < 0.01$). Our estimations showed that influence and susceptibility increase with age; relaxed people are 96% more influential than stressed people ($P < 0.001$); obese people are 23% more self-motivated ($P < 0.001$); socially active people are 29% more influential ($P < 0.001$); those who self-characterize as "keep-to-themselves" people have a 79% greater susceptibility ($P < 0.001$). Relaxed people exert the most influence on non-stressed peers at 109% more than baseline ($P < 0.001$). Our findings could enable new and effective personalized behavioral interventions to spread healthy behaviors in next-generation digital therapeutics.

Keywords: Overweight · Obesity · Social and motivation factors · Physical activities

1 Introduction

The US healthcare system is currently transforming from a transaction-based industry into a value-based care model [25]. With this transformation, there

© Springer Nature Switzerland AG 2021
D. Mohaisen and R. Jin (Eds.): CSoNet 2021, LNCS 13116, pp. 184–196, 2021.
https://doi.org/10.1007/978-3-030-91434-9_17

is a growing dependence on leveraging innovative technologies to deliberately impact the health expression of populations [25,34]. A new discipline, Digital Therapeutics, is characterized by online or mobile programs that actually adjust health risk, treat conditions, and measurably improve health outcomes [25,34]. Companies such as Omada Health, Cala Health, Telcare and Welldoc are proving successful in impacting health and thus opening up an era in which digital therapeutics may well yield results that dwarf the outcomes achieved when our narrow therapeutic armamentarium consisted mainly of in-person visits coupled with pharmaceuticals [34].

The Diabetes Prevention Program [17] was a successful analog (in-person) intervention, resulting in a 58% reduction in the incidence rate of diabetes, among its cohort of 1,079 participants. Converting this analog model to a digital therapeutics program would not only provide the ability to scale across larger and larger populations [25], but also would avail a secondary opportunity: leveraging the digital exhaust of users to uncover the multitude of conditions that combine to create the most therapeutic outcomes. Today, early digital therapeutics approaches offer relatively similar programs to most participants [10,16,24,25,32,33,35]. In addition, the programs offered by these approaches usually do not take into account the impact of social networks, which have been demonstrated to be important in spreading healthy behaviors, e.g., physical activities [15,27,30,31]. Harvesting high-definition insights from the digital transactions of users and their social effects will enable significant personalization. Therefore, a more data-driven customization of the intervention is highly expected to yield, by design, ever improving health outcomes. In particular, data-driven approaches may dramatically improve a program's ability to arrange participants into preferred groupings, with the aim to optimize social accountability or socially-dependent impact of the digital therapeutics program.

2   YesiWell Study

We have proposed a system which comprehensively combines physical activities, social activities, biomarkers, and biometric measures in a fine-grained time scale, for identifying self-motivated, influential, and susceptible participants to influence toward increasing physical activities in a health social network, utilizing lifestyle monitoring and online communities. Our study demonstrates the viability and significance of our proposed system. The YesiWell study was conducted for 335 days in 2010–2011, as collaboration among PeaceHealth Laboratories, SK Telecom Americas, and the University of Oregon. We recorded daily physical activities, social activities (i.e., text messages, social games, meet-up events, competitions, etc.), and biomarkers and biometric measures (i.e., cholesterol, triglyceride, BMI, etc.) for a group of 254 individuals who formed a health social network. Physical activities were reported every minute via a mobile device carried by each user. All users enrolled in an online social network application, allowing them to befriend and communicate with each other via public postings, replies, comments, and private messages. Users were able to organize competitions, social games, and meet-up events among various teams. Biomarkers and biometric measures were recorded via quarterly laboratory tests, including at baseline, and wireless digital scales connected to the backend server, respectively. We also collected data from 100 survey items, which covered diverse aspects of mental health, food consumption, wellness, stress level, social activity level, sleep disorder, etc.

Overall, we have 7 million data points for physical activities; 10,000 records for BMI and wellness score[1] [31]; 3,101 instances of participation in competitions; 1,765 instances of participation within 278 social games; 2,656 messages sent; 1,828 friend connections; 1,300 goals set; 14,138 survey answers, etc. Users volunteered to join the study; therefore, they were not under any pressure to exercise more or less. After 6 months, the YesiWell users increased leisure walking minutes by 164% on average (i.e., from 129.2 min/week to 341 min/week), compared with 47% among the control participants, who did not use the Yesi-Well social network ($P < 0.05$ performed by t-test) [15]. The YesiWell users also lost more weight than the controls (5.2 pounds compared with 1.5 pounds) ($P < 0.01$ performed by t-test) [15]. The system and our study provide strong evidence that interventions using online social networks can successfully promote physical activity increase and weight loss. Such systems may increasingly be refined over a greater number of users to provide a higher-granularity understanding of the dynamics of the spread of healthy behaviors (i.e., self-motivation, influence, and susceptibility), and to inform next-generation digital therapeutics.

3   Self-motivation, Influence, and Susceptibility

Identifying the effects of social and motivational factors is critical to understanding how healthy behaviors, i.e., physical activities, spread in digital therapeutics programs. The correlations between social influence and self-motivation are empirically elusive within social science. Scholars in disciplines as diverse as economics, sociology, psychology, finance, and management are interested in understanding such correlations as whether happiness, obesity, and smoking are "contagious" [8,9,13]; and whether risky behaviors and information spread via peer influence [5,11,12,14,18,19,22,23,26,28]. *To what degree is a given behavior change or contagion as a consequence of a social epidemic* [1,4,5]*, and/or self-motivation* [2,6,21,36]*?* Comprehending estimates and data toward answering to that question is critical to policy decision-making, as the success of personalized intervention strategies in these domains depends on such analyses.

Due to limited resources, early digital therapeutics methods and studies have been designed and conducted based on insufficient observable factors, i.e., either lacking physical world factors (e.g., physical activities, biomarkers, social events, etc.) or lacking online world factors (e.g., online social networks, e-mail, instant messaging, mobile phone communications, etc.) [10,16,24,25,32,33,35]. Those insufficiencies pose limitations to the completeness with which researchers may perceive the effects of self-motivation, influence, and susceptibility on human

---

[1] Wellness score is a composite score of one's health based on lifestyle parameters, biometrics, and biomarkers.

Table 1. Examples of influential messages

Join me for the 50,000 steps M-F next week!
Wow! I believe we are about half way to our Valentine Day goal Take a look at your total and see if you can double it by then... plus about 10%... We can do it!
You don't really have to try - except just get out and walk What fun especially with our warmer weather
Getting close to summer! Join me in the final leg of the school year Earn "extra credit" steps with Wellness game points, too!
You're on, Mark! See you Monday at the flag poles
Last day for Peacehealth Oregon to "step it up". We have those long hallways in our buildings to walk on our breaks and lunch time Come on let's get her done!!! We can do it!!!
How about a connect 5 games for Turkey month? Let's keep motivated as the holidays kick off!

Table 2. Descriptive statistics of user and peers

	# influence trials	# messages sent	# activations	avg (# messages sent)	avg (# activations) message)	avg (# activations)/ trial)
Age: 18–39	172	226	127	1.3140	0.5619	0.7384
Age: 40–49	175	228	137	1.3029	0.6009	0.7829
Age: 50–59	482	621	372	1.2884	0.5990	0.7718
Age: 60+	179	208	156	1.1620	0.75	0.8715
Overweight (BMI: 25–29.9)	781	1024	604	1.3111	0.5898	0.7735
Obesity (BMI: 30+)	227	259	188	1.1410	0.7259	0.8282
Relax	129	164	110	1.2713	0.6707	0.8527
Non-stress	586	747	451	1.2747	0.6037	0.7696
Stress	293	372	231	1.2696	0.6209	0.7884
Active in hanging out	200	247	166	1.2350	0.6721	0.83
Share to friends	479	628	362	1.3111	0.5764	0.7557
Keep to themselves	133	160	105	1.2030	0.6562	0.7894
Common number of friends	316	362	258	1.1456	0.7127	0.8165
Many friends	263	356	212	1.3536	0.5955	0.8060
Too many friends	429	565	322	1.3170	0.5699	0.7506

behavior change, especially the spread of healthy behaviors, such as exercise, in digital therapeutics programs. Understanding whether self-motivation, influence, susceptibility, or a combination of the three drives social contagions, and accurately identifying self-motivated, influential, and susceptible individuals in networks, could enable new personalized behavioral interventions to spread healthy behaviors, e.g., exercise, and to impact lifestyle-induced health problems in a cost-effective, scalable manner. In this study, we have examined the quantity and efficacy of influential messages sent from individuals to other users in the network, in order to identify the participants who were self-motivated, influential, and susceptible participants with regard to the spreading of physical activities. Conceptually, we monitored the influence of each sender upon each recipient, across 7 days, to see whether the recipient would increase physical activities

Table 3. Descriptive statistics of peers in terms of competitions, social games, and goals set

	# influence	# competitions	# social games	# goals set	avg (# competitions)	avg (# social games)	avg (# goals set)
Age: 18–39	254	1,175	142	597	4.6260	0.5591	2.3504
Age: 40–49	188	857	72	567	4.5585	0.3830	3.0160
Age: 50–59	438	2,830	274	1,147	6.4612	0.6256	2.6187
Age: 60+	128	401	47	428	3.1328	0.3672	3.3438
Overweight (BMI: 25–29.9)	781	4,084	431	2,137	5.2292	0.5519	2.7362
Obesity (BMI: 30+)	227	1,179	104	602	5.1938	0.4581	2.6520
Relax	102	355	46	428	3.4804	0.4510	4.1961
Non-stress	543	3,539	324	1,369	6.5175	0.5967	2.5212
Stress	363	1,369	165	942	3.7713	0.4545	2.5950
Active in hanging out	195	597	65	450	3.0615	0.3333	2.3077
Share to friends	394	2,981	268	1,148	7.5660	0.6802	2.9137
Keep to themselves	246	799	105	763	3.2480	0.4268	3.1016
Common number of friends	289	570	53	640	1.9723	0.1834	2.2145
Many friends	304	1,169	130	913	3.8454	0.4276	3.0033
Too many friends	415	3,524	352	1,186	8.4916	0.8482	2.8578

in comparison to the previous week's total by at least 2,500 steps, i.e., by the average number of steps/day of sedentary US adults [7]. If the recipient met that threshold, then the sender was considered to have successfully activated the recipient in that particular influence trial.

More formally: First, we identify two sets of users in and between whom there were message connections. Second, we open an observation window of 7 days starting from the first-message timestamp from to in order to investigate the level and directionality of influence. By choosing 7-day windows, we can avoid unbalance given different windows; since, they contain the same week-days and a weekend. An influential user-to-user message generally refers to any communication between peers that could conduct influence, such as invitations, encouragement, follow-up, competitions, progress reports, fitness, goal, notification, etc. (Tables 1, 2 and 3, Figs. 1 and 2). In our 10-month study, an individual can activate another individual multiple times and can activate multiple individuals at the same time. Self-motivation, influence, and susceptibility were estimated, from modeling time to peer activation, as a function of the peer's treatment status– whether influential messages had been received, and if so, how many, and in how many competitions, social games, and meet-up events users had participated prior to the time of activation. Note that users tended to exercise more; they set their goals and joined competitions, social games, and meet-up events. Throughout the 335 days, users tried to activate their peers to increase physical activities 1,008 times, by sending 2,656 messages. This resulted in 792 unique peer activations, or a 78% increasing physical activities by at least 2,500 steps (Tables 1, 2 and 3, Figs. 1 and 2).

Our statistical approach uses hazard modeling, which is the standard technique for estimating social contagions in sociology, economics, and marketing (e.g., [20]). However, we improved existing approaches by distinguishing self-motivation and two types of peer-based influence on increasing physical activ-

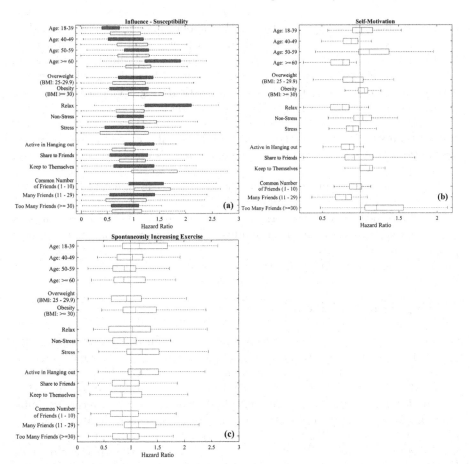

Fig. 1. Effects of age, BMI, stress level, social activity level, and #friend connections. Influence (dark gray) and susceptibility to influence (white) in fig. (a), self- motivation (white) in fig. (b), and spontaneous (white) in fig. (c) are shown with SEs (boxes), 95% confidence intervals (whiskers). The figure presents hazard ratios (HRs) representing the percent increase ($HR > 1$) or decrease ($HR < 1$) in adoption hazards associated with each attribute. Estimates are shown relative to the baseline case for each attribute, which is the average for all individuals given that attribute.

ities: (1) self-motivation-driven physical activity increasing, which occurs in response to participating in competitions, social games, and meet-up events; (2) spontaneous physical activity increasing[2], which occurs in the absence of self-motivation and influence; and (3) influence-driven physical activity increasing, which occurs in response to influential messages. This improvement is significant; since, in the absence of self-motivation, human behavior outcomes among peers can be a consequence of influence, homophily, assortativity, simultaneity, and

---

[2] Spontaneity can be considered a form of "intrinsic self-motivation.".

correlated effects [3,5,29]. To estimate the moderating effects of an individual i's attributes on the influence exerted by i on peer j (and to distinguish them from the moderating effects of j's attributes on j's susceptibility to influence), we use a continuous-time single-failure proportional hazards model. Moreover, we extend the model by adding the effect of participating in competitions, social games, and meet-up events to model self-motivation in increasing physical activities of peer j. Survival models provide information about how quickly peers react (rather than simply whether they react), and they also correct for censoring of peer actions that may occur beyond the experiment's observation window, i.e., a week.

Models of dyadic (two-party) relationships between influential individuals and potentially susceptible individuals test whether influence depends on self-motivation of peers and characteristics of the relationship between a given pair; e.g., whether relaxed people are more influential on stressed people than stressed people are on relaxed people. To estimate the effect of dyadic relationships, we use a continuous-time single-failure proportional hazards model.

4   Experimental Results

On average, in our findings, susceptibility and influence increase with age (Fig. 1a). Meanwhile, spontaneity decreases with age (Fig. 1c). People under the age of 40 are the least susceptible to influence, with the least likelihood of influencing their peers to increase physical activities. Relative to the baseline, they have a 15% lower hazard of increasing physical activities upon receiving influential messages ($P < 0.05$; the statistical significance of all estimates is derived from χ^2 tests), and a 71% lower likelihood of influencing their peers via sending influential messages ($P < 0.001$). However, people in the same age quartile (18–39) can spontaneously increase physical activities. They have a 20% higher likelihood of spontaneously increasing physical activities.

Relative to people younger than 40, people with the age of 60+ are significantly more influential and susceptible; they have a 143% greater likelihood of influencing their peers to increase physical activities ($P < 0.001$), and an 18% higher hazard of increasing physical activities ($P < 0.01$). However, people with the age of 60+ do not spontaneously increase physical activities; relative to people younger than 40, they have a 20% lower likelihood of spontaneously increasing physical activities ($P < 0.01$). In addition, people in the same age quartile (60+) are not self-motivated in increasing physical activities; they have a 24% lower self-motivation hazard of increasing physical activities when they participate in social games, competitions, and meet-up events ($P < 0.01$).

Meanwhile, people in the age quartile 50–59 are strongly self-motivated in increasing physical activities; relative to aging adults (60+), they have a 41% stronger likelihood of self-motivation in increasing physical activities ($P < 0.001$).

Overweight people are 16% more influential than obese people ($P < 0.001$). However, obese people are 23% more susceptible ($P < 0.001$), 11% more

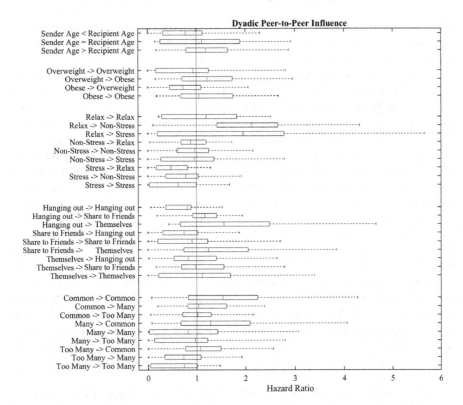

Fig. 2. Dyadic influence models involving age, BMI, stress level, social activity level, and number of friend connections. The results include the relative age, BMI, stress level, social activity level, and number of friend connections of senders and recipients, with SEs (boxes) with 95% CI (whiskers). The figure presents hazard ratios (HRs) representing the percent increase ($HR > 1$) or decrease ($HR < 1$) associated with each attribute. The baseline case represents dyads in which the attribute being examined is unreported in the individual data in both peers. "Common", "Many", and "Too Many" refer to # of friends.

self-motivated ($P < 0.05$), and 13% more spontaneous ($P < 0.01$) in increasing physical activities than overweight people.

Relaxed people are significantly more influential; they have a 78% greater likelihood of influencing their peers to increase physical activities ($P < 0.001$). However, relaxed people are the least self-motivated in increasing physical activities; they have a 10% lower self-motivation hazard ($P < 0.05$).

People with no stress are 17% more susceptible ($P < 0.05$). In addition, people with no stress are strongly self-motivated; relative to relaxed people, people with no stress have a 21% greater likelihood of self-motivation in increasing physical activities ($P < 0.001$).

Stressed people are significantly less effective in influencing their peers to increase physical activities; relative to relaxed people, they are 96% less influential ($P < 0.001$). However, stressed people can spontaneously increase physical activities; relative to people with no stress, stressed people are 20% more spontaneous ($P < 0.001$).

Analyzing the correlations among social activity level, the number of friend connections, self-motivation, influence, and susceptibility reveals the potential roles that different social relationships play in the spread of physical activities. Influence increases with the level of social activity from the "keep-to-themselves" people to the "active-in-hanging-out" people. On the other hand, susceptibility is negatively correlated with social activity. "Active-in-hanging-out" people are 29% more influential ($P < 0.001$), but 31% less susceptible to influence ($P < 0.001$) than the baseline. Those who self-characterize as "keep-to-themselves" are susceptible to influence, but have the least likelihood of influencing their peers to increase physical activities. Relative to the "active-in-hanging-out" people, those who self-characterize as "keep-to-themselves" people have a 33% lower influence, and a 79% greater susceptibility ($P < 0.001$). Interestingly, the "keep-to- themselves" people have a significantly stronger self-motivation to increase physical activities; meanwhile, the "active-in-hanging-out" people are not self-motivated. Relative to the "active-in-hanging-out" people, the "keep-to-themselves" people have a 34% greater self-motivation ($P < 0.001$). In spite of low self-motivation, the "active-in-hanging-out" people can spontaneously increase physical activities significantly more than the others. In fact, the "active-in-hanging-out" people have a 29% greater spontaneous hazard in increasing physical activities ($P < 0.001$).

Unlike general online social networks, such as Facebook and Twitter, having many friend connections within our YesiWell social network does not guarantee a strong likelihood of influencing their peers to increase physical activities. In fact, influence and susceptibility are negatively correlated with the number of friend connections. People with too many friend connections exert the least influence on their peers in increasing physical activities; they have a 44% lower likelihood of influencing their peers ($P < 0.001$). However, people with too many friend connections have a significantly higher self-motivation in increasing physical activities when participating in social events; they have a 51% greater self-motivation ($P < 0.001$). In addition, people with too many friends are among the least susceptible individuals to influence and the least likely to spontaneously increase physical activities. Relative to people with 1–10 friends, those with too many friends have a 57% lower susceptibility ($P < 0.001$). People with too many friends also have a 20% lower spontaneous hazard in increasing physical activities compared with those with many friends (11–29) ($P < 0.001$). These results suggest that people who make too many friend connections increase physical activities because of their strong self-motivation, and not because of social influence, susceptibility, or spontaneous hazard.

On the other hand, people with 1–10 friends are the most susceptible to influence, and have the strongest likelihood of influencing their peers; i.e., 57% more susceptible and 51% more influential than those with 30+ friends ($P < 0.001$). These results suggest that high-quality social relationships, i.e., frequently interacting with friends, actively hanging out with them, is more important in the spread of physical activities, than just having a high number of friend connections. This observation is strengthened by our previous finding that declaring friend connections has some benefit, but that such benefit is marginal compared with that from actual social interactions, e.g., hanging out and exchanging messages among users, in terms of spreading physical activities.

People exert the most influence on peers of the same or younger age, i.e., 15% more influence than baseline ($P < 0.05$) (Fig. 2). In non-dyadic models, we found that overweight people were more influential than obese people (Fig. 1). Dyadic models (Fig. 2) further revealed that overweight people exert 26% more influence over obese people than over other overweight people ($P < 0.001$). In addition, obese people exert 37% more influence over other obese people than over overweight people ($P < 0.01$). With regards to stress level, dyadic models discover that relaxed people exert the most influence on non-stressed peers at 109% more than the baseline ($P < 0.001$), and 72% more than relaxed peers ($P < 0.001$), while also influencing stressed peers by 95% more than the baseline ($P < 0.001$), and 61% more than relaxed peers ($P < 0.001$).

Interestingly, people of different social activity levels, including those who actively hang out, who are not stressed, and who keep to themselves, share a similar behavior pattern in influencing their peers; that is, they exert the most influence over keep-to-themselves peers and the least influence over actively-hanging-out peers ($P < 0.05$ hold for all tests). For instance, actively-hanging-out people exert more influence over keep-to-themselves peers than over the baseline (34%, $P < 0.05$), over non-stressed peers (33%, $P < 0.05$), and over actively-hanging-out peers (83%, $P < 0.01$).

Finally, dyadic models further reveal that people of different numbers of friend connections, including a common number of friends (1–10), many friends (11–29), and too many friends (30+), share a similar behavior pattern in influencing their peers; that is, they exert the most influence over those with 1–10 friends ($P < 0.01$ holds for all tests). For instance, people with 1–10 friends exert more influence over people in the same quartile than over the baseline (68%, $P < 0.001$), over those with many friends (58%, $P < 0.001$), and over those with too many friends (63%, $P < 0.001$).

These results have implications for policies designed to promote the spread of physical activities and healthy behaviors. They show the general utility of our approaches for informing intervention strategies, for targeting healthcare-oriented social events/programs, and for policy-making toward healthy communities.

5   Conclusions

Our system, which combines social activities, physical activities, and biometrics/biomarkers for a group of 254 individuals over 10 months, continuously measures self-motivation, influence, and susceptibility to being influenced into increasing physical activities. By applying online and mobile programs, our system can be refined and can be scalable to a larger number of users, to provide a higher-granularity understanding of the dynamics of the spread of healthy behaviors and to inform next-generation digital therapeutics. This will enable significant personalization. In fact, by understanding individual social and psychological factors that govern self-motivation, susceptibility, and influence, we can personalize intervention strategies, policies, and nudges.

Acknowledgments. This work is supported by the NIH grant R01GM103309 to the SMASH project. The work was done when Dr. Phan was at the University of Oregon.

References

1. Creating social contagion through viral product design: A randomized trial of peer influence in networks. Manage. Sci. **57**(9), 1623–1639 (2011). https://doi.org/10.1287/mnsc.1110.1421
2. Amrai, K., Motlagh, S.E., Zalani, H.A., Parhon, H.: 3rd world conference on educational sciences - 2011 the relationship between academic motivation and academic achievement students. Proc. Soc. Behav. Sci. **15**, 399–402 (2011). https://doi.org/10.1016/j.sbspro.2011.03.111
3. Aral, S., Muchnik, L., Sundararajan, A.: Distinguishing influence-based contagion from homophily-driven diffusion in dynamic networks. Proc. Natl. Acad. Sci. **106**(51), 21544–21549 (2009). https://doi.org/10.1073/pnas.0908800106
4. Aral, S., Walker, D.: Identifying social influence in networks using randomized experiments. IEEE Intell. Syst. **26**(5), 91–96 (2011)
5. Aral, S., Walker, D.: Identifying influential and susceptible members of social networks. Science **337**(6092), 337–341 (2012). https://doi.org/10.1126/science.1215842
6. Bellemare, C., Lepage, P., Shearer, B.: Peer pressure, incentives, and gender: an experimental analysis of motivation in the workplace. Labour Econ. **17**(1), 276–283 (2010). https://doi.org/10.1016/j.labeco.2009.07.004
7. Tudor-Locke, C., Johnson, W.D., Katzmarzyk, P.T.: Accelerometer-determined steps per day in us adults. Med. Sci. Sports Exerc. **41**(7), 1384–91 (2009)
8. Christakis, N.A., Fowler, J.H.: The spread of obesity in a large social network over 32 years. N. Engl. J. Med. **357**(4), 370–379 (2007). https://doi.org/10.1056/NEJMsa066082
9. Christakis, N.A., Fowler, J.H.: The collective dynamics of smoking in a large social network. N. Engl. J. Med. **358**(21), 2249–2258 (2008). https://doi.org/10.1056/NEJMsa0706154
10. Renders, C.M., Valk, G.D., Griffin, S.J., Wagner, E., van Eijk, J.T., Assendelft, W.J.: Interventions to improve the management of diabetes mellitus in primary care, outpatient and community settings. Cochrane Database Syst. Rev. **4** (2000). https://doi.org/10.1002/14651858.CD001481
11. Eagle, N., Macy, M., Claxton, R.: Network diversity and economic development. Science **328**(5981), 1029–1031 (2010). https://doi.org/10.1126/science.1186605
12. Eisenberg, D., Golberstein, E., Whitlock, J.L.: Peer effects on risky behaviors: New evidence from college roommate assignments. J. Health Econ. **33**, 126–138 (2014). https://doi.org/10.1016/j.jhealeco.2013.11.006
13. Fowler, J.H., Christakis, N.A.: Dynamic spread of happiness in a large social network: longitudinal analysis over 20 years in the Framingham heart study. BMJ **337** (2008). https://doi.org/10.1136/bmj.a2338

14. Golder, S.A., Macy, M.W.: Diurnal and seasonal mood vary with work, sleep, and daylength across diverse cultures. Science **333**(6051), 1878–1881 (2011). https://doi.org/10.1126/science.1202775
15. Greene, J., Sacks, R., Piniewski, B., Kil, D., Hahn, J.S.: The impact of an online social network with wireless monitoring devices on physical activity and weight loss. J. Primary Care Commun. Health **4**(3), 189–194 (2013). https://doi.org/10.1177/2150131912469546
16. Diabetes Prevention Program Research Group: Reduction in the incidence of type 2 diabetes with lifestyle intervention or metformin. N. Engl. J. Med. **346**(6), 393–403 (2002)
17. The Diabetes Prevention Program (DPP) Research Group: The diabetes prevention program (DPP). Diabetes Care **25**(12), 2165–2171 (2002). https://doi.org/10.2337/diacare.25.12.2165
18. Guille, A., Hacid, H., Favre, C., Zighed, D.A.: Information diffusion in online social networks: a survey. SIGMOD Rec. **42**(2), 17–28 (2013). https://doi.org/10.1145/2503792.2503797
19. Henry, D.B., Schoeny, M.E., Deptula, D.P., Slavick, J.T.: Peer selection and socialization effects on adolescent intercourse without a condom and attitudes about the costs of sex. Child Dev. **78**(3), 825–838 (2007)
20. Iyengar, R., den Bulte, C.V., Valente, T.W.: Opinion leadership and social contagion in new product diffusion. Mark. Sci. **30**(2), 195–212 (2011)
21. Jelas, Z.M., et al.: International conference on learner diversity 2010 effects of FLEP on self-motivation and aspiration to learn among low-achieving students: An experimental study across gender. Procedia. Soc. Behav. Sci. **7**, 122–129 (2010). https://doi.org/10.1016/j.sbspro.2010.10.018
22. Kempe, D., Kleinberg, J., Tardos, E.: Maximizing the spread of influence through a social network. In: Proceedings of the Ninth ACM SIGKDD International Conference on Knowledge Discovery and Data Mining, KDD 2003, pp. 137–146 (2003)
23. Kossinets, G., Watts, D.J.: Empirical analysis of an evolving social network. Science **311**(5757), 88–90 (2006). https://doi.org/10.1126/science.1116869
24. Kulshreshtha, A., Kvedar, J.C., Goyal, A., Halpern, E.F., Watson, A.J.: Use of remote monitoring to improve outcomes in patients with heart failure: a pilot trial. Int. J. Telemed. Appl. **2010** (2010). https://doi.org/10.1155/2010/870959
25. Kvedar, J.C., Fogel, A.L., Elenko, E., Zohar, D.: Digital medicine's march on chronic disease. Nat. Biotechnol. **34**, 239–246 (2016)
26. Lazer, D., et al.: Computational social science. Science **323**(5915), 721–723 (2009). https://doi.org/10.1126/science.1167742
27. Leahey, T.M., Kumar, R., Weinberg, B.M., Wing, R.R.: Teammates and social influence affect weight loss outcomes in a team-based weight loss competition. Obesity **20**(7), 1413–1418 (2012)
28. Leskovec, J., Adamic, L.A., Huberman, B.A.: The dynamics of viral marketing. ACM Trans. Web **1**(1) (2007). https://doi.org/10.1145/1232722.1232727
29. Manski, C.F.: Identification problems in the social sciences. Sociol. Methodol. **23**, 1–56 (1993)
30. Phan, N., Ebrahimi, J., Kil, D., Piniewski, B., Dou, D.: Topic-aware physical activity propagation in a health social network. IEEE Intell. Syst. **31**(1), 5–14 (2016). https://doi.org/10.1109/MIS.2015.92
31. Phan, N., Dou, D., Xiao, X., Piniewski, B., Kil, D.: Analysis of physical activity propagation in a health social network. In: Proceedings of the 23rd ACM International Conference on Conference on Information and Knowledge Management, pp. 1329–1338 (2014). https://doi.org/10.1145/2661829.2662025

32. Rice, J.B., Desai, U., Cummings, A.K.G., Birnbaum, H.G., Skornicki, M., Parsons, N.B.: Burden of diabetic foot ulcers for medicare and private insurers. Diabetes Care **37**(3), 651–658 (2014). https://doi.org/10.2337/dc13-2176

33. Sepah, C.S., Jiang, L., Peters, L.A.: Long-term outcomes of a web-based diabetes prevention program: 2-year results of a single-arm longitudinal study. J. Med. Internet Res. **17**(4), e92 (2015). https://doi.org/10.2196/jmir.4052

34. Steinberg, J.D., Horwitz, G., Zohar, D.: Building a business model in digital medicine. Nat. Biotechnol. **33**, 910–920 (2015)

35. Watson, A.J., Kvedar, J.C., Rahman, B., Pelletier, A.C., Salber, G., Grant, R.W.: Diabetes connected health: a pilot study of a patient- and provider-shared glucose monitoring web application. J. Diabetes Sci. Technol. **3**(2), 345–352 (2009). https://doi.org/10.1177/193229680900300216

36. Wentzel, K.R.: Social relationships and motivation in middle school: the role of parents, teachers, and peers. J. Educ. Psychol. **90**(2), 202–209 (1998)

Understanding the Issues Surrounding COVID-19 Vaccine Roll Out via User Tweets

Jose Esparza[1](✉), Gissella Bejarano[2], Arti Ramesh[3], and Anand Seetharam[3]

[1] Universidad Privada del Norte, Trujillo, Peru
joseesparza@tutanota.com
[2] Baylor University, Waco, TX, USA
gissella_bejaranonic@baylor.edu
[3] Binghamton University, Binghamton, NY, USA
{artir,aseethar}@binghamton.edu

Abstract. Vaccinations have emerged as one of the key tools to combat the COVID-19 pandemic, reduce infections and to enable safe re-opening of societies. Vaccinating the entire world population is a challenging undertaking and with demand far exceeding supply in the world, it is expected that topics surrounding vaccinations generate a wide array of discussions. Therefore, in this paper, we collect data from Twitter during the early days of the COVID-19 vaccination program and adopt a linguistic approach to better understand and appreciate peoples' concerns and opinions with regards to the roll out of the vaccines. We begin by studying the term frequencies (i.e., unigrams and bigrams) and observe discussions around *vaccination doses, receiving doses, vaccine supply, scheduling appointments and wearing masks* as the vaccination efforts get underway. We then adopt a seeded topic modeling approach to automatically identify the main topics of discussion in the tweets and the main issues being discussed in each topic. We observe that our dataset has nine distinct topics. For example, we observe topics related to *vaccine distribution, eligibility, scheduling* and *COVID variants*. We then study the sentiment of the tweets with respect to each of the nine topics and observe that the overall sentiment is negative for most of the topics. We only observe a higher percentage of positive sentiment for topics related to obtaining *information* and *schools*. Our research lays the foundation to conduct a more fine-grained analysis of the various issues faced by the people as the pandemic recedes over the course of the next few years.

1 Introduction

Vaccines are one of the key tools to control the COVID-19 pandemic that has caused immense suffering to people around the world. Though the first vaccines were authorized in countries around the world in late 2020, the majority of the world population is still unvaccinated and demand far exceeds supply. Vaccinating the entire world population to bring an end to the pandemic is a challenging task and economic conditions, inequity, logistics, variants and misinformation have compounded the difficulty of this endeavor. Therefore, understanding the

© Springer Nature Switzerland AG 2021
D. Mohaisen and R. Jin (Eds.): CSoNet 2021, LNCS 13116, pp. 197–205, 2021.
https://doi.org/10.1007/978-3-030-91434-9_18

issues faced by the people and addressing their concerns with regards to the COVID-19 vaccines is key to getting shots in the arms of people.

In this work, we collect and analyze approximately 650K user communications from Twitter in the United States between January and February 2021 related to COVID-19 vaccines to better understand and appreciate peoples' concerns and issues. Though Twitter communications can be biased [7], they present us the opportunity to conduct a large scale analysis of the opinions of the people as self-expressed by them. Therefore, in this paper, we conduct a linguistic study to unearth the fine-grained topics of discussion and the sentiment of the people with regards to the mass COVID-19 vaccination programs. We begin our linguistic analysis by investigating term frequencies (i.e., unigrams and bigrams). We observe from the term frequencies that large number of user communications are centered around *receiving the vaccine, vaccine distribution, appointment and availability.* We also investigate the bigrams that do not contain the word vaccine as one of the words and observe that users also express their opinions on *wearing masks, health workers* and *nursing homes* while discussing the availability, access and distribution of vaccines.

We then design a seeded LDA model to identify the key topics of discussion in our data. We observe that there are nine main topics of discussion, namely *scheduling, information, dose, distribution, eligibility, COVID cases, COVID variants, trials, and schools* with regards to the vaccination efforts. For example, we observe that words such as *county, clinic, site, resident, and state* highlight the main constraints with regards to scheduling appointments (*Topic: Scheduling*). Similarly, we observe that with regards to the topic *distribution,* words such as *million, speed, president, and deliver* underscore the difference between the demand and supply and the hurdles associated with the vaccine distribution process. With respect to the topic *COVID variants,* the presence of words such as *mask, effective, safe, and immunity* demonstrate peoples' concern on whether the vaccines will be effective against existing and new variants.

Finally, we conduct sentiment analysis of the tweets to identify the sentiment associated with the various topics. We observe that the overall sentiment is negative for seven of the nine topics. We observe that for the topics *information* and *school,* the overall sentiment is positive. Additionally, we observe that the overall sentiment is particularly negative (higher than 85%) for the topics *eligibility* and *distribution.* Our research unearths the main topics of discussion and concerns of the people with regards to the COVID-19 vaccine roll out and paves the way for further fine-grained analysis. By identifying the main talking points, our study arms government officials with tools that can be used when rolling out booster doses of the vaccines so as to mitigate some of the issues faced by the public during the first phase of the vaccination program.

2   Data and Methods

We collect tweets every day for a period of one month from January 4th, 2021 to February 4th, 2021. As vaccine availability and accessibility, as well as political

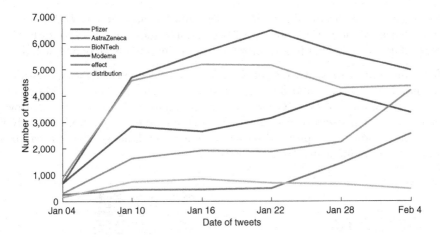

Fig. 1. Occurrence of criteria words in tweets

and social situations vary significantly between countries, we geographically limit our data collection to English language tweets from the United States. Our script runs every thirty seconds to ensure that we remain within the limits imposed by the Twitter data collection API. We collect a total of **640,311** tweets using specific search criteria.

Search Criteria: As our main goal is to understand peoples' opinions about the vaccines, our initial search criteria included words such as *vaccine* and the three most popular drug companies developing vaccines at that time, namely *Pfizer*, *Moderna*, and *AztraZeneca*. We note that during our data collection period, the Johnson & Johnson vaccine trial results had not yet been submitted for emergency authorization and therefore, it is not part of our search terms. After a few iterations we added more words related to the vaccine roll out such as *distribution* and *effect*. Figure 1 shows the number of tweets collected for each search keyword in the aforementioned time period, excluding the two most popular words: *COVID* and *vaccine*. We observe that the number of tweets mentioning Pfizer is the highest, followed by Moderna and then AstraZeneca. This is because only the Pfizer and Moderna vaccines are authorized in the US, while AstraZeneca is yet to receive emergency authorization. Additionally, between Pfizer and Moderna, the Pfizer vaccine has been more widely available, which also explains the higher number of tweets. An increase in the number of tweets related to AstraZeneca towards the end of January could be due to that fact that it is the more widely available vaccine in the rest of the world.

Preprocessing: To enable us to obtain meaningful results, we pass each tweet through a pipeline that returns a simplified form to be used by our algorithms. We remove hyperlinks, special characters, emoticons, non-English words and words with length less than three. Additionally, wherever applicable, we use full forms of the words (i.e., transform *can't* to *can not*), remove stop words and use lemmatization to obtain the simplest common form of words. Following [1], in our work, we do not use stemming because it makes some operations harder, particularly when trying to interpret the results obtained from our topic models.

Bias and Gaps in Data Collection: We acknowledge that Twitter data can be biased due to spam and bots, as shown in previous studies [5]. Additionally, due to the nature of the COVID-19 pandemic and the long time period over which the vaccines will be distributed and administered to people, our study is only able to capture peoples' opinions in the month of January when the vaccines were first being made available to the public.

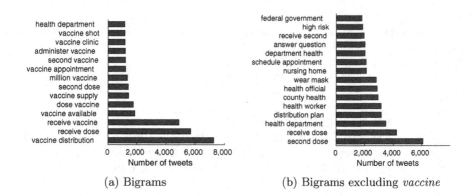

(a) Bigrams   (b) Bigrams excluding *vaccine*

Fig. 2. Frequency of unigrams and bigrams

3   Linguistic Analysis

In this section, we conduct a linguistic analysis to unearth the main concerns of the people and the issues faced by them. We first investigate the term frequencies (i.e., unigrams and bigrams) to identify the primary points of discussion surrounding vaccinations. We then leverage the term frequencies to design a seeded topic model that helps us identify and investigate the key topics of discussion in the tweets. Our topic model helps us automatically group tweets into separate topics and sheds light on the the sub-topics of discussion in each topic. We conduct a sentiment analysis of the tweets to appreciate the sentiment of the people with respect to each individual topic identified by our topic model.

3.1   Term Frequencies—Unigrams and Bigrams

We first evaluate the unigram and bigram frequencies to understand the key talking points in the tweets. We found that the most frequent unigram is *vaccine*, with 300,000 repetitions; while *receive, health, distribution, dose, state, vaccination, county, shot* and *work* range from 30,000 to 15,000 approximately in that specific order. In other words, the rest of the top 10 unigrams corresponds to less than 10% of the most frequent unigram. Words such as *health, receive* and *dose* appears as top words as they are closely related to vaccinations. Figure 2a shows the top bigrams and we observe word pairs such as *vaccine distribution* and *vaccine appointment* that highlight the main concerns of the people. We observe that majority of the bigrams contain vaccine as one of the words. As vaccine is our main search keyword, we also study the bigrams that do not contain the word vaccine (Fig. 2b). We observe from Fig. 2b that people are talking about receiving their dose (*receive dose*). Given the time period of our data collection, we know that only a small fraction of the population had obtained their vaccinations, and even they were yet to receive their second dose. Therefore, we observe people anxiously discussing about receiving the second dose of the vaccine (*second dose* and *receive second*). We observe word pairs such as *wear mask* and *high risk*, which indicates users urging others to wear masks to limit the spread of the virus. We also observe discussions surrounding *health care workers* and the difficulty in *scheduling appointments*.

3.2 Topic Modeling

In this subsection, we design a Seeded LDA model [4] to discern the underlying topics of discussion in the tweets. Seeded LDA is a seeded version of the Latent Dirichlet Allocation (LDA) topic model [2] that uses seeded words to guide the topic discovery. The model works with two main parameters α and β that represent the mixture of topics for any given document and the distribution of words per topic, respectively. With values of $\alpha = 0.001$, $\beta = 0.1$ and 2000 iterations, we obtain a high sparsity of topics for our dataset.

By carefully perusing through the data, we determine that there are nine distinct topics of discussion in our dataset and choose the seed words appropriately to automatically group the tweets. Table 1 provides an overview of the different topics in the dataset, the seed words used to guide the tweets to the various topics. Table 1 also shows the number of tweets assigned to each topic by our Seeded LDA model.

Table 1. Topics, seeded words and frequency of tweets

Topic	Seeded words	Frequency
Distribution	Distribution, plan, stock, supply	47,777
Scheduling	Appointment, schedule, available, register	44,546
COVID variants	Variant, spread, mutation, discover	37,054
Information	Question, discus, inform, expert	34,391
Schools	School, teacher, reopen	33,818
Eligibility	Eligible, phase, adult, group	32,905
COVID cases	Death, case, test, rate, patient	29,478
Dose	Dose, shot, reaction, sore	23,315
Trials	Trial, research, response, efficacy	19,235

In Table 2 we show the top words for each topic excluding the seeded ones discovered by the model. We observe from the table that words such as *county, clinic, site, resident, state* underscore the primary constraints related to scheduling appointment (*Topic: Scheduling*). Similarly, words such as *million, federal, president, administration, speed, deliver* highlight the logistical hurdles associated with the distribution of vaccines to people around the country (*Topic: Distribution*). We observe that vaccine eligibility is an important topic of discussion as demonstrated by words such as *worker, priority, staff, home, state*.

We also observe that some words such as *health, vaccination, receive and state* occur across multiple topics. However, it is important to note that the context in which these words are used can differ across as well as within topics. For example, the word *receive* in the topic *scheduling* is primarily contained in tweets where people are discussing when they will be able to receive the vaccine. In comparison, the word *receive* in the topic *eligibility* mainly refers to the age group that is currently eligible to receive the vaccine.

Table 2. Top ten words for each topic using SeededLDA

Topic	Words
Scheduling	County, health, vaccination, site, clinic, receive, department, resident, information, state
Information	Join, health, answer, community, link, support, late, information, pharmacy, meeting
Dose	Second, receive, effect, single, allergic, yesterday, symptom, injection, severe, fever
Distribution	State, administration, million, government, federal, vaccination, president, official, speed, delivery
Eligibility	Worker, receive, health, resident, home, state, vaccinate, vaccination, priority, staff
COVID cases	Receive, report, administer, state, million, positive, data, number, vaccination, high
COVID variants	Effective, mask, work, study, wear, south, social, safe, strain, immunity
Trials	Company, data, approve, clinical, country, emergency, drug, produce, million, development
Schools	Receive, work, family, vaccinate, staff, nurse, student, parent, friend, able

4   Sentiment Analysis

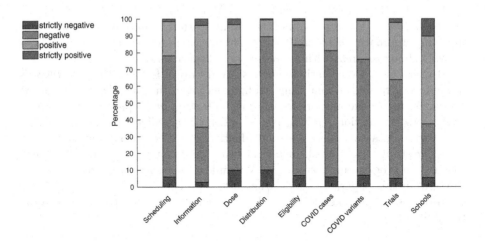

Fig. 3. Sentiment distribution in each topic

For each topic, we perform sentiment analysis using a pre-trained BERT-based model called RoBERTa, which achieves a 95% accuracy on the Stanford SST test dataset. Before training the sentiment classification model, we pre-process the data (e.g., remove hyperlinks, special characters, emoticons, non-English words). We keep the original words that can be used by the transformer attention step to get a better insight using certain words such as connectors. We observe from Fig. 3 that the overall sentiment with respect to most topics is negative. The key reason behind the negative sentiment is the lack of availability of vaccines to the general public in the beginning of 2021. For topics such as *information* and *schools*, we observe that the overall sentiment in the tweets is positive. We hypothesize that most people engage positively with *information* related to vaccines on social media.

5   Related Work

Since the beginning of the COVID-19 pandemic, researchers have studied a variety of different types of data including social network data to understand the impact of the pandemic on peoples' lives. Even before the outbreak of COVID-19, researchers have investigated Twitter data to understand the concerns of the public during the H1N1 influenza pandemic [11]. In this section, we provide an overview of the some of the recent and most relevant research in this space.

Shanthakumar et al. adopt linguistic models to investigate the societal impact of the COVID-19 pandemic [9]. They build upon their research and extend this linguistic analysis to understand the impact of the pandemic on deep societal issues such as mental health, addiction and unemployment [10]. Sarker et al. examine tweets to analyze COVID-19 symptoms as self-reported by users and use them to create a symptom lexicon [8]. Similarly, the authors in [14] also analyze COVID-19 related discussions, concerns, and sentiments using tweets posted by Twitter users. In another recent work, the authors design a topic modeling approach and identify 45 different topics related to concerns about COVID-19 across areas with socioeconomic disparities [12].

Because global vaccination efforts started only a few months ago, there is limited work related to analyzing public reaction to the COVID-19 vaccination programs. In [3], the authors measure the intention of people to participate in COVID-19 vaccination trials. Quyen et al. apply machine learning approaches on Twitter data to understand and identify anti-vaccination content [13]. In a similar study, Kwok et al. investigate the sentiment towards the COVID-19 vaccination program in Australia [6]. We build on existing work, but in contrast to them, in this paper, we adopt a linguistic approach to investigate the issues faced by the people when the COVID-19 vaccination program got underway in the United States.

6   Conclusions

In this paper, we investigated the opinions, reactions and concerns of the people with respect to the COVID-19 vaccine roll out in the United States by analyzing approximately 650K vaccine-related user communications from Twitter. We studied the unigram and bigram term frequencies and observed that users were discussing about *scheduling appointments*, *vaccine supply* and *wearing masks*. To obtain a deeper understanding of the underlying topics of discussion, we designed a seeded LDA topic model and identified that there are nine main topics of discussion in our data. Some example topics of discussion are *vaccine distribution* and *appointment scheduling*. Another topic of discussion was about the efficacy of the vaccines against the different COVID variants. We then conducted sentiment analysis of the tweets with respect to the different topics and observed an overall negative sentiment for most of the topics, except discussions related to topics of *vaccine information* and *schools*. We continue to collect user tweets to understand how the pandemic continues to affect various aspects of social life including vaccinations.

References

1. Alexandra Schofield, Måns Magnusson, L.T., Mimno, D.: Understanding text pre-processing for latent Dirichlet allocation. ACL Workshop for Women in NLP (2017). https://www.cs.cornell.edu/~xanda/winlp2017.pdf
2. Blei, D.M., Ng, A.Y., Jordan, M.I.: Latent Dirichlet allocation. J. Mach. Learn. Res. **3**, 993–1022 (2003)
3. Detoc, M., Bruel, S., Frappe, P., Tardy, B., Botelho-Nevers, E., Gagneux-Brunon, A.: Intention to participate in a COVID-19 vaccine clinical trial and to get vaccinated against COVID-19 in France during the pandemic. Vaccine **38**, 7002–7006 (2020)

4. Jagadesh Jagarlamudi, H.D., Udupa, R.: Incorporating lexical priors into topic models, pp. 204–213 (2012). https://www.aclweb.org/anthology/E12-1021
5. Kwak, H., Lee, C., Park, H., Moon, S.: What is Twitter, a social network or a news media? In: Proceedings of the 19th International Conference on World Wide Web, pp. 591–600 (2010)
6. Kwok, S.W.H., Vadde, S.K., Wang, G.: Tweet topics and sentiments relating to COVID-19 vaccination among Australian twitter users: machine learning analysis. J. Med. Internet Res. **23**(5), e26953 (2021)
7. Morstatter, F., Liu, H.: Discovering, assessing, and mitigating data bias in social media. J. Online Soc. Netw. Media **1**, 1–13 (2017)
8. Sarker, A., Lakamana, S., Hogg-Bremer, W., Xie, A., Al-Garadi, M.A., Yang, Y.C.: Self-reported COVID-19 symptoms on Twitter: an analysis and a research resource. J. Am. Med. Inf. Associ. **27**(8), 1310–1315 (2020). https://doi.org/10.1093/jamia/ocaa116
9. Shanthakumar, S.G., Seetharam, A., Ramesh, A.: Analyzing societal impact of COVID-19: a study during the early days of the pandemic. In: 2020 IEEE International Conference on Parallel & Distributed Processing with Applications, Big Data & Cloud Computing, Sustainable Computing & Communications, Social Computing & Networking (ISPA/BDCloud/SocialCom/SustainCom), pp. 852–859. IEEE (2020)
10. Shanthakumar, S.G., Seetharam, A., Ramesh, A.: Understanding the societal disruption due to COVID-19 via user tweets. In: IEEE Smartcomp 2021. IEEE (2021)
11. Signorini, A., Segre, A.M., Polgreen, P.M.: The use of twitter to track levels of disease activity and public concern in the U.S. during the influenza a H1N1 pandemic. PLoS One **6**, e19467 (2011)
12. Su, Y., Venkat, A., Yadav, Y., Puglisi, L.B., Fodeh, S.J.: Twitter-based analysis reveals differential COVID-19 concerns across areas with socioeconomic disparities. Comput. Biol. Med. **132**, 104336 (2021)
13. To, Q.G., et al.: Applying machine learning to identify anti-vaccination tweets during the COVID-19 pandemic. Int. J. Environ. Res. Public Health **18**, 4069 (2021)
14. Xue, J., et al.: Twitter discussions and emotions about the COVID-19 pandemic: machine learning approach. J. Med. Internet Res. **22**(11), e20550 (2020)

Complex Networks Analytics

Minimize Travel Time with Traffic Flow Density Equilibrium on Road Network

Qinghua Tang[1,2], Demin Li[1,2(✉)], Shuang Zhou[1,2], and Yue Fu[1,2]

[1] College of Information Science and Technology, Donghua University,
Shanghai 201620, China
`deminli@dhu.edu.cn`
[2] Engineering Research Center of Digitized Textile and Apparel Technology,
Ministry of Education, Shanghai 201620, China

Abstract. With the complexity of the urban road network and the increase of the automatic electric vehicles (AEVs) on the road, the time spend on traffic is gradually increasing. Therefore, how to find the travel path with the minimize time in the complex road network to make balanced use of road resources remains a challenge. In is paper, we first propose a minimum travel time optimization model for AEV, which takes into account the constraints of traffic flow density. One of the reasons that affect the travel time cost is the choice of travel path, the other is the traffic flow density of the road section. When a large number of vehicles rush into the same road at the same time, which will increase the traffic flow density on the road, and then increase the travel cost of AEV. Therefore, second, we further consider the equilibrium of the traffic flow density of the road section, and obtain the AEV minimum travel time optimization model based on the equilibrium of the traffic flow density on road network, which can reduces the occurrence of traffic congestion and the travel cost of AEV. Finally, we propose optimal path planning (OPP) algorithm to solve the optimization problem and some real scenes of Songjiang District in Shanghai are used to verify the feasibility of the proposed model.

Keywords: Automatic electric vehicle · Traffic flow density · Minimum travel time · Equilibrium

1 Introduction

In recent years, with the improvement of the domestic economic level, the average vehicle ownership of each family has gradually increased [1], leading to a gradual increase in traffic volume and a sharp rise in traffic pressure [2]. This situation

This work is supported by NSF of China under Grant No. 61772130, No. 71171045, and No. 61901104; the Innovation Program of Shanghai Municipal Education Commission under Grant No. 14YZ130; and the International S&T Cooperation Program of Shanghai Science and Technology Commission under Grant No. 15220710600.

ⓒ Springer Nature Switzerland AG 2021
D. Mohaisen and R. Jin (Eds.): CSoNet 2021, LNCS 13116, pp. 209–217, 2021.
https://doi.org/10.1007/978-3-030-91434-9_19

directly aggravates the environmental pollution and traffic pressure. Therefore, in order to alleviate the development of this trend, AEV has become the focus of research. AEV is a kind of intelligent vehicle, which belongs to higher level electric vehicle [3]. Private cars and AEVs also spend more and more time on the road, which will not only cause traffic congestion, but also increase the cost of travel. Therefore, optimal route planning should be developed to alleviate traffic congestion and reduce travel time [4].

For path planning in vehicle travel, Wang, et al. in [5] introduce effective travel time, which uses the k-shortest path algorithm to generate the path of the iterative set, and then determines the optimal path. For the common shortest path algorithm, A* algorithm [6], the key to determine the shortest path is the evaluation function, which determines the path according to the cost of each road section. However, there is a lack of research on the balanced utilization of each road section. In fact, traffic flow imbalance may lead to traffic congestion [7]. In [8], an improved ant colony algorithm for finding the best solution in local search is proposed. Ants use pheromone tracking mechanism and directional guidance mechanism to find the shortest path. However, the intelligent algorithm will fall into local optimization with great probability, resulting in inaccurate results and additional cost. Therefore, we select the travel path by calculating the minimum travel time with consider the impact of traffic flow, and can flexibly adjust the traffic flow density threshold of road sections under different driving environments to realize the shortest time path planning with less probability of congestion. For the travel time minimization model considering the influence of traffic flow density equilibrium and the uncertainty of ordinary vehicle driving on AEV travel is rarely discussed. The contributions of this paper are as follows,

– First, an AEV minimum travel time optimization model considering the constraints of traffic flow density is proposed, which involves the influence of real-time overstocked vehicle on road section.
– Second, considering the equilibrium of traffic flow density, the optimization model is simplified by Jain's fairness index equation, which can make balanced use of road resources and reduces the occurrence of traffic congestion.
– Finally, for the optimization model in this paper, we propose an optimal path planning (OPP) algorithm based on network traffic conservation, and the simulation results show that the path with lower travel time can be obtained through the path optimization model proposed in this paper.

2   Road Network

In order to better analyze the problems raised in this paper. We have given some definitions of the road network in this section. Among them, the relevant definitions of road network, origin-destination pair and accessible path are all referred to in [9]. We define the road network $W(P, M)$, where P is the intersection node set and M is the road section set, $|P| = p, |M| = m$. $\forall i, j \in P$, $(i, j) \neq (j, i)$, if $(i, j) \in M$, which can be expressed as $(1, -1)$. And number the road sections in a certain order, expressed as $M = \{1, 2, \cdots, m\}$. In order to facilitate the following discussion, we express the road network W as a matrix in which rows are the number of nodes and columns are the number of sections. The origin-destination (O-D) of the h-th AEV is represented by vector b_h, the origin node and destination node are 1 and -1 respectively, and other nodes are 0. In [9], a network flow conservation equation is proposed, which ensure the vehicle is able to reach the respective destination. So, we can obtain some available routes from origin to destination by the equation of $Wx_h = b_h$, x_h is a vector to denote the path of the h-th AEV.

Our work is based on real-time traffic information. For better analysis, we make the following basic assumptions:

- Each AEV can obtain the real-time road section loads.
- Each AEV can receive traffic accident information to dynamically adjust the path.
- Each AEV can receive the travel path selected by other AEVs and the estimated travel time.
- AEV mission destinations are deployed with charging stations.

3 Problem Formulation and Algorithm Design

In this paper, we consider a group of AEVs which start their journey from the central station when they are fully charged. Each AEV has its own mission destination, in order to ensure that each AEV can arrive at the destination smoothly and complete the corresponding loading service. We consider minimizing the travel time of AEV on the road network with the traffic flow density equilibrium.

3.1 Formula Description

In general, there are multiple reachable paths from the original node to the destination node in the road network, but for a group of AEVs, different path assignments lead to different travel times. Therefore, considering the influence of other AEV route selection, the total number of vehicles on section g can be expressed as follows,

$$q_g = \sum_{h=1}^{N} x_{g,h} + r_g \quad g \in M \tag{1}$$

Here, N represents the number of AEVs in this group, $x_{g,h}$ represents the selection status of the h-th AEV for road section g. if the h-th AEV is assigned to road section g, $x_{g,h} = 1$, otherwise $x_{g,h} = 0$. $\sum_{h=1}^{N} x_{g,h}$ represents the total number of AEVs assigned to road section g, and r_g represents the number of vehicles overstocked on road section g. Based on the definition formula of vehicle average speed proposed in [10], We consider the speed under the real-time road loads is expressed as follows,

$$\widetilde{V_g} = \frac{V_g}{[1 + \alpha(\frac{q_g}{q_g^{max}})^\beta]} \quad g \in M \tag{2}$$

Where, q_g^{max} is the maximum capacity of the road section g. V_g represent the free-flow speed. $\frac{q_g}{q_g^{max}}$ represents the traffic flow density on road segment g. α and β are adjustive parameters.

In order to avoid the congestion caused by the unbalanced utilization of road resources in the road network, we consider the traffic flow density in the routing process of AEV, and the range of traffic flow density on the road section can be adjusted according to its own environment and the time period. Therefore, the traffic flow density constraint on the road section can be expressed as follows,

$$0 \le \frac{q_g}{q_g^{max}} \le Q_g \quad g \in M \tag{3}$$

Here, Q_g is the adjustment parameter, which can be adjusted according to the different environment on the road section. Therefore, the AEV minimum travel time routing problem for a group of trips with tasks can be expressed as follows,

$$\min_{x_{g,h}} \sum_{g=1}^{m} \frac{L_g}{\widetilde{V_g}} x_{g,h} \tag{4}$$

$$s.t. \quad W x_h = b_h \tag{5}$$

$$\widetilde{V_g} = \frac{V_g}{[1 + \alpha(\frac{q_g}{q_g^{max}})^\beta]} \tag{2}$$

$$0 \le \frac{q_g}{q_g^{max}} \le Q_g \tag{3}$$

$$x_{g,h} = 0 \, or \, 1 \tag{6}$$

where, m is the number of road section on road network, and L_g is the length of the road section g. We denote $a_g = \frac{q_g}{q_g^{max}}$. And based on the Jain's fairness index equation proposed in [11], we define equality constraints (7) to achieve road network equilibrium as follow,

$$\frac{(\sum_{g=1}^{m} a_g)^2}{m \sum_{g=1}^{m} (a_g)^2} = 1 \tag{7}$$

Proposition 1. *For $\forall m \ge 2, m \in N^*$, if $\frac{(\sum_{g=1}^{m} a_g)^2}{m \sum_{g=1}^{m} (a_g)^2} = 1$, then $a_1 = a_2 = \cdots = a_m$.*

Proof. Here, N^* is the set of non-zero positive integers. We prove this proposition by mathematical induction.

First, when $m = 2$, if $\frac{(\sum_{g=1}^{2} a_g)^2}{2 \sum_{g=1}^{2} (a_g)^2} = 1$, which may be transformed to obtain

$$(a_1 - a_2)^2 = 0 \tag{8}$$

Because, $a_1 \geq 0, a_2 \geq 0$,so $a_1 = a_2$. The proposition holds.

Second, suppose that when $m = k$, and $k \geq 2, k \in N^*$, the proposition is established, that is, if $\frac{(\sum_{g=1}^{k} a_g)^2}{k \sum_{g=1}^{k}(a_g)^2} = 1$, then $a_1 = a_2 = \cdots = a_k$. Then when $m = k + 1$, and $k + 1 \geq 2, k + 1 \in N^*$, if $\frac{(\sum_{g=1}^{k+1} a_g)^2}{(k+1) \sum_{g=1}^{k+1}(a_g)^2} = 1$, which may be transformed to obtain

$$\frac{(\sum_{g=1}^{k} a_g)^2}{k \sum_{g=1}^{k}(a_g)^2} + \frac{a_{k+1}(a_{k+1} + 2 \sum_{g=1}^{k} a_g)}{k \sum_{g=1}^{k}(a_g)^2} = 1 + \frac{a_{k+1}^2(k + 1)}{k \sum_{g=1}^{k}(a_g)^2} + \frac{1}{k} \tag{9}$$

Since the proposition holds when $m = k$, we express $a_1 = a_2 = \cdots = a_k = \gamma$, so Eq. (9) can be simplified as

$$k(a_{k+1} - \gamma)^2 = 0 \tag{10}$$

Because, $a_{k+1} \geq 0$, so $a_{k+1} = \gamma$. Therefore, when $m = k + 1$, the proposition also holds.

In conclusion, the proposition holds for $\forall m \geq 2, m \in N^*$.

According to Proposition 1, when considering road network equilibrium, the traffic flow density of each section should tend to be equal. For example, the most ideal state is $a_1 = a_2 = \cdots = a_m = \gamma \leq \lambda$, where, λ is the equalization threshold that can be adjusted. Because complete equalization is difficult, AEV preferentially selects the road section within the equalization threshold during path planning. However, when there are many vehicles in the road network and close to the maximum load capacity, the above road network equilibrium may be realized within a certain error range.

3.2 Algorithm Design

In this paper, the optimization model is a typical 0–1 integer programming problem, we propose the optimization algorithm for this problem.

As shown in Algorithm 1, and we outline the method of solving the optimal path, which explains how to choose a path that minimizes travel time within the constraints of traffic flow density on road network. And the equilibrium threshold λ will affect the existence and uniqueness of the path solution. If λ is too large, there may be multiple optimal solutions, and if λ is too small, there may be no solutions. Therefore, the adjustment of the λ value is the key to the solution process.

Theorem 1. *The complexity of the proposed OPP algorithm with traffic flow equilibrium is $O(pm + n + c)$.*

Proof. According to the road network W, the complexity of the feasible path from the origin node to the destination node is $O(pm)$, where p and m are the number of nodes and sections in the road network respectively. The complexity of the optimal path under the traffic flow density constraint is $O(n + c)$, where

---

Algorithm 1 : The optimal path planning (OPP) algorithm

1: BEGIN
2: /*Initialization*/
3: Input the matrix of road network W
4: Input origin-destination pairs b_h
5: Input traffic flow density equilibrium parameter λ
6: Load the current-time traffic information
7: **while** $O \neq D$ **do**
8:     Calculate Eq. (5)
9:     Obtain all available paths.
10:     **for** each available path **do**
11:         **if** $0 \leq a_g \leq \lambda$ **then**
12:             Calculates the $\sum_{g=1}^{m} \frac{L_g}{V_g} x_{g,h}$
13:             **if** the sum is minimum **then**
14:                 Output the optimal path x_h
15:             **end if**
16:         **end if**
17:     **end for**
18: **end while**
19: END

---

n is the number of inequality constraints in the model and c is the number of iterations. Therefore, the complexity of the proposed OPP algorithm is $O(pm + n + c)$.

4   Simulation Result

We use simulation to evaluate the feasibility of the optimization model and OPP algorithm proposed in this paper, and this section describes and analyzes the simulation environment and results.

4.1   Simulation Setup

In order to verify the correctness and feasibility of the road network optimization model proposed in this paper, we carried out python simulation experiment on it. We considered a realistic urban scene in Songjiang District, Shanghai, China, as shown in Fig. 1, and used Java OpenStreetMap (JOSM) [12] to extract traffic road information and number nodes, as shown in Fig. 2. The road network consists of 13 nodes and 36 road sections, with two-way traffic between them, which is represented as a matrix W with 13 rows and 36 columns. The parameters in the model are set as follows: $\alpha = 0.5$, $\beta = 6$, the AEV's total number is 400, and free running speed is set to 25km/h, and use Baidu map to obtain the length of each section.

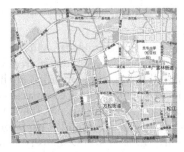 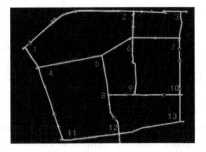

Fig. 1. Simulation scenario: Songjiang town

Fig. 2. Simplified road network

4.2 Simulation Analysis

Take the original node 1 and the destination node 10 as an example, we compare the path planning results of OPP algorithm and A* algorithm, we know that there are multiple possible paths from node 1 to node 10. When the threshold of traffic flow density is set to 0.8, the path planning results of OPP algorithm and A* algorithm are 1→4→5→8→9→10 and 1→2→6→7→10 respectively. Then, by simulating different original-destination pairs of travelling AEVs, we compare the travel time of the path obtained by the OPP algorithm proposed in this paper with the travel time of the path obtained by the A* algorithm, as follow Fig. 3,

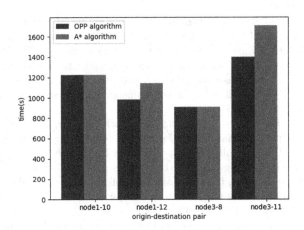

Fig. 3. Travel time of AEV with different original-destination pairs

As shown above, the travel time of the path obtained by the OPP algorithm proposed in this paper is always less than or equal to the travel time of the path obtained by the A* algorithm. This is because, without considering the balance of the road network, AEV often only consider their own travel needs when choosing travel routes, which will lead to road congestion, and increases the travel time.

5   Conclusion

In this paper, we propose an optimization model for the minimize travel time of AEV based on the road network. In this model, in order to use road resources in a balanced way, avoid a large number of AEVs flooding into the same road at the same time causing traffic congestion, we consider traffic flow density constraints in the AEV routing process, and use the road section as evenly as possible. We also consider the real-time overstocked vehicles on road section, and considering the road network equilibrium, we further simplify the traffic flow density constraint. For the solution of the model, we propose OPP algorithm, and through the real map environment simulation. The experimental simulation results are consistent with the theoretical analysis and prove the correctness and feasibility of the proposed model.

In future work, we will discuss the communication architecture of real-time traffic information acquisition in detail, and discuss the charging problem of AEV when the power is insufficient during driving, as well as the power consumption cost of AEV during driving.

References

1. Zhu, M., Liu, X., Tang, F.: Public vehicles for future urban transportation. IEEE Trans. Intell. Transp. Syst. **17**(12), 3344–3353 (2016). https://doi.org/10.1109/TITS.2016.2543263
2. He, Z., Cao, J., Li, T.: A real-time traffic estimation based vehicular path planning solution using VANETs. In: International Conference on Connected Vehicles and Expo.: Institute of Electrical and Electronics Engineers, pp. 172–178 (2012)
3. Sethuraman, G., Reddy Ragavareddy, S.S., Ongel, A., Lienkamp, M., Raksincharoensak, P.: Impact assessment of autonomous electric vehicles in public transportation system. In: IEEE Intelligent Transportation Systems Conference, ITSC 2019, art. no. 8917256, pp. 213–219 (2019)
4. Artmeier, A., Haselmayr, J., Leucker, M., Sachenbacher, M.: The shortest path problem revisited: optimal routing for electric vehicles. In: Dillmann, R., Beyerer, J., Hanebeck, U.D., Schultz, T. (eds.) KI 2010. LNCS (LNAI), vol. 6359, pp. 309–316. Springer, Heidelberg (2010). https://doi.org/10.1007/978-3-642-16111-7_35
5. Wang, S., Shao, H., Tao, L.: Travel time reliability-based optimal path finding. In: Third International Joint Conference on Computational Science and Optimization, vol. 2, pp. 531–534. IEEE (2010)
6. Guruji, A.K., Agarwal, H., Parsediya, D.K.: Time-efficient A* algorithm for robot path planning. Proc. Technol. **23**, 144–149 (2016)
7. Souza, A.M.D., Yokoyama, R.S., Maia, G.: Real-time path planning to prevent traffic jam through an intelligent transportation system. In: Computers and Communication, pp. 726–731 (2016)
8. Zhang, S., Liu, X., Wang, M.: A novel Ant colony optimization algorithm for the shortest-path problem in traffic networks. Filomat **32**(5), 1619–1628 (2018)
9. Guo, H., Cao, Z., Seshadri, M., Zhang, J., Niyato, D., Fastenrath, U.: Routing multiple vehicles cooperatively: minimizing road network breakdown probability. IEEE Trans. Emerg. Top. Comput. Intell. **1**(2), 112–124 (2017)

10. Alexander, S., Dowling, R.: Improved speed-flow relationships for planning applications. Transp. Res. Rec.: J. Transp. Res. Board **1572**, 18–23 (1997)

11. Jain, R., Chiu, D., Hawe, W.: A quantitative measure of fairness and discrimination for resource allocation in shared computer systems (1984). http://www.cse.wustl. edu/jain/papers/fairness.htm. Accessed 30 Apr 2014

12. Huber, S., Rust, C.: osrmtime: calculate travel time and distance with OpenStreetMap data using the open source routing machine (OSRM). Soc. Sci. Electron. Publ. **16**(2), 416–423 (2016)

Network Based Framework to Compare Vaccination Strategies

Rishi Ranjan Singh[✉], Amit Kumar Dhar, Arzad Alam Kherani,
Naveen Varghese Jacob, Ashitabh Misra, and Devansh Bajpai

Department of Electrical Engineering and Computer Science,
Indian Institute of Technology, Bhilai, Raipur, Chhattisgarh, India
{rishi,amitkdhar,arzad.alam,naveenjacob,ashitabhm,
devanshb}@iitbhilai.ac.in

Abstract. We propose a network based framework to model spread of disease. We study the evolution and control of spread of virus using the standard SIR-like rules while incorporating the various available models for social interaction. The dynamics of the framework has been compared with the real-world data of COVID-19 spread in India. This framework is further used to compare vaccination strategies.

Keywords: Network based framework · Vaccination strategies · Disease spread modelling

1 Introduction

Pandemics are rapidly spreading diseases which are results of disorders caused by various germs. COVID-19, the ongoing pandemic is due to a novel coronavirus named *Severe Acute Respiratory Syndrome Coronavirus 2 (SARS-CoV-2)*. Modeling of infection spread, prediction of disease spread by inferring data at early stages of endemic, quantifying and forecasting the spread of infectious disease, are some of the basic direction of study when a disease outbreak occurs. These studies help to understand and identify various mechanisms to slow down the spread and flatten the curve of spread until antivirals/vaccines are developed. For some diseases, it takes quite a large amount of time to establish a certified antiviral medication or develop a vaccination. In the meantime, if the basic reproduction number of disease is not small enough (≤ 1), it doesn't die on its own. The basic reproduction number of COVID-19 has been estimated in early studies [11,14] to be around 1.8–2.7. Therefore, even if one infected case remains on the planet, there will be a chance of next wave of outbreak until a major fraction of the global population is immunized. Once some certified vaccine has been developed, the next challenge is to deliver it to the people to slow down

A. Misra—Currently working at University of Illinois at Urbana-Champaign, USA,
D. Bajpai—Currently working at Goldman Sachs Services Private Limited, Bengaluru, India.

the spread and further make the disease disappear. While there are several constraints involved in producing a vaccine on a large scale for the world population, countries with a higher population, like India, faces the same challenge severely. In the beginning, due to scarcity and limited availability of vaccines, an optimal strategy to vaccinate people plays a crucial role.

In this paper, we first propose a network based modular framework for modeling spread of COVID-19. Then we use real-world data in the context of India to fine tune various input parameters in the model to simulate similar spread dynamics as in the real-world. Afterwards, we apply this framework to compare performance of various existing immunization strategies. The paper is organized in 6 sections after the introduction. Section 2 discusses a few related works. Next, in Sect. 3, we discuss the design of framework. Section 4 mentions considered vaccination strategies. Simulation results are compiled in Sect. 5. We conclude and discuss future direction in Sect. 6.

2   Related Work

In a similar study, Yang et al. [16] used a network developed from empirical social contact data and applied fixed choice designs to identify contacts. They considered conventional SIR model where infection seeding started from 1% of the population. Kherani et al. [7] recently proposed a queuing model to study the spread of an infection. Vaccination is a mechanism to provide immunity to a person. It reduces the spreading rate by reducing the susceptible population which may also increase the likelihood that the disease dies out. Random immunization strategy may demand to vaccinate a large portion of the overall population, at times almost the whole population, to control a disease outbreak [1]. Therefore, due to scarcity in the availability of vaccine soon after the invention of the vaccine, it is not feasible to follow random immunization. Few target immunization strategies exist that suggest vaccinating highly connected people in the population but due to the unavailability of related data, these approaches might not be practical [4]. Several evolutionary algorithms have been designed to find optimal vaccination strategies [3,5,10,12] some of which rely on genetic paradigm.

3   Network Based Framework

In this section, we propose a network based modular framework to model spread of disease in context of pandemic and use it further to compare various vaccinations strategies. Several challenges are involved while designing such a framework. Few are mentioned next. There is a huge uncertainty in the number of parameters involved. The germs may mutate over a period of time and it may affect spreading rate, severity etc. The disease might end in a single wave of spread or there might be multiple waves of the infection. For spreading diseases due to novel viruses, there is an uncertainty in the availability of vaccination. In the early days, there is non-availability of sufficient clinical proofs regarding reinfection and quantification of the volume of reinfection. The duration after taking the vaccination shot when vaccine becomes effective and the duration for which the immunity due to vaccine or recovery holds is not well established. One of the major challenge is the unavailability of accurate contact network in real world. In this paper, we generate synthetic contact network based on the following three standard network generation models:

- Small World Network Model by Watts-Strogatz (WS)
- Scale Free Network Model by Barabasi-Alberts (BA)
- Random Network Model by Erdos-Renyi (ER)

In the proposed framework, social interaction among individuals in a population is assumed to be similar to a real-world social network. Each individual is considered as a node of a network and the link of the network represent the interactions among people. Nodes are classified as Susceptible(S), Exposed(E), Infected(I), Recovered(R), and, Vaccinated(V) based on SEIRV model in order to mimic the various state of a person during disease propagation as shown in Fig. 1. Every node is initially classified as susceptible. The disease starts at seed nodes by changing the status to exposed. In this paper, exposed nodes denote those people who have caught the infection post exposure but yet have not been detected as infected in the system based on testing. A person would be only in any one of these states at a time. In order to control the spread of the epidemic, different strategies like lockdown, containment, and immunization enforced by government authorities are also incorporated into our system. All these process are incorporated as modules.

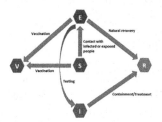

Fig. 1. SEIRV MODEL

Model: Any framework requires to capture following functionalities to show the dynamics of disease propagation similar to the real-world. It should capture the availability and scarcity of testing kits in the initial days of spread. It should have the provision to start vaccination only after a counter day in simulation that captures the start of vaccination drive and in parallel the daily availability of vaccination should be taken into account. Mutation of virus is a common phenomenon and due to it after the initial days of spread, spreading virus exist in the form several strains/variants. The mutation may cause change in the transmission rate which needs to be captured. An exposed person doesn't recover immediately, the duration of recovery varies from person to person. A right mechanism to capture the duration of natural recovery when the exposed person remains undetected or the duration of medicated recovery after testing positive is required. Lockdown and unlock are a globally known modus operandi in the early days of pandemic when there is scarcity of testing kits and yet there is no sign/scarcity of vaccine. During COVID-19 it has been used widely by various countries and states to contain as well as control the spread. It is very important to capture the restrictions imposed during lockdown and versions of unlocking periods. The possibility of developing symptoms or remain asymptomatic during infection and the capacity/duration for being active to transmit infection is another essential feature which is required in any such framework. In the proposed framework, all of the above discussed requirements are encapsulated as modules. Being modular in nature, the framework has the flexibility to easily modify a module representing a functionality based on the real-world requirement while disturbing other functionalities. There is not sufficient clinical studies exist that shed light on the reinfection post vaccination or post immunity after recovery in the case of COVID-19. Although, this framework easily allows to add the possibility of limited time immunity due to vaccination/ natural recovery, we have left it for future work. Algorithm 1 portrays the skeleton of the proposed framework where the modules are present in an abstraction form.

Algorithm 1: Modular Network-based Framework for COVID-19 Spread

---

1: INPUT: Parameters, Network Type Choice and Vaccination Strategy Choice
2: Generate Synthetic Interaction Network.
3: $Contact\_Profile()$ ▷ Initialise contact probability on the links
4: $Seeding\_Profile()$ ▷ Choose seed nodes to start disease propagation.
5: Set $Counter = 0$ ▷ Day of seeding counts as the zeroth day.
6: **while** Number of Infected/Exposed nodes at the end of $Counter$ day $\neq 0$ **do**
7:     $Mutation\_Profile()$ ▷ Change in transmission rate due to mutation.
8:     $Lockdown\_Profile()$ ▷ Change in interaction probability due to lockdown.
9:     **for** Each Infected/Exposed node **do**
10:       **if** Node is Exposed **then**
11:         $Natural\_Recovery\_Profile()$ ▷ Does node recovers today naturally?
12:         **if** Naturally Recovered Today **then**
13:           Change State of this node from $Counter + 1$ day : $E \rightarrow R$
14:         **else**   ▷ Node did not recover yet and is still exposed.
15:           $Symptomatic\_Profile()$ ▷ Does node develop Symptoms?
16:           $Testing\_profile()$ ▷ Does testing option available?
17:           **if** Tested Positive **then**
18:             Change State pf this node from $Counter + 1$ day: $E \rightarrow I$
19:           **else**
20:             **for** Each Susceptible Neighbor **do**
21:               $Transmission\_Profile()$ ▷ Does transmission happen?
22:               **if** Disease transmits to that Neighbor **then**
23:                 Change state of neighbor from $Counter + 1$ day: $S \rightarrow E$
24:               **end if**
25:             **end for**
26:           **end if**
27:         **end if**
28:       **else**   ▷ Node is Marked Infected
29:         $Medicated\_Recovery\_Profile()$ ▷ Does node recovers today?
30:         **if** Recovered Today **then**
31:           Change State of this node from $Counter + 1$ day: $I \rightarrow R$
32:         **end if**
33:       **end if**
34:     **end for**
35:     $Vaccination\_Profile()$ ▷ Vaccinate Susceptible nodes as per availability.
36:     Counter=Counter+1
37: **end while**

Input Parameters. Following parameters are required before starting the algorithm: number of nodes denoting the population (N), choise of synthetic network and parameters to generate it, ratio of population to be made as seed nodes (r), initial testing capacity and weekly incremental value, threshold day for recovery (after which an E or I node definitely recovers), scaling factor in case of medicated recovery, days when lock down and versions of unlock is placed, scaling factor for transmission probability when virus mutates, days when mutations occur, scaling down factor for meeting probability during lockdown and various versions of unlock, containment threshold, that checks whether the no.of infection are sufficient to declare containment zone, choice of vaccination strategy, parameters for vaccination (starting day of vaccination, initially available quantity and daily increment factor), and fraction of overall population to be classified into various age category.

Modules. The modules present in abstraction in Algorithm 1 representing various functionalities are described below:

– **Contact_Profile():** This module is to set daily contact probability as weights on the links. In the real-world, people make contact with other people during their daily activities which happens in different forms, for different duration, and in different environment set-ups. These interactions would be different for each individual based on the role they play in a society like a law-enforcer, or a health professional might have a lot of interaction than a common man. Also this social connection varies with the nature of the population like urban, semi-urban and rural. As the number of contact increases, a person is more likely to spread the influenza and might be called a super spreader[2]. In this paper, we have assigned uniformly random interaction probability values on the links. The values are generated based on Gaussian random number distribution.

– **Seeding_Profile():** This module chooses the seed nodes to start disease propagation. The fraction of the nodes to be initially made exposed to the disease for further transmission needs to be given as input to the framework. The approximate number of initial reported cases in the beginning of spread could be given as inputs in order to simulate for an actual condition prevailing in a region under study. In this paper, we have considered uniformly random selection for choosing seed nodes among the population to start the disease propagation.

– **Mutation_Profile():** The virus for spreading infectious disease mutates over time and its spreading rate as well as severity due to infection changes over time post mutation. This framework is mainly designed to understand the performance of vaccinations strategies in controlling propagation of a pandemic. Therefore, we consider a simplified version of mutation where the transmission probability is scaled up/down by a tuning factor. In this assumed model, effect of mutation is activated when the day counter in simulation reaches to the preset days given as input in the framework for mutation.

- **Lockdown_Profile():** Infectious disease often traverse from human to human if people interact within a distance limit for some duration. There might be undetected yet exposed people, therefore, staying away from close interaction is the best available solution to reduce the virus attack [6]. During COVID-19 spread, different lockdown restrictions have been imposed to reduce the interaction among people. In this paper, we have modelled a simplified version of real-world lockdown imposition. This version scales down contact probability of every link by a multiplicative tuning factor $0 \leq LF \leq 1$. $LF \in [0, 1]$ denotes the scale of lockdown restriction. $LF = 0$ means that all interactions are severed while $LF = 1$ means that no lock down restrictions, i.e., contacts are made with the initial assigned probability values. Lockdown restrictions and unlock relaxations are activated when the day counter in simulation reaches to the preset days given as input in the framework for applying restrictions and relaxations.
- **Natural_Recovery_Profile():** This module is to check if an exposed node naturally recovers on a counter day based on the number of days since when it has been exposed. We computes the conditional probability that a node recovers on Xth day since exposure, given that it had not recovered on previous days. Based on this computed probability value, this module decides if the exposed node under consideration recovers on this counter day or not. Naturally recovering probability is assigned based on a exponential function.
- **Medicated_Recovery_Profile():** This module is to check if an infected node on medication recovers on a counter day based on the number of days since when it has been exposed and since when it has been detected as infected. The decision is made based on a conditional probability that a node recovers on Xth day since exposure and Yth day since infected, given that it had not recovered on previous days. Based on this computed probability value, this module decides if the infected node under consideration recovers on this counter day or not. The medicated recovery probability is computed similarly to the probability value for natural recovery while taking into account the speed-up effect in recovery due to medication after detection. We have assumed to consider a linear scaling of natural recovering probability value based on the day since when this exposed node got detected and started using medication.
- **Symptomatic_Profile():** This module aims to bring in the functionality of symptomatic and asymptomatic case. It considers the factor that every exposed node does not develop symptoms due to varying immunity across the population. In this paper, we implement this functionality with the help of symptomatic probability and consider that only the nodes that develop symptoms get tested if testing kits are available. We assume that probability of developing symptoms after x counter days since exposure is generated similarly as the transmission probability, i.e., the symptomatic probability value is also assigned based on a normal distribution.
- **Testing_Profile():** This module incorporates the capacity of testing. In the early days of COVID-19, due to scarcity of testing kits/ long procedure and limited testing centers, not anyone could be tested for the disease. One of the essential condition was to have the established symptoms. Once sufficient

many centers were established and capacity of testing was increased sufficiently, then this requirement was relaxed. We try to model the capacity of testing in the form of the chance of getting tested given the symptoms are present. Only symptomatic exposed cases are tested based on the availability of testing. The capacity of testing on the current day counter is computed based on the initial capacity of testing and daily increment factor given as input. In this paper, we assume that the capacity of testing is increased linearly and after a threshold counter day, the capacity reaches sufficient that any one can be tested.

- **Transmission_Profile()**: This module checks if an exposed node transmits disease successfully to a susceptible neighbor. The success of transmission depends on the contact probability value on the link between these two nodes and the transmission probability of the exposed node representing the capability of spreading the disease. The transmission probability of an exposed node is computed based on the current day counter and the day counter when that node was moved to exposed category. In this paper, we have assumed that the transmission probability is assigned based on a normal distribution.
- **Vaccination_Profile()**: This module applies considered vaccination strategies which are discussed in next section. Before starting the vaccination it checks whether the current day counter is higher than the day counter when the vaccination drive starts. It also computes the number of vaccine vials available on the current day counter. This quantity is computed based on the initial vaccination quantity and daily increment factor given as input. In this paper, we assume that the production of vaccine as well as the availability of vaccine vials is increased linearly after the start of vaccination. This module vaccinates Susceptible/Exposed nodes as per the chosen vaccination strategy and availability of vaccine. Change state of these chosen nodes from $Counter + t$ day: $\boldsymbol{S} \rightarrow \boldsymbol{V}$ if these nodes remain susceptible for the next t days, where t is the number of days after which vaccine becomes effective.

Let N_{SV}^t, N_{EV}^t, N_{SE}^t, N_{ER}^t, N_{IR}^t be the number of susceptible nodes vaccinated, exposed nodes vaccinated, susceptible nodes exposed to disease, exposed nodes showed symptom and tested positive, exposed nodes naturally recovered, and infected nodes recovered with medicines respectively at the end of counter day t. Similarly, let N_S^t, N_E^t, N_I^t, N_R^t, and N_V^t be the number of, susceptible, exposed nodes, infected, recovered, and vaccinated nodes in the beginning of counter day t. As per the proposed model and algorithm, these values are related as follows: $N_S^1 = N - SEED$, $N_E^1 = SEED$, $N_I^1 = N_R^1 = N_V^1 = 0$ where N denotes the population size and $SEED$ denote the number of seed infections

$$N_S^{t+1} = N_S^t - N_{SV}^t - N_{SE}^t$$
$$N_E^{t+1} = N_E^t + N_{SE}^t - N_{EI}^t - N_{ER}^t - N_{EV}^t$$
$$N_I^{t+1} = N_I^t + N_{EI}^t - N_{IR}^t$$
$$N_R^{t+1} = N_R^t + N_{ER}^t + N_{IR}^t$$
$$N_V^{t+1} = N_V^t + N_{SV}^t + N_{EV}^t$$

4   Considered Vaccination Strategies

Vaccines actually simulate the bodies natural immune system to create anti-bodies that could effectively fight against pathogens. The distribution of vaccine doses during a pandemic needs to be optimized when the demand is high and supply is less especially in a highly populated country like India. The proposed system would vaccinate the susceptible and exposed people based on the availability of doses and type of vaccination strategy. In the current framework, vaccination begins with an initial quantity that is calculated from the average of first week vaccines distributed in India, collected from https://www.covid19india.org/. The increment in the number of available dose is assumed to follow a linear curve. In this paper, we compare following vaccination strategies.

Random Vaccination: Individuals are randomly selected from the subject population, based on the quantity of vaccine doses available. In this approach, no information about the network structure is required, hence nodes are selected without any knowledge about its position [16]. Thus, no preference is given to any of the nodes.

Age-Based Vaccination: Government of India initially vaccinated all COVID-19 Front-line Warriors followed by elder population above 60 and people with comorbidities [8]. In subsequent phases, the target population were people above 45 and 18 years of age. Although to reduce the mortality rate the elderly should be given initial preference, it would be better to vaccinate the youth responsible for spreading the virus [9]. In the system, nodes are randomly classified into four age groups, namely, 65 and above (6.72%), 55–64 (7.91%), 18–54 (59.07%) and below 18 age groups. In order to mimic the actual age based vaccination followed in India the elder group is vaccinated first, followed by the 55–64 and 18–54 groups.

Ring vaccination: As per medical definition, Ring vaccination is a type of vaccination that vaccinates all susceptible individuals in the demarcated area of an epidemic outbreak. If an individual gets infected, the probability of infection getting transmitted to all the neighbouring nodes is high. Hence, all these neighbours need to be vaccinated immediately. A ring of protected people would act like a buffer to cover the infected person and stops further spreading of the virus [15]. This is a traditional vaccination strategy that had been used to control small pox.

Acquaintance Vaccination: This vaccination is applied in networks with large heterogeneity. First, a node is randomly selected like in the case of random vaccination and then, from its connected neighbours a new random node is selected for vaccination. This is based on the assumption that the acquaintance node might be more exposed to infections [15]. In actual scenario, this vaccination is implemented by asking a random person to nominate one of his friend for vaccination. Hence, it is also called as Nominated vaccination [16].

High Degree Vaccination: The number of contact a person have depends on the nature of his occupation and social behaviour. If the interaction is more,

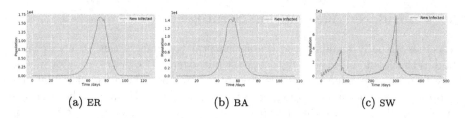

(a) ER   (b) BA   (c) SW

Fig. 2. Comparison of daily New Infection in ER, BA, SW synthetic networks with Random Vaccination starting from day:300 and the real-world data of Indian COVID19 daily new confirmed cases.

there is more chance to get exposed from an infected person and spread to others. In this vaccination strategy, all the nodes having degree above a threshold value, say average degree, are tracked and immunized to break the viral chain [16].

Top Degree Centrality Vaccination: The rate of transmission of pandemic could be reduced by identify and vaccinating those people with the highest number of contacts. These super-spreader would, otherwise spread the virus to a large group of people. Conventionally it selects some fraction of highest degree nodes for vaccination. [16]. In this paper, we implement it by choosing the highest degree nodes, vaccinate them if vaccines are available and, follow the same for next lower highest degree. Similar strategies based on other type of centrality measures [13] can be implemented in future.

5   Results and Discussion

In this section, we compile and discuss the acquired simulation results. The simulation results achieved using the proposed framework scales linearly.

5.1   Comparison with the Real-World Data

In this section, first we plot simulation results for the three considered synthetic network generators and compare to identify the one that shows most similar disease dissemination as in the real-world. Figure 2 contains plots for ER, BA and WS network generator models. The networks were generated with same number of nodes and similar average degree, i.e., similar number of links. It is evident from the plots that results on synthetic small world networks generated based on WS model deliver most similar disease diffusion as observed in the real-world data in context of India given in Fig. 3a. In this simulation, the average degree has been considered 16 across all models. For higher average degree, even network generated based on WS model also exhibits a single wave due to faster disease dissemination in denser network.

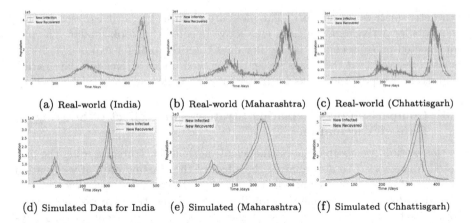

(a) Real-world (India)   (b) Real-world (Maharashtra)   (c) Real-world (Chhattisgarh)

(d) Simulated Data for India   (e) Simulated (Maharashtra)   (f) Simulated (Chhattisgarh)

Fig. 3. Comparison of real-world new daily infection and recovery count of India, Maharastra and Chhattisgarh data with simulated data

Next, we show the plots for daily new infection and daily new recovery for India as well as two states (Maharashtra and Chhattisgarh) of India where the proposed framework using WS small world network has exhibited similar dynamics for disease spread as happened in the real-world. The plots are compiled in Fig. 3. Apart from capturing the two waves of the real-world data, it is also observed that the daily new recovery curve (detected cases) follows the daily infected curve by few counter days similar to how it occurs in the real-world data. The real-world data has been curated from https://www.covid19india.org/. The detail list of input parameters for these simulations have been given in the full-version paper available on www.iitbhilai.ac.in/index.php?pid=csonet2021.

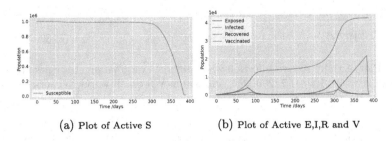

(a) Plot of Active S   (b) Plot of Active E,I,R and V

Fig. 4. Plot of SEIRV model in WS small world network with random vaccination starting from day:300

5.2   Disease Spread

In this section we show plots depicting the changes in the number of nodes belonging to various class in SEIRV model in a simulation on WS model based synthetic small world network. The plots are given in Fig. 4. It is observed that the plots for active exposed cases and active infected case look similar. The plot for infected nodes seems to lag by few days than the plot for exposed nodes. It is also observed that the peaks for active exposed nodes arrives few days earlier than the corresponding peaks for active infected nodes.

5.3   Comparison of Vaccination Strategies and Effect of Starting Date of Vaccination

In this section, we compare performance of vaccination strategies using the proposed framework. We also analyse the effect of the starting date of vaccination on the performance of vaccination strategies and the containment power against spread of disease. The plots are compiled in Fig. 5 and the expected number of total infected nodes for various strategies applied from different starting days has been compiled in Table 1.

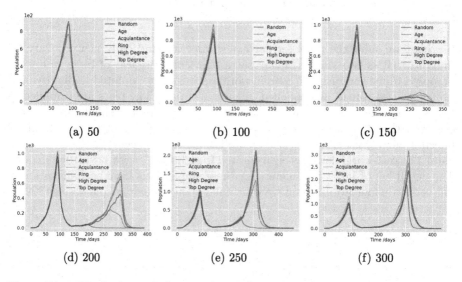

Fig. 5. Plot of Daily Active Infection when different vaccination strategies are applied from day X = 50, 100, 150, 200, 250, and 300 in WS Small world Network

The expected results are generated by averaging the results over 5 runs of each simulation. It is evident from the plots that starting vaccination at early stage helps contain the disease quickly and in most of the cases there occurs a single wave of disease spread. But as we delay the start of vaccination process, the chance as well as the size of second wave increases. Ring vaccination has turn out to be most efficient among the considered strategies but it is not a realistic strategy to implement in most of the cases. The high degree vaccination strategy comes next in minimizing the expected number of total infection cases which is a more realistic approach. In India, the front-line workers were vaccinated first which is similar to vaccinating high degree nodes.

Table 1. Expected number of infected nodes when various vaccination strategies are applied from day X = 50, 100, 150, 200, 250, and 300

Type of vaccination	Day 50	Day 100	Day 150	Day 200	Day 250	Day 300
Random	31871.92	32688.17	43847.83	60724.67	130430.42	134503
Age	32799.22	33621.10	43979.57	73105.28	116993.64	127114
Acquaintance	31956.15	32376.68	45000.64	74050.65	113310.40	145875
Ring	7140	29634.84	31809.37	36775.42	53171.70	103298.92
High degree	32541.70	38092.75	42786.47	75622.65	124645.70	150917.08
Top degree	28787.8	34831.84	39820.14	56640.85	99415.98	142957.00

6   Conclusion and Future Directions

We proposed a network based framework to compare performance of vaccination strategies during a pandemic. It has been observed that the results for the dynamics of disease spread on small-world networks were the most similar to the real-world spread among the considered synthetic network generators. Although, Ring vaccination has turnout to be most efficient among the considered strategies, top-degree seems most realistic yet efficient approach to consider. Simulation results show that if the vaccination starts early, it may contain the dissemination of disease quickly. The dynamics of the spread scales linearly in the proposed model. Considering more accurate probability generation functions based on real-world behaviour is one of the future directions. A synthetic network generator that can depict the real-world physical social relationships better than the general synthetic generators considered in this paper is desired.

Acknowledgements. This work is supported by a research grant under a special call under the MATRICS scheme by SERB, India (MSC/2020/000374). This works is partially supported by Research Initiation Grant from IIT Bhilai (2004800).

References

1. Anderson, R.M., Anderson, B., May, R.M.: Infectious Diseases of Humans: Dynamics and Control. Oxford University Press, Oxford (1992)
2. Barabási, A.L., Albert, R.: Emergence of scaling in random networks. Science **286**(5439), 509–512 (1999)
3. Calonaci, C., Chiacchio, F., Pappalardo, F.: Optimal vaccination schedule search using genetic algorithm over MPI technology. BMC Med. Inform. Decis. Mak. **12**(1), 129 (2012)
4. Cohen, R., Havlin, S., Ben-Avraham, D.: Efficient immunization strategies for computer networks and populations. Phys. Rev. Lett. **91**(24), 247901 (2003)
5. Hu, X.-M., Zhang, J., Chen, H.: Optimal vaccine distribution strategy for different age groups of population: a differential evolution algorithm approach. Math. Probl. Eng. **2014**, 7 (2014). Article ID 702973. https://doi.org/10.1155/2014/702973
6. Johansson, M.A., et al.: SARS-CoV-2 transmission from people without COVID-19 symptoms. JAMA Netw. Open **4**(1), e2035057–e2035057 (2021)

7. Kherani, A.A., Kherani, N.A., Singh, R.R., Dhar, A.K., Manjunath, D., et al.: On modeling of interaction-based spread of communicable diseases. In: Gervasi, O. (ed.) ICCSA 2021. LNCS, vol. 12949, pp. 576–591. Springer, Cham (2021). https://doi.org/10.1007/978-3-030-86653-2_42

8. Kumar, V.M., Pandi-Perumal, S.R., Trakht, I., Thyagarajan, S.P.: Strategy for COVID-19 vaccination in India: the country with the second highest population and number of cases. NPJ Vaccines 6(1), 1–7 (2021)

9. Matrajt, L., Eaton, J., Leung, T., Brown, E.R.: Vaccine optimization for COVID-19: who to vaccinate first? medRxiv (2020)

10. Patel, R., Longini, I.M., Jr., Halloran, M.E.: Finding optimal vaccination strategies for pandemic influenza using genetic algorithms. J. Theor. Biol. **234**(2), 201–212 (2005)

11. Sanche, S., Lin, Y., Xu, C., Romero-Severson, E., Hengartner, N., Ke, R.: High contagiousness and rapid spread of severe acute respiratory syndrome coronavirus 2. Emerg. Infect. Dis. **26**(7), 1470 (2020)

12. Sanders, L., Woolley-Meza, O.: Optimal vaccination of a general population network via genetic algorithms. BioRxiv, p. 227116 (2018)

13. Singh, R.R.: Centrality measures: a tool to identify key actors in social networks. In: Biswas, A., Patgiri, R., Biswas, B. (eds.) Principles of Social Networking. SIST, vol. 246, pp. 1–27. Springer, Singapore (2022). https://doi.org/10.1007/978-981-16-3398-0_1

14. Wu, J.T., Leung, K., Bushman, M., et al.: Estimating clinical severity of COVID-19 from the transmission dynamics in Wuhan, China. Nat. Med. **26**, 506–510 (2020). https://doi.org/10.1038/s41591-020-0822-7

15. Xu, Z., Zu, Z., Zheng, T., Zhang, W., Xu, Q., Liu, J.: Comparative analysis of the effectiveness of three immunization strategies in controlling disease outbreaks in realistic social networks. PLoS One **9**(5), e95911 (2014)

16. Yang, Y., McKhann, A., Chen, S., Harling, G., Onnela, J.P.: Efficient vaccination strategies for epidemic control using network information. Epidemics **27**, 115–122 (2019)

Groups Influence with Minimum Cost in Social Networks

Phuong N. H. Pham[1], Canh V. Pham[2(✉)], Hieu V. Duong[2],
Trung Thanh Nguyen[2], and My T. Thai[3]

[1] Faculty of Information Technology, Ho Chi Minh city University of Food Industry,
Ho Chi Minh, Vietnam
phuongpnh@hufi.edu.vn

[2] ORlab, Faculty of Computer Science, Phenikaa University, Hanoi 12116, Vietnam
{canh.phamvan,thanh.nguyentrung}@phenikaa-uni.edu.vn,
dvhieubg95@gmail.com

[3] Department of Computer and Information Science and Engineering,
University of Florida, Gainesville, USA
mythai@cise.ufl.edu

Abstract. This paper studies a Group Influence with Minimum cost which aims to find a seed set with smallest cost that can influence all target groups, where each user is associated with a cost and a group is influenced if the total score of the influenced users belonging to the group is at least a certain threshold. As the group-influence function is neither submodular nor supermodular, theoretical bounds on the quality of solutions returned by the well-known greedy approach may not be guaranteed. To address this challenge, we propose a bi-criteria polynomial-time approximation algorithm with high certainty. At the heart of the algorithm is a novel group reachable reverse sample concept, which helps speed up the estimation of the group influence function. Finally, extensive experiments conducted on real social networks show that our proposed algorithm outperform the state-of-the-art algorithms in terms of the objective value and the running time.

Keywords: Viral marketing · Group influence · Approximation algorithm · Online social network

1 Introduction

Information diffusion in Online Social Networks (OSNs) is a central research topic due to its tremendous commercial value. By leveraging the "word of mouth" effect, companies and organizations have used social networks as an effective mean of communication to promote products, spread opinion and renovation, persuade voters, etc. In a seminal work [10] published almost twenty years ago, Kempe *et al.* introduced the Influence Maximization (IM) problem, which aims to find a set of k users (called *seed set*) in a social network to initiate a propagation process that can influence a largest possible number of users, under some

© Springer Nature Switzerland AG 2021
D. Mohaisen and R. Jin (Eds.): CSoNet 2021, LNCS 13116, pp. 231–242, 2021.
https://doi.org/10.1007/978-3-030-91434-9_21

predefined propagation model. Since then, this problem and its notable variants have demonstrated theirs significant role in various real-world problems, not only in viral marketing [12,15], but also in other fields such as epidemics control in social network [13,20], social network monitoring [25], recommendation system [24], etc. As in many realistic scenarios, users decision and behavior tends to be dependent on his/her group and most of important decisions or works, which would affect to many individuals, are done by a group of key persons. Therefore, creating an impact on groups or communities would be able to bring more benefits than individuals, and deserves a special consideration.

Motivated by aforementioned phenomenon, recent studies have been carried out on a general version of IM, whose objective is to maximize the number of influenced groups of users instead of the number of influenced individuals, by choosing some seed set of at most k users (see, e.g. [7,17,23,27,28]). Also, one can consider a dual problem of this problem by asking for the minimum number of seed nodes to influence a given number of groups. Along this line, this paper investigates a slightly general problem, named *Groups Influence with Minimum cost* (GIM), which aims to find a seed set of minimum cost nodes to influence all the groups in the network. Different from existing works, we consider the role of each user in a group by assigning a score to him/her, and each group admits a *threshold* representing how difficult it is to be influenced. Specifically, a group is influenced if the total score of the influenced members reaches its threshold.

One can easily seen that GIM subsumes IM and its dual version as special cases and thus it is **NP**-hard to solve, not only by the combinatorial structure of the problem, but also by the #**P**-hardness of the calculation of the group influence function (denoted by $\sigma(\cdot)$). Another challenge is that $\sigma(\cdot)$ is neither a submodular nor suppermodular, implying that the classical greedy algorithms when being applied to GIM may not result in any approximation guarantee.

In this work, we address the above challenges and our contributions can be summarized as follows. Assume that $G = (V, E)$ is a social network under a diffusion model, $\mathcal{C} = \{C_1, C_2, \ldots, C_K\}$ is a set of target groups, each group C_i has a threshold t_i, and each node u has a cost $c(u)$ and a score $b(u)$.

- We first show that the group influence function, denoted by $\sigma(\cdot)$, is neither submodular nor suppermodular, and then develop a novel sampling technique, named *Group Reachable Reverse* (GRR), to estimate $\sigma(\cdot)$. This technique plays an importance role in our proposed algorithm.
- We devise a bi-criteria approximation algorithm, named *Groups Influence Approximation* (GIA) by proposing an algorithmic framework for generating multiple candidate solutions with theoretical bounds. Specifically, GIA is a $(O(\ln K + \ln(n \ln n)), 1 - \epsilon)$-bicriteria approximation with high probability, that is, GIA runs in polynomial time and returns a solution S satisfying $c(S) \leq O(\ln K + \ln(n \ln n))$OPT, and $\sigma(S) \geq (1 - \epsilon)K$ with high probability (w.h.p), where OPT is the total cost of an optimal solution, and ϵ is any fixed positive constant.
- We conduct extensive experiments on real social networks to demonstrate the effectiveness and scalability of our proposed algorithms. It is shown that our algorithms provide significantly higher quality solutions than existing

methods, while their running time is several times faster than that of the state-of-the art algorithms.

Organization. The rest of our paper is organized as follows. We give a short literature review in Sect. 2. In Sect. 3, we introduce the information diffusion, problem formulation and sampling method to estimate group influence function. Section 4 presents our main algorithm. The experiments are presented in Sect. 6 and the conclusion is given in Sect. 5.

2   Related Work

Kempe *et al.* [10] first introduced the Influence Maximization (IM) problem as a discrete optimization problem under two classical information diffusion models: Independent Cascade (IC) and Linear Threshold (LT). There are two main challenges: (1) IM is **NP**-hard and it cannot be approximated within a ratio of $1 - 1/e + \epsilon$ for any $\epsilon > 0$; (2) calculating influence spread of a seed node is #**P**-hard [3]. This work has inspired a vast mount of studies on developing efficient algorithms for IM [4,14,21,22] as well as its variants [9,15]. By utilizing monotonic and submodularity of the influence spread function, [10] also proposed a naive greedy algorithm providing an approximation ratio of $1 - 1/e$. Several fast heuristic algorithms were proposed for the large-scale networks [3], without any theoretical guarantees. In another work, Borgs *et al.* [2] introduced the concept of Reverse Influence Sampling (RIS) which paves the way to the development of a linear time $(1 - 1/e - \epsilon)$-approximation algorithm (w.h.p). Subsequent works have been considered for reducing the sampling complexity and running time by improving the RIS algorithm [14,21,22]. In other direction, several works have extended IM to different variants, such as, budget constraint [16], topic queries [1], competitive influence [18], and misinformation detection [19,26].

The groups influence (or community) maximization is one of IM variations, which has gained much attention recently. In this context, every user on social media usually belongs to a particular group and his/her behavior is influenced by those groups, making the creation of an influence on groups of users that reap more benefits than individuals. Nguyen *et al.* [17] aimed to find a seed set of k nodes that influences the largest number of communities and show that the group influence function is neither submodular nor supermodular, and developed several approximation algorithms to solve this problem. The authors in [27,28] investigated the problem of Group Influence Maximization, which is to select k seed users such that the number of eventually activated groups is maximized, in which, a group becomes active if β percent of nodes in this group are influenced. Later, Tsang *et al.* [23] considered the influence maximization problem with several fairness constraints and propose an algorithmic framework which utilizes monotonic and submodular multi-objective functions techniques to give the approximate solutions. More recently, [7] proposed the mix integer programming approach for seeking the exact solution of this problem but it only applies to a specific set of sample graphs due to its high complexity. These studies focused on the problem of maximizing the influence of groups with limited budgets, which is different from our problem. Therefore, the existing algorithms cannot readily be applied to our problem.

3   Diffusion Models and Problem Definition

3.1   Independent Cascade Model

We model an OSN as a directed graph $G = (V, E)$ where V is the set of nodes and E is the set of edges with $|V| = n$ and $|E| = m$. Let $N_{in}(v)$ and $N_{out}(v)$ be the set of in-neighbors and out-neighbor of node v, respectively. Given a seed set (initial influenced nodes) S, an information diffusion process happens in the network and hence more nodes can be activated. In this paper we focus on the IC model but our approach can be modified to handle the LT model as well. In the IC model, each edge $e = (u, v) \in E$ has a propagation probability $p(e) \in [0, 1]$ representing the information transmission from a node u to a node v. The diffusion process from S happens in discrete time steps $t = 1, 2, \ldots$, as follows. At round $t = 0$, all nodes in S are *active* and other nodes in $V \backslash S$ are *inactive*. At step $t \geq 1$, for each node u activated at step $t - 1$, it has a single chance to activate each currently inactive node $v \in N_{out}(u)$ with a successful probability $p(e)$. If a node is activated it remains active till the end of the diffusion process. The propagation process ends at step t if there is no new node is activated in this step.

The IC model is equivalent to a *live-edge* model defined as follows. From the graph $G = (V, E)$, we generate a random sample graph g by selecting edge $e \in E$ with probability $p(e)$ and not selecting e with probability $1 - p(e)$. We refer to g as a sample of G and write $g \sim G$. The probability that g is generated from G is $\Pr[g \sim G] = \prod_{e \in E(g)} p(e) \prod_{e \notin E(g)} (1 - p(e))$, where $E(g)$ is the set of edges in the graph g. The influence spread from a set node S to a node u is $\mathbb{I}(S, u) = \sum_{g \sim G} \Pr[g \sim G] \cdot r(S, u)$, where $r(S, u) = 1$ if u is reachable from S in g and $r(S, u) = 0$ otherwise. The influence spread of S in network G is: $\mathbb{I}(S) = \sum_{u \in V} \mathbb{I}(S, u)$.

3.2   Problem Definition

We are given a social network $G = (V, E)$ under the IC model and a collection of K disjoint groups $\mathcal{C} = \{C_1, C_2, \ldots, C_K\}$ (called target groups), where $C_i \subseteq V, C_i \cap C_j = \emptyset$, for every pair of nodes (i, j) with $i \neq j$. Denote by $C(u)$ the group that contains node u. To determine a group is influenced or not, we extend the group influence model in [17] by scoring each node in the group based on the fact that each user has a different role in his/her group. Thus, each node $u \in V$ has a *cost* $c(u)$ and a *score* $b(u)$. We denote $b_{max} = \max_{v \in V} b(v)$ and $b_{min} = \min_{v \in V} b(v)$. The weight $c(u)$ measures the cost or the price of the node u that has to pay if u is chosen as a seed node. The node score $b(u) > 0$ ($b(u)$ is an integer) indicates the role of node u in group $C(u)$. Each group C_i is assigned a threshold f t_i ($t_i > 0$), which reflects the minimum total score that we must reach if we want to influence group C_i. We say that group C_i is influenced iff the total score of influenced nodes in C_i is at least t_i. We define a *cost function* $c : 2^V \rightarrow \mathbb{R}_+$ and a *group influence function* $\sigma : 2^V \rightarrow \mathbb{R}_+$ as follows. For a given seed set $S \subseteq V$, define $c(S) = \sum_{u \in S} c(u)$ is the total cost of S and $\sigma(S)$ is the (expected) number of groups in $\mathcal{C}$ are influenced by the seed set S when the diffusion process ends, that is,

$$\sigma(S) = |\{C_i : \sum_{v \in C_i} \mathbb{I}(S, v) b(v) \geq t_i, C_i \in \mathcal{C}\}| \tag{1}$$

In the special case where each group C_i has only one node, the group influence function $\sigma(\cdot)$ above becomes the influence spread function $\mathbb{I}(\cdot)$ of the IM problem. As a consequence, computing $\sigma(\cdot)$ is #**P**-hard. On the other hand, one can easily verify that the function $\sigma(\cdot)$ is neither submodular nor supermodular. The function $\sigma(\cdot)$ is submodular if for every pair of subsets $A, B \subseteq V$ it holds that $\sigma(A) + \sigma(B) \geq \sigma(A \cup B) + \sigma(A \cap B)$. If the inequality holds in the reversed direction we call $\sigma(\cdot)$ a supermodular function. Due to our group influence process is an extended version of [17], the function $\sigma(\cdot)$ is also neither submodular nor supermodular. We now formally define the *Groups Influence with Minimal cost* (GIM) problem as follows.

Definition 1 (GIM). *An instance of* GIM *is given by* $(G, \mathcal{C})$, *where* $G = (V, E)$ *is a social network under* IC *model, and* $\mathcal{C}$ *is a collection of disjoint target groups* $\{C_1, C_2, \ldots, C_K\}$, $C_i \cap C_j = \emptyset$. *The objective is to find a seed set* $S \subseteq V$ *of a minimum total cost that influences all the groups in* $\mathcal{C}$.

It is not hard to prove the inappximibility of GIM problem, states in Theorem 1, by reducing from the classical Set Cover problem [8].

Theorem 1. GIM *has no polynomial-time algorithm attaining an approximation ratio of* $(1 - \epsilon) \ln n$ *for any* $\epsilon > 0$, *unless* **NP** $\subset$ **DTIME**$(n^{O(\log \log n)})$.

3.3   An Estimator of Group Influence

In this section, we first introduce the concept of *Group Reverse Reachable (GRR)* sample, by extending existing sampling methods in [2,17].

Definition 2 (GRR sample). *Given an instance of* GIM *problem* $(G, \mathcal{C})$, *a GRR sample is generated by the following four steps:*

1. *Randomly select a group* C_i *with probability* $\frac{1}{K}$ *(call* C_i *a source group).*
2. *Generate a sample graph* g *according to the live-edge model under* IC *model.*
3. *For each node* $u \in C_i$, *return a node set* $R_g(u)$ *that is reachable from* u *in* g. *We call* u *is a source node.*
4. *Return a GRR sample* $R_g = \{R_g(u) | \forall u \in C_i\}$, *we also refer to* $C(R_g)$ *as the source group of* R_g *and* $t(R_g)$ *as the threshold of* $C(R_g)$.

Our GRR sample is a nature extended version of the RR sample [10] by combining RR samples with the source node belongs to the source group. Moreover, our GRR sample is also an extended version of Reverse Influence Community (RIC) [17] which uses to estimate the group influence with the sore of each node is equal to 1. The main differences between ours and RIC are: (1) the definition of GRR sample specifically determines whether or not a group is influenced via the total score of influenced nodes but this is not well defined in the RIC even when the score is equal to 1, (2) storing the reachable influence set for each node in GRR sample can help us exploit some important properties that used for analyzing approximation ratio of proposed algorithms in the next sections.

For a set $S \subseteq V$ and a GRR sample R_g, and for $R_g(u) \in R_g$, if $R_g(u) \cap S \neq \emptyset$, we say that S *covers* node u, define:

$$\text{cover-score}(S, R_g(u)) = b(u) \cdot \min\{|S \cap R_g(u)|, 1\} \qquad (2)$$

is *score* of source node u *covered* by S in R_g, we denote following random variable:

$$X_g(S) = \begin{cases} 1, & \text{if } \sum_{R_g(u) \in R_g} \text{cover-score}(S, R_g) \geq t_i \\ 0, & \text{otherwise} \end{cases} \qquad (3)$$

The variable $X_g(S)$ indicates that the total score of nodes that are covered by S is greater than threshold t_i or not? When $X_g(S) = 1$, C_i is influenced by S in sample graph g. We also say that a sample R_g is influenced by S. The probability of generating a sample R_g is: $\Pr[R_g] = \frac{1}{K} \sum_{g \sim G: Re(u,g) = R_g(u), \forall u \in C(R_g)} \Pr[g \sim G]$, where $Re(u, g)$ is the set of nodes that can reach to u in g. We now show that we can estimate the value of $\sigma(S)$ by the expectation of $X_g(S)$, is a key property of GRR sample that helps us devise the algorithms.

Lemma 1. *For any set $S \subseteq V$, we have $\sigma(S, C) = K \cdot \mathbb{E}[X_g(S)]$ where the expectation is taken over the randomness of g.*

Due to the space limitation, we omit some lemmas and proofs. From Lemma 1, we have an estimation of group influence function over a collection of GRR sets $\mathcal{R}$ is: $\hat{\sigma}(S) = \frac{K}{|\mathcal{R}|} \cdot \sum_{R_g \in \mathcal{R}} X_g(S)$.

4   Proposed Algorithm

In this section, we introduce our proposed algorithms for the GIM problem. From the analysis in Sect. 3, we can use $\hat{\sigma}(S)$ to closely estimate $\sigma(S)$ if the number of samples $|\mathcal{R}|$ is sufficiently large. Therefore, instead of solving GIM directly, we find the solution of the following problem:

Definition 3 (Samples Influence with Minimal Cost (SIM) problem). *Given a set of GRR samples $\mathcal{R}$. The problem asks to find a seed set $S \subseteq V$ with minimal total cost so that $\hat{\sigma}(S) = K$, i.e., find $S = \arg\min_{S' \subseteq V: \hat{\sigma}(S) = K} c(S')$.*

The idea behind of our algorithms is that we propose a algorithm for solving the SIM problem and use them as a core in our framework, which creates multiple candidate solutions and select a final solution. We prove the approximation guarantees by utilizing martingale theory [5].

An Approximation Algorithm for SIM. First of all, it is not hard to see that SIM problem also is NP-hard and $\hat{\sigma}_{\mathcal{R}}(\cdot)$ is non submodular and suppermodular. Therefore, similar to GIM, it does not admit a naive greedy algorithm with any approximation ratio. We handle the challenges by introducing a lower bounded function F of $\hat{\sigma}_{\mathcal{R}}(\cdot)$ and exploit its properties to devise an approximation algorithm. Define $f(S, R_g)$ is the total score of all source nodes in R_g which are influenced by set S in sample graph g: $f(S, R_g) = \sum_{u \in C(R_g)} \text{cover-score}(S, R_g(u))$. We can see that $f(S, R_g)$ is a non negative and monotonic set function respect to $S \subseteq V$. Define $g(S, R_g) = \min\{1, f(S, R_g)/t(R_g)\}$, we have following Lemma.

Lemma 2. *For all $S \subseteq T \subseteq V$ and $v \notin T$, $\Delta_v g(S, R_g) \geq \frac{b_{min}}{b_{max}} \cdot \Delta_v g(T, R_g)$*

Denote $\Delta_T f(S, R_g) = f(S \cup T, R_g) - f(S, R_g)$. In order to influence all samples in $\mathcal{R}$, we need to find S such that $g(S, R_g) = 1, \forall R_g \in \mathcal{R}$. Therefore, we find S with a minimal total cost such that $F(S, \mathcal{R}) = \frac{K}{T} \sum_{R_g \in \mathcal{R}} g(S, R_g) = \hat{\sigma}(S) = K$. Since $F(S, \mathcal{R})$ is a linear combination of $g(S, R_g)$, it is easy to show that $\Delta_v F(S, \mathcal{R}) \geq \frac{b_{min}}{b_{max}} \cdot \Delta_v F(T, \mathcal{R})$, for all $S \subseteq T \subseteq V$ and $v \notin V \backslash T$. $F(\cdot)$ is a lower bounded function of $\hat{\sigma}_{\mathcal{R}}(\cdot)$ and they have the same value at set S which can influence all samples in $\mathcal{R}$. We propose Modified Greedy (MoGreedy) algorithm

---

Algorithm 1: MoGreedy$(\mathcal{R}, \mathcal{C})$

Input: A set of GRR samples $\mathcal{R}$, set of groups $\mathcal{C} = \{C_1, C_2, \ldots, C_K\}$

1. $S \leftarrow \emptyset$
2. **while** $F(S, \mathcal{R}) < K$ **do**
3. $\quad \lfloor \quad v_{max} \leftarrow \arg\max_{v \in V \backslash S}(\min\{K, F(S \cup \{v\}, \mathcal{R})\} - F(S, \mathcal{R}))/c(v),$
   $\quad\quad\quad S \leftarrow S \cup \{v_{max}\}$
4. **return** S

---

which utilizes the above characteristic of $F(\cdot, \mathcal{R})$. The pseudocode is presented in Algorithm 1. The general idea is that: we iteratively add a node v into the current solution S, which maximizes the marginal gain per its cost $\Delta_v(S)/c(v) = \min\{K, F(S \cup \{v\}, \mathcal{R})\} - F(S, \mathcal{R})/c(v)$ until the value of $F(S)$ achieves K.

Theorem 2. *Algorithm 1 provides a $\frac{b_{max}}{b_{min}}(1 + \ln(|\mathcal{R}|t_{max}))$-approximation solution for SIM problem.*

Complexity. At each iteration, MoGreedy scans at most n nodes and calculate marginal gain value of F'. Therefore, it takes $O(|S|n)$ time complexity.

Main Proposed Algorithm. We now present Groups Influence Approximation (GIA) algorithm, a $(1 - \epsilon, O(\ln K + \ln \ln n))$-bi criteria approximation algorithm w.h.p for GIM problem. Our algorithm is inspired by the idea of Stop-and-Stare framework for IM problem [14], which devises a stopping condition to check the quality of candidate solutions. Due to the different between GIM and IM, we introduce another stopping condition to check the candidate solutions and establish the number of required samples that ensure the theoretical bounds of the final solution. GIA algorithm operates in multiple iterations and finds

---

Algorithm 2: GIA algorithm

Input: Graph $G = (V, E)$, groups $\mathcal{C} = \{C_1, C_2, \ldots, C_K\}$, $\epsilon, \delta \in (0, 1)$.

1. $N_{max} = (2 + \frac{2}{3}\epsilon)\frac{K}{\epsilon^2}\ln(2\binom{n}{k_{max}}/\delta)$, $N_1 \leftarrow (2 + \frac{2}{3}\epsilon)\frac{1}{\epsilon^2}\ln(1/\delta)$
2. $i_{max} \leftarrow \lceil \log_2(N_{max}/N_1)\rceil$, $\delta_1 \leftarrow \delta/(2i_{max})$
3. Generate a set of N_1 samples $\mathcal{R}_1$
4. **for** $i = 1$ to i_{max} **do**
5. $\quad$ $S_i \leftarrow \mathsf{MoGreedy}(\mathcal{R}_i, \mathcal{C})$ and calculate $F_l(S, \mathcal{R}_i, \epsilon, \delta_1)$ by Lemma 3
6. $\quad$ **if** $F_l(S, \mathcal{R}_i, \epsilon, \delta_1) \geq K - \epsilon K$ or $i = i_{max}$ **then**
7. $\quad\quad |$ **break**
8. $\quad$ **else**
9. $\quad\quad |$ Double size of $\mathcal{R}_i$ by generating $|\mathcal{R}_i|$ samples and adding them into $\mathcal{R}_i$
10. $\quad\quad \lfloor$ $\mathcal{R}_{i+1} \leftarrow \mathcal{R}_i$

11. **return** S

---

a candidate solution at each iteration by leveraging MoGreedy algorithm and checks the quality of these solutions based on static evidences. Denote $k_{max} = \arg\max_{k=1\ldots n}\binom{n}{k}$, the algorithm needs at most $N_{max} = (2 + \frac{2}{3}\epsilon)\frac{K}{\epsilon^2}\ln(2\binom{n}{k_{max}}/\delta)$ samples and operates in at most $i_{max} = \lceil \log_2(N_{max}/N_1)\rceil$ iterations, where $N_1 = (2 + \frac{2}{3}\epsilon)\frac{1}{\epsilon^2}\ln(\frac{n}{\delta})$. We then show that N_{max} is the number of samples required that can ensure the approximation ratio by Theorem 3. At iteration i, the algorithm generates a set of $(2 + \frac{2}{3}\epsilon)\frac{1}{\epsilon^2}\ln(\frac{1}{\delta})2^{i-1}$ samples $\mathcal{R}_i$ and finds a candidate solution S_i by utilizing MoGreedy algorithm (line 5). We devise an stopping condition and check the quality of S_i in line 7. Note that, we do not reuse the stopping condition in [14], which is used in a recent work [17]. Our stopping condition is based on a lower bound of function $F_l(S, \mathcal{R}, \epsilon, \delta)$ of f, defined in Lemma 3. We show that F_l gives a lower bound value of f w.h.p in Lemma 3. The algorithm then checks termination condition in line 6. This condition also helps us prove the approximation ratio more succinctly than Stop and Stare. If the condition is true, the algorithm returns S_i as a final solution. Otherwise, it doubles size of $\mathcal{R}_i$ and moves to the next iteration. The details of the algorithm described in Algorithm 2.

Theoretical Analysis. We now analyze the performance of our algorithm.

Lemma 3. *Given $\epsilon, \delta \in (0, 1)$, for any set $S \subseteq V$ and a set of samples $\mathcal{R}$, denote $c = \ln(1/\delta)$, $T = |\mathcal{R}|$ we have $\Pr[\sigma(S) \geq F_l(S, \mathcal{R}, \epsilon, \delta)] \geq 1 - \delta$, where*

$$F_l(S, \mathcal{R}, \epsilon, \delta) = \min\{\hat{\sigma}_{\mathcal{R}}(S) - \frac{Kc}{3T}, \hat{\sigma}_{\mathcal{R}}(S) + \frac{K}{T}(\frac{2c}{3} - \sqrt{\frac{4c^2}{9} + 2Tc\frac{\hat{\sigma}_{\mathcal{R}}(S)}{K}})\}.$$

Lemma 4. *For any set of GRR samples $\mathcal{R}$, we have:* $\hat{\sigma}_{\mathcal{R}}(S^*) = K$.

Theorem 3 (Approximation ratio). *For any input parameters $\epsilon, \delta \in (0,1)$, GIA algorithm returns a solution S satisfying* $\Pr[\sigma(S) \geq K - \epsilon K] \geq 1 - \delta$ *and* $c(S) \leq \frac{b_{max}}{b_{min}} \left(1 + \ln\left((2 + \frac{2}{3}\epsilon)\epsilon^{-2}\right) + \ln K + \ln(nt_{max}\ln(n/\delta))\right)$ OPT.

Theorem 4 (Complexity). GIA *algorithm has* $O\left((n\ln n + \ln(\frac{1}{\delta})\epsilon^{-2})|C|\eta + n^2)\log n\right)$ *time complexity, where $\rho = |\bigcup_i C_i|$ and η is the expectation of influence spread of a node.*

5 Experiments

In this section, we conduct some experiments illustrating the performance of our GIA algorithm as compared with the current state-of-the-art algorithms on three metrics: the total cost of seed nodes, the time efficiency, and the value of group influence function on various network datasets.

5.1 Experimental Settings

Dataset. We use three networks in recent work [17]: Facebook with 747 nodes and 60.05K edges, Wiki with 7.1K nodes and 103.6K edges and Epinions with 76K nodes and 508.8K edges.

Algorithms Compared. To our knowledge, there is no existing algorithm can be adopted to solve the GIM problem directly. Therefore, we compare our GIA and EGI algorithms with the state-of-the-art algorithms for the closest problem: Influence Maximization at Community level (IMC) [17]. Also, we adapt High Degree, a common baseline algorithm for related problem on information diffusion [3,4,10]. These algorithms are described in detail as follows. **UBG (Upper Bound Greedy)** [17]: This the best performance algorithm for the Influence Maximization at Community level (IMC) problem, which finds a set seed of k nodes that can influence largest the number of groups while GIM problem requires to find the set of nodes with minimal cost that can influence all target groups. Therefore, we adapt UBG algorithm with some modifications as follows. We first initialize an empty candidate solution S. We then sequentially use UBG with k from 1 to n to find the best influence node then add it into S until the objective is at least K. **MAF** [17]: This is also an algorithm for the IMC problem. We also modify this algorithm as for UBG to adapt for the GIM problem. **High Degree (HD):** We repeatedly select a node with highest degree until the current solution influences all target groups. For all the above algorithms, we use the Monte-Carlo method in [6] to obtain an (ϵ, δ)-approximation for estimating influence group function. For each algorithm, we run 10 times to get the average results.

Parameters Setting. All experiments are under the IC model with edge probabilities set to $p(u,v) = 1/|N_{in}(v)|$ as in prior works [10,14,17]. We set parameters $\epsilon = 0.1$ and $\delta = 1/n$ as the default setting. For the purpose of comparing

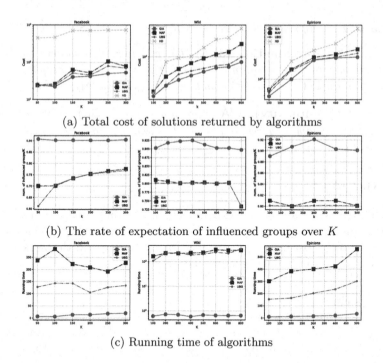

(a) Total cost of solutions returned by algorithms

(b) The rate of expectation of influenced groups over K

(c) Running time of algorithms

Fig. 1. Performance of algorithms

our algorithm with current algorithms for the group influence problem, we set $s(u) = 1, \forall u \in U$, and the thresholds $t_i = \sum_{u \in C_i} s(u)/2$ for $i = 1 \ldots K$ according to the setting in [17]. Each node has its cost calculated under Normalized Linear model with the support $(0, 1]$ according to recent works [11, 16].

5.2 Experiment Results

We first compare the quality of algorithms, measured by the total cost of seed set (Fig. 1(a)). In general, GIA always provides the best solutions and outperforms other algorithms by a considerable gap. The total cost of solutions obtained by our algorithm is up to 2.12 and 1.6 times lower than those of MAF and UBG, respectively. We further report the ratio of number of influenced groups over K of algorithms in Fig. 1(b). It can be seen that GIA can output solutions with the group influence that is above $(1 - \epsilon)K$ and outperforms MAF and UBG. These results show that the proposed algorithm is more efficient than the other algorithms. They do not only select the smaller-cost set of nodes but also ensure that the number of influenced group is above $(1 - \epsilon)K$. Figure 1(c) shows the running time of algorithms. We do not report the running time of HD because it a simple heuristic algorithm and can finish within one second. GIA is several times faster than MAF and UBG. This is because the mechanisms of MAF and UBG consist of many iterations to find the seed set that can reach to the terminal condition. In contrast, GIA follows the mechanism of our framework which can finds the final solution after a few loops.

6 Conclusion

In this paper, we studied the GIM problem, arising from the goal of reaping the benefits of influencing user groups on social networks under a more realistic scenario. Solving the problem is challenging because of its hardness and inapproxibility and the properties of group influence function. We developed a bi-criteria approximation algorithm, called GIA, to solve GIM. The experiment results demonstrate that our algorithm outperforms the state-of-the-art ones both on the cost and on the running time.

Acknowledgements. This paper is partially supported by Vietnam National Foundation for Science and Technology Development (NAFOSTED) under Grant No. 102.01-2020.21.

References

1. Barbieri, N., Bonchi, F., Manco, G.: Topic-aware social influence propagation models. Knowl. Inf. Syst. **37**(3), 555–584 (2013)
2. Borodin, A., Filmus, Y., Oren, J.: Threshold models for competitive influence in social networks. In: Saberi, A. (ed.) WINE 2010. LNCS, vol. 6484, pp. 539–550. Springer, Heidelberg (2010). https://doi.org/10.1007/978-3-642-17572-5_48
3. Chen, W., Lakshmanan, L.V.S., Castillo, C.: Information and Influence Propagation in Social Networks. Synthesis Lectures on Data Management, Morgan & Claypool Publishers, San Rafael (2013)
4. Chen, W., Yuan, Y., Zhang, L.: Scalable influence maximization in social networks under the linear threshold model. In: ICDM 2010, The 10th IEEE International Conference on Data Mining, pp. 88–97 (2010)
5. Chung, F.R.K., Lu, L.: Survey: concentration inequalities and martingale inequalities: a survey. Internet Math. **3**(1), 79–127 (2006)
6. Dagum, P., Karp, R.M., Luby, M., Ross, S.M.: An optimal algorithm for Monte Carlo estimation. SIAM J. Comput. **29**(5), 1484–1496 (2000)
7. Farnadi, G., Babaki, B., Gendreau, M.: A unifying framework for fairness-aware influence maximization. In: Companion of The 2020 Web Conference 2020, Taipei, Taiwan, 20–24 April 2020, pp. 714–722 (2020)
8. Feige, U.: A threshold of ln n for approximating set cover. J. ACM **45**(4), 634–652 (1998)
9. Goyal, A., Bonchi, F., Lakshmanan, L.V.S., Venkatasubramanian, S.: On minimizing budget and time in influence propagation over social networks. Soc. Netw. Anal. Min. **3**(2), 179–192 (2013)
10. Kempe, D., Kleinberg, J.M., Tardos, É.: Maximizing the spread of influence through a social network. In: Proceedings of the Ninth ACM SIGKDD International Conference on Knowledge Discovery and Data Mining, pp. 137–146 (2003)
11. Li, X., Smith, J.D., Dinh, T.N., Thai, M.T.: TipTop: (almost) exact solutions for influence maximization in billion-scale networks. IEEE/ACM Trans. Netw. **27**(2), 649–661 (2019)

12. Li, Y., Zhang, D., Tan, K.: Targeted influence maximization for online advertisements. PVLDB **8**(10), 1070–1081 (2015)
13. Nguyen, H.T., Cano, A., Vu, T., Dinh, T.N.: Blocking self-avoiding walks stops cyber-epidemics: a scalable GPU-based approach. IEEE Trans. Knowl. Data Eng. **32**(7), 1263–1275 (2020)
14. Nguyen, H.T., Thai, M.T., Dinh, T.N.: Stop-and-stare: optimal sampling algorithms for viral marketing in billion-scale networks. In: International Conference on Management of Data, SIGMOD, pp. 695–710 (2016)
15. Nguyen, H.T., Thai, M.T., Dinh, T.N.: A billion-scale approximation algorithm for maximizing benefit in viral marketing. IEEE/ACM Trans. Netw. **25**(4), 2419–2429 (2017)
16. Nguyen, H., Zheng, R.: On budgeted influence maximization in social networks. IEEE J. Sel. Areas Commun. **31**(6), 1084–1094 (2013)
17. Nguyen, L.N., Zhou, K., Thai, M.T.: Influence maximization at community level: a new challenge with non-submodularity. In: 39th IEEE International Conference on Distributed Computing Systems, ICDCS 2019, pp. 327–337 (2019)
18. Pham, C.V., Duong, H.V., Bui, B.Q., Thai, M.T.: Budgeted competitive influence maximization on online social networks. In: Chen, X., Sen, A., Li, W.W., Thai, M.T. (eds.) CSoNet 2018. LNCS, vol. 11280, pp. 13–24. Springer, Cham (2018). https://doi.org/10.1007/978-3-030-04648-4_2
19. Pham, C.V., Pham, D.V., Bui, B.Q., Nguyen, A.V.: Minimum budget for misinformation detection in online social networks with provable guarantees. Optim. Lett. 1–30 (2021)
20. Pham, C.V., Phu, Q.V., Hoang, H.X., Pei, J., Thai, M.T.: Minimum budget for misinformation blocking in online social networks. J. Comb. Optim. **38**(4), 1101–1127 (2019)
21. Tang, J., Tang, X., Xiao, X., Yuan, J.: Online processing algorithms for influence maximization. In: Proceedings of the 2018 International Conference on Management of Data, pp. 991–1005. Association for Computing Machinery (2018)
22. Tang, Y., Shi, Y., Xiao, X.: Influence maximization in near-linear time: a martingale approach. In: Proceedings of the 2015 ACM SIGMOD International Conference on Management of Data, pp. 1539–1554 (2015)
23. Tsang, A., Wilder, B., Rice, E., Tambe, M., Zick, Y.: Group-fairness in influence maximization. In: Proceedings of the Twenty-Eighth International Joint Conference on Artificial Intelligence, pp. 5997–6005 (2019)
24. Ye, M., Liu, X., Lee, W.: Exploring social influence for recommendation: a generative model approach. In: International ACM Conference on research and development in Information Retrieval, SIGIR 2012, pp. 671–680 (2012)
25. Zhang, H., Alim, M.A., Li, X., Thai, M.T., Nguyen, H.T.: Misinformation in online social networks: detect them all with a limited budget. ACM Trans. Inf. Syst. **34**(3), 18:1–18:24 (2016)
26. Zhang, H., Zhang, H., Li, X., Thai, M.T.: Limiting the spread of misinformation while effectively raising awareness in social networks. In: Thai, M.T., Nguyen, N.P., Shen, H. (eds.) CSoNet 2015. LNCS, vol. 9197, pp. 35–47. Springer, Cham (2015). https://doi.org/10.1007/978-3-319-21786-4_4
27. Zhu, J., Ghosh, S., Wu, W.: Group influence maximization problem in social networks. IEEE Trans. Comput. Soc. Syst. **6**(6), 1156–1164 (2019)
28. Zhu, J., Ghosh, S., Wu, W., Gao, C.: Profit maximization under group influence model in social networks. In: Tagarelli, A., Tong, H. (eds.) CSoNet 2019. LNCS, vol. 11917, pp. 108–119. Springer, Cham (2019). https://doi.org/10.1007/978-3-030-34980-6_13

Recovering Communities in Temporal Networks Using Persistent Edges

Konstantin Avrachenkov[1(✉)], Maximilien Dreveton[1], and Lasse Leskelä[2]

[1] Inria Sophia Antipolis, 2004 Route des Lucioles, 06902 Valbonne, France
{k.avrachenkov,maximilien.dreveton}@inria.fr
[2] Department of Mathematics and Systems Analysis, Aalto University,
Otakaari 1, 02150 Espoo, Finland
lasse.leskela@aalto.fi

Abstract. This article studies the recovery of static communities in a temporal network. We introduce a temporal stochastic block model where dynamic interaction patterns between node pairs follow a Markov chain. We render this model versatile by adding degree correction parameters, describing the tendency of each node to start new interactions. We show that in some cases the likelihood of this model is approximated by the regularized modularity of a time-aggregated graph. This time-aggregated graph involves a trade-off between new edges and persistent edges. A continuous relaxation reduces the regularized modularity maximization to a normalized spectral clustering. We illustrate by numerical experiments the importance of edge persistence, both on simulated and real data sets.

Keywords: Graph clustering · Temporal networks · Spectral methods · Stochastic block model

1 Introduction

Complex networks are commonly used to describe and analyze interactions between entities. A natural problem arising consists in identifying meaningful structures within the complex system. Community recovery, i.e., partitioning the set of nodes of a network into *communities* based on some common properties of the vertices, is now a well established area [6].

In many situations, interactions between node pairs vary over time, and classical graph-based models are replaced by temporal networks models [10]. Examples include communication, interaction, and transportation networks. The longitudinal dimension of data raises new challenges to traditional clustering algorithms. Previous research has focused on evolving communities [22], for example by generalizing belief-propagation methods [8], developing variational EM algorithms [15] and introducing new spectral methods [5,13] or modularity-based

This work has been done within the project of Inria - Nokia Bell Labs "Distributed Learning and Control for Network Analysis" and was partially supported by COSTNET Cost Action CA15109.

© Springer Nature Switzerland AG 2021
D. Mohaisen and R. Jin (Eds.): CSoNet 2021, LNCS 13116, pp. 243–254, 2021.
https://doi.org/10.1007/978-3-030-91434-9_22

methods [16,20]. Nevertheless, all of the aforementioned works focus on evolving communities for which the interactions between nodes are re-sampled at every time step. One can then treat each layer independently by applying static community detection and smoothing the community predictions. When the communities are non-evolving, the extra longitudinal dimension brings new information, and each additional snapshot makes the clustering easier. The recovery bounds established in [1] highlight this fact. Nonetheless, simple temporal aggregation of the data might lose important features such as temporal patterns. As such, [1] proposes an online algorithm tailored for Markov edge evolution dynamics, while [2] studies a spectral algorithm using the squared adjacency matrix.

In this work, we introduce a temporal extension of the degree-corrected stochastic block model [9,11], in which the community labeling is fixed and the interactions between node pairs follow a Markov evolution which only depends on the community labeling and on the degree correction parameters. To the best of our knowledge, we are the first to introduce degree-corrected parameters into temporal network models with edge persistency. We show that the maximum likelihood inference reduces to the maximization of the regularized modularity of a time-aggregated graph, in the limit of a large number of snapshots and sparse interactions. This graph is not simply the sum of the adjacency matrices over all snapshots. Instead, it involves a trade-off between the newly formed edges and the persistent ones, and this trade-off depends on the difference between the edge-persistence between intra-community and inter-community node pairs. A continuous relaxation then leads to a normalized spectral clustering algorithm. Finally, we validate the importance of taking into account the persistent edges in simulated and real data sets.

Notations. Throughout this article, matrices are represented by capital letters (A, W, etc.), and the corresponding matrix elements by A_{ij}, W_{ij}, etc. Tr A denotes the trace of a square matrix A, and $A \odot B$ denotes the entrywise product of two matrices. Finally, 1_n is the n-by-1 vector of all ones, and the indicator of an event B is denoted by $1(B)$. The Kronecker delta is denoted by $\delta(x, y)$, so that $\delta(x, y) = 1(x = y)$.

By a slight abuse of notations, graphs are represented by their adjacency matrix A. We will assume that graphs are undirected but potentially weighted, hence A is symmetric, with non-negative entries. Clustering a graph with N nodes into K clusters accounts to assigning to each node i a label $Z_i \in [K]$.

2   Degree-Corrected Temporal Network Model with Markov Edge Dynamics

Consider a population of N nodes partitioned into K static communities such that node i belongs to community $Z_i \in [K]$. We write $A_{ij}^t = 1$ if nodes i and j interact at time t, and $A_{ij}^t = 0$ otherwise. We investigate methods of recovering the community structure $Z = (Z_1, \ldots, Z_N)$ from an observed adjacency tensor $A = (A_{ij}^t)$. The following section describes a versatile statistical model for this setting.

2.1   Model Description

A degree-corrected temporal stochastic block model with N nodes, K blocks and T snapshots is a probability distribution

$$\mathbb{P}(A \,|\, Z, F, \theta) \;=\; \prod_{1 \le i < j \le N} F_{Z_i Z_j}^{\theta_i \theta_j}\left(A_{ij}^1, \ldots, A_{ij}^T\right) \tag{1}$$

of a symmetric adjacency tensor $A \in \{0,1\}^{N \times N \times T}$ with zero diagonal entries, where $Z = (Z_1, \ldots, Z_N)$ is a community assignment with $Z_i \in \{1, \ldots, K\}$ indicating the community of node i, $F = (F_{k\ell}^{xy})$ is a collection of probability distributions over $\{0,1\}^T$, and $\theta = (\theta_1, \ldots, \theta_N)$ is a vector of node-specific degree correction parameters, with $0 \le \theta_i < \infty$.

In the following, we will restrict ourselves to homogeneous inter-block interactions with Markov edge dynamics, for which the nodes' static community labellings are sampled uniformly at random from the set $[K]$ of all node labellings, and

$$F_{Z_i Z_j}^{\theta_i \theta_j}(x) \;=\; \begin{cases} \mu_{x_1}^{\theta_i \theta_j} \prod_{t=2}^{T} P_{x_{t-1}, x_t}^{\theta_i \theta_j} & \text{if } Z_i = Z_j, \\ \nu_{x_1}^{\theta_i \theta_j} \prod_{t=2}^{T} Q_{x_{t-1}, x_t}^{\theta_i \theta_j} & \text{otherwise,} \end{cases} \tag{2}$$

with initial distributions

$$\mu^{\theta_i \theta_j} = \begin{pmatrix} 1 - \theta_i \theta_j \mu_1 \\ \theta_i \theta_j \mu_1 \end{pmatrix} \quad \text{and} \quad \nu^{\theta_i \theta_j} = \begin{pmatrix} 1 - \theta_i \theta_j \nu_1 \\ \theta_i \theta_j \nu_1 \end{pmatrix}, \tag{3}$$

and transition probability matrices

$$P^{\theta_i \theta_j} = \begin{pmatrix} 1 - \theta_i \theta_j P_{01} & \theta_i \theta_j P_{01} \\ 1 - P_{11} & P_{11} \end{pmatrix} \quad \text{and} \quad Q^{\theta_i \theta_j} = \begin{pmatrix} 1 - \theta_i \theta_j Q_{01} & \theta_i \theta_j Q_{01} \\ 1 - Q_{11} & Q_{11} \end{pmatrix}. \tag{4}$$

The parameters θ_i account for the fact that some nodes might be more inclined than others to start new connections, similarly to the degree-corrected block model of [11]. To keep the model simple, we do not add degree correction parameters in front of P_{11}; hence once a connection started, the probability to keep it active is simply P_{11} or Q_{11}. Moreover, we assume that $\min_{i,j}\{\theta_i \theta_j \delta\} \le 1$, where $\delta = \max\{\mu_1, \nu_1, P_{01}, Q_{01}\}$. Finally, we normalise the degree correction parameters so that $\sum_i 1(Z_i = k)\theta_i = \sum_i 1(Z_i = k)$ for all k.

2.2   Maximum Likelihood Estimator

Proposition 1. *A maximum likelihood estimator for the Markov block model defined by (1)–(2) is any community assignment $Z \in [K]^N$ that maximizes*

$$\sum_{i,j} \delta(Z_i, Z_j) \left\{ A_{ij}^1 \left(\rho_1^{\theta_i \theta_j} - \rho_0^{\theta_i \theta_j} \right) + \rho_0^{\theta_i \theta_j} + \left(A_{ij}^1 - A_{ij}^T \right) \ell_{10}^{\theta_i \theta_j} \right\}$$

$$+ \sum_{i,j} \delta(Z_i, Z_j) \sum_{t=2}^{T} \left\{ \left(\ell_{01}^{\theta_i \theta_j} + \ell_{10}^{\theta_i \theta_j} \right) \left(A_{ij}^t - A_{ij}^{t-1} A_{ij}^t \right) + \ell_{11}^{\theta_i \theta_j} A_{ij}^{t-1} A_{ij}^t - \log \frac{Q_{00}^{\theta_i \theta_j}}{P_{00}^{\theta_i \theta_j}} \right\}$$

where $\rho_a^{\theta_i \theta_j} = \log \frac{\mu_a^{\theta_i \theta_j}}{\nu_a^{\theta_i \theta_j}}$ *and* $\ell_{ab}^{\theta_i \theta_j} = \log \frac{P_{ab}^{\theta_i \theta_j}}{Q_{ab}^{\theta_i \theta_j}} - \log \frac{P_{00}^{\theta_i \theta_j}}{Q_{00}^{\theta_i \theta_j}}.$

The proof of Proposition 1 is presented in Appendix A.1. The MLE derived in Proposition 1 is more complex that summing all snapshots independently. In particular, the terms $A_{ij}^{t-1} A_{ij}^t$ account for *persistent edges* over two consecutive snapshots. Denote by $A_{\text{pers}}^t = A^{t-1} \odot A^t$ the entrywise product of adjacency matrices A^{t-1} and A^t. Then A_{pers}^t is the adjacency matrix of the graph containing the persistent edges between $t-1$ and t, and $A_{\text{new}}^t = A^t - A_{\text{pers}}^t$ corresponds to the graph containing the edges freshly appearing at time t.

Assuming that the number of snapshots T is large, we can ignore the boundary terms, and the MLE expressed in Proposition 1 reduces to maximizing

$$\sum_{t=2}^T \sum_{ij: Z_i = Z_j} \left(\left(\ell_{01}^{\theta_i \theta_j} + \ell_{10}^{\theta_i \theta_j} \right) \left(A_{ij}^t - A_{ij}^{t-1} A_{ij}^t \right) + \ell_{11}^{\theta_i \theta_j} A_{ij}^{t-1} A_{ij}^t - \log \frac{Q_{00}^{\theta_i \theta_j}}{P_{00}^{\theta_i \theta_j}} \right).$$

By utilising (3)–(4), we can further simplify it to express this as a modularity. Recall given a weighted graph W, a partition Z and a resolution parameter γ, the regularized modularity is defined as [19, 21]

$$\mathcal{M}(W, Z, \gamma) = \sum_{i,j} \delta(Z_i, Z_j) \left(W_{ij} - \gamma \frac{d_i d_j}{2m} \right)$$

where $d_i = \sum_j W_{ij}$ and $m = \sum_i d_i$. Hence, suppose that $P^{\theta_i \theta_j}$ and $Q^{\theta_i \theta_j}$ are nondegenerate, and $\mu^{\theta_i \theta_j}$ (resp. $\nu^{\theta_i \theta_j}$) is the stationary distribution of $P^{\theta_i \theta_j}$ (resp. $Q^{\theta_i \theta_j}$). In a sparse setting, P_{01} and Q_{01} are small, and after a Taylor expansion (see Appendix A.2 for the full derivations) the previous expression is approximately equal to $\mathcal{M}(W, Z, \gamma)$, where W is defined by

$$W = \sum_{t=2}^T \left(\alpha A_{\text{new}}^t + \beta A_{\text{pers}}^t \right) \tag{5}$$

with

$$\alpha = \log \frac{P_{01}}{Q_{01}} + \log \frac{1 - P_{11}}{1 - Q_{11}} \quad \text{and} \quad \beta = \log \frac{P_{11}}{Q_{11}}, \tag{6}$$

and $\gamma = (P_{01} - Q_{01}) \frac{\alpha(\mu_1 + (K-1)\nu_1) + (\beta - \alpha)(\mu_1 P_{11} + (K-1)\nu_1 Q_{11})}{K}$.

Comparison with Previous Work. Correspondence between maximum likelihood estimator and modularity maximization are long known in static block models [18]. Analogously to the single-layer case, the modularity of a temporal network, with possibly time-dependent community structure, was previously defined in [16, 20] by

$$\sum_{t=1}^T \mathcal{M}(A^t, Z^t, \gamma_t) + \sum_{t=1}^T \sum_{s \neq t} \sum_i \omega_i^{st} \delta \left(Z_i^s, Z_i^t \right) \tag{7}$$

where γ_t is the resolution parameter for layer t, Z_i^t is the community membership of node i at time step t, and w_i^{st} denotes a coupling between time instants s and t. For a static community structure, the second term in (7) is irrelevant. When the resolution is constant over time, the relevant term in (7) can be written as

$$\sum_{t=1}^{T} \mathcal{M}(A^t, Z, \gamma) = \mathcal{M}(A^{\mathrm{agg}}, Z, \gamma),$$

where $A^{\mathrm{agg}} = \sum_{t=1}^{T} A^t$ is the weighted adjacency matrix of the time-aggregated data. In contrast, the matrix W in (5) involves a trade-off between new edges and persistent edges. We notice that $W = A^{\mathrm{agg}}$ only if $\alpha = \beta = 1$.

2.3 Temporal Spectral Clustering Combining New and Persistent Edges

Following our analysis in Sect. 2.2, the community prediction should verify

$$\hat{Z} = \underset{Z \in [K]^N}{\arg\max} \mathcal{M}(W, Z, \gamma)$$

where W is defined in Eq. (5) and γ is a proper resolution parameter. This optimisation problem is NP-complete in general [4], but can be approximately solved by continuous relaxation. We can choose the relaxation so that the optimization problem reduces to normalized spectral clustering algorithm on the weighted graph W (we refer to [17] and to the Appendix for the full computations). We note that in order to compute the normalized Laplacian of W, we should restrict $\alpha, \beta \geq 0$, which is not necessarily guaranteed by Formula (6). We summarize this in Algorithm 1.

Algorithm 1: Spectral clustering for temporal networks with Markov edge dynamics and static node labeling.

Input: Adjacency matrices $A^1, \ldots, A^T$, number of clusters K, parameter α, β.
Output: Predicted membership matrix $\hat{Z} \in \hat{Z}_{N,K}$

1   **Process:**
     – Let $W = \sum_{t=2}^{T} \alpha A_{\mathrm{new}}^t + \beta A_{\mathrm{pers}}^t$ where $A_{\mathrm{new}}^t = A^t - A^{t-1} \odot A^t$ and $A_{\mathrm{pers}}^t = A^{t-1} \odot A^t$;
     – Compute $\mathcal{L} = I_n - D^{-1/2} W D^{-1/2}$ where $D = \mathrm{diag}(W 1_n)$;
     – Compute $\hat{X} \in \mathbb{R}^{N \times K}$ whose columns consist of the K orthonormal eigenvectors of $\mathcal{L}$ associated to the K smallest eigenvalues.

2   **Return** $z \leftarrow \mathrm{kmeans}\left(D^{-1/2} \hat{X}, K\right)$.

---

3   Numerical Experiments

The Python source code for reproducing our results is available online[1].

3.1   Synthetic Data

Effect of Persistent Edges. We first examine the effect of the choice of the parameters α and β in Algorithm 1. For this, we let $\alpha = 1$ and we plot in Fig. 1 the averaged accuracy obtained on 25 realizations of stochastic block models with Markov edge dynamics for various β. While spectral clustering on the time-aggregated graph (corresponding to $\beta = 1$) works well, it is striking to notice that other values of β give better results. The choice of β depends on the probabilities of persistent interactions. For example, if $P_{11} > Q_{11}$ (Fig. 1a), then $\beta > 1$ are preferred, while if $P_{11} < Q_{11}$ (Fig. 1b) large choice of β are penalized. This is in accordance to the values of α, β derived in Formula (6) (albeit in Formula (6), α and β could be negative).

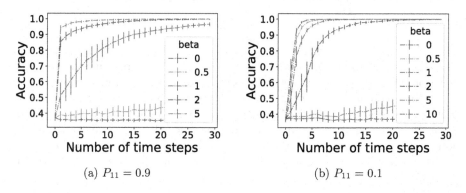

(a) $P_{11} = 0.9$ (b) $P_{11} = 0.1$

Fig. 1. Accuracy of Algorithm 1 on a SBM with 300 nodes in $K = 3$ blocks, degree correction parameters $\theta_1 = \cdots = \theta_n = 1$, and a stationary Markov edge evolution $\mu_1 = 0.04$, $\nu_1 = 0.02$ and $Q_{11} = 0.3$. The results are averaged over 25 synthetic graphs, and error bars show the standard deviation.

Effect of Degree Correction Parameters. We show the robustness of Algorithm 1 on the degree correction parameters in Fig. 2. More precisely:

- Figure 2a generates θ_i according to $|\mathcal{N}(0, \sigma^2)| + 1 - \sigma\sqrt{2/\pi}$ where $|\mathcal{N}(0, \sigma^2)|$ denotes the absolute value of a normal random variable with mean 0 and variance σ^2. We choose $\sigma = 0.25$.
- Figure 2b generates the θ_i from a Pareto distribution with density function $f(x) = \frac{am^a}{x^{a+1}}\mathbb{1}(x \geq m)$ with $a = 3$ and $m = 2/3$ (chosen such that $\mathbb{E}\theta_i = 1$).

Note that the sampling of the θ_i's enforces $\mathbb{E}\theta_i = 1$ in both settings. We notice that in both cases, letting $\beta \neq 1$ improves the performance of Algorithm 1.

---

[1] https://github.com/mdreveton/Spectral-clustering-with-persistent-edges.

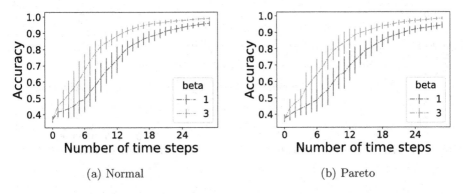

(a) Normal   (b) Pareto

Fig. 2. Accuracy of Algorithm 1 with $\alpha = 1$ and different β, on a SBM with 300 nodes and $K = 3$ blocks (with uniform prior), and a stationary Markov edge evolution $\mu_1 = 0.06$, $\nu_1 = 0.03$, $P_{11} = 0.7$ and $Q_{11} = 0.4$, for different generation of the degree correction parameters θ. The results are averaged over 25 synthetic graphs, and error bars show the standard deviation.

3.2 Social Networks of High School Students

We investigate three data sets collected during three consecutive years from a high school Lyceé Thiers in Marseilles, France [7,14]. Nodes correspond to students, interactions to close-proximity encounters, and communities to classes, with dimensions given in Table 1.

Table 1. Dimensions of three data sets of interacting high school students.

Year	N	K	T
2011	118	3	5609
2012	180	5	11273
2013	327	9	7375

We make a hypothesis that the temporal characteristics of the interactions are similar each year. We then use the 2011 data set to estimate the transition probability matrices P and Q, and use these for clustering the 2012 and 2013 data sets. We assume that $\theta_i = 1$ (no degree correction). A standard estimator of Markov chain transition probability matrices [3] gives

$$\widehat{P} = \begin{pmatrix} 0.9992 & 0.0008 \\ 0.37 & 0.63 \end{pmatrix} \quad \text{and} \quad \widehat{Q} = \begin{pmatrix} 0.999967 & 3.3 \times 10^{-5} \\ 0.48 & 0.52 \end{pmatrix}.$$

Using (6), leads to $\hat{\alpha} = 2.9$ and $\hat{\beta} = 0.18$. We observe in Fig. 3b that this choice of parameters gives a better accuracy on the 2013 data set than simply applying spectral clustering on the time-aggregated graph ($\alpha = \beta = 1$). For the 2012 data set (Fig. 3a), this improvement is not so clearly visible.

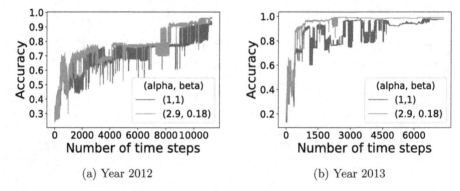

(a) Year 2012   (b) Year 2013

Fig. 3. Accuracy of Algorithm 1 on the 2012 and 2013 high school datasets, using uniform $\alpha = \beta = 1$ (blue) and adjusted α, β predicted using 2011 data (orange). (Color figure online)

To understand why Algorithm 1 performs better for 2013 than for 2012, we have listed in Table 2 temporal transition probabilities and clustering weights $\hat{\alpha}, \hat{\beta}$ estimated separately for each dataset. For year 2012, the difference between intra-community edge persistence $\hat{P}_{11}$ and inter-community edge persistence $\hat{Q}_{11}$ is small, implying that persistent edges do not add much extra information for distinguishing communities ($\hat{\beta} \approx 0$). For years 2011 and 2013, this difference is larger, manifesting that edge persistence contains information that can be employed to recover communities with a higher accuracy.

Table 2. Markov chain transition probabilities and adjusted clustering weights estimated separately for each dataset.

Dataset	$\hat{P}_{01}$	$\hat{Q}_{01}$	$\hat{P}_{11}$	$\hat{Q}_{11}$	$\hat{\alpha}$	$\hat{\beta}$	$\hat{\beta}/\hat{\alpha}$
2011	0.00080	0.000033	0.63	0.52	2.9	0.58	0.060
2012	0.00050	0.000011	0.57	0.56	3.8	0.01	0.003
2013	0.00150	0.000014	0.64	0.40	4.5	0.07	0.015

A   Proofs of Main Statements

A.1   Maximum Likelihood Computations (Proposition 1)

Proof (Proof of Proposition 1). By the temporal Markov property, the log-likelihood of the model can be written as $\log \mathbb{P}(A \mid Z, \theta) = \log \mathbb{P}(A^1 \mid Z, \theta) + \sum_{t=2}^{T} \mathbb{P}(A^t \mid A^{t-1}, Z, \theta)$. By denoting $\rho_a^{\theta_i \theta_j} = \log \frac{\mu_a^{\theta_i \theta_j}}{\nu_a^{\theta_i \theta_j}}$, we find that

$$\log \mathbb{P}(A^1 \mid Z, \theta) = \frac{1}{2} \sum_{i,j} \sum_a \delta(A_{ij}^1, a)\Big(\delta(Z_i, Z_j)\rho_a^{\theta_i\theta_j} + \log \nu_a^{\theta_i\theta_j}\Big)$$

$$= \frac{1}{2} \sum_{i,j} \delta(Z_i, Z_j) \sum_a \delta(A_{ij}^1, a)\rho_a^{\theta_i\theta_j} + c_1(A),$$

where $c_1(A) = \frac{1}{2}\sum_{i,j}\sum_a \delta(A_{ij}^1, a)\log \nu_a^{\theta_i\theta_j}$ does not depend on the community structure. Similarly, by denoting $R_{ab}^{\theta_i\theta_j} = \log \frac{P_{ab}^{\theta_i\theta_j}}{Q_{ab}^{\theta_i\theta_j}}$ we find that

$$\log \mathbb{P}(A^t \mid A^{t-1}, Z, \theta) = \frac{1}{2} \sum_{i,j} \sum_{a,b} \delta(A_{ij}^{t-1}, a)\delta(A_{ij}^t, b)\Big(\delta(Z_i, Z_j)R_{ab}^{\theta_i\theta_j} + \log Q_{ab}^{\theta_i\theta_j}\Big)$$

$$= \frac{1}{2} \sum_{i,j} \delta(Z_i, Z_j) \sum_{a,b} \delta(A_{ij}^{t-1}, a)\delta(A_{ij}^t, b)R_{ab}^{\theta_i\theta_j} + c_t(A),$$

where $c_t(A) = \frac{1}{2}\sum_{i,j}\sum_{a,b}\delta(A_{ij}^{t-1}, a)\delta(A_{ij}^t, b)\log Q_{ab}^{\theta_i\theta_j}$ does not depend on the community structure. Simple computations show that

$$\sum_a \delta(A_{ij}^1, a)\rho_a^{\theta_i\theta_j} = A_{ij}^1(\rho_1^{\theta_i\theta_j} - \rho_0^{\theta_i\theta_j}) + \rho_0^{\theta_i\theta_j}$$

and

$$\sum_{a,b} \delta(A_{ij}^{t-1}, a)\delta(A_{ij}^t, b)R_{ab}^{\theta_i\theta_j} = R_{00}^{\theta_i\theta_j} + A_{ij}^{t-1}(R_{10}^{\theta_i\theta_j} - R_{00}^{\theta_i\theta_j}) + A_{ij}^t(R_{01}^{\theta_i\theta_j} - R_{00}^{\theta_i\theta_j})$$

$$+ A_{ij}^{t-1}A_{ij}^t(R_{11}^{\theta_i\theta_j} - R_{01}^{\theta_i\theta_j} - R_{10}^{\theta_i\theta_j} + R_{00}^{\theta_i\theta_j})$$

$$= R_{00}^{\theta_i\theta_j} + A_{ij}^{t-1}\ell_{10}^{\theta_i\theta_j} + A_{ij}^t\ell_{01}^{\theta_i\theta_j} + A_{ij}^{t-1}A_{ij}^t(\ell_{11}^{\theta_i\theta_j} - \ell_{01}^{\theta_i\theta_j} - \ell_{10}^{\theta_i\theta_j}).$$

By collecting the above observations, we now find that $\log \mathbb{P}(A \mid Z, \theta)$ equals

$$c(A) + \frac{1}{2} \sum_{i,j} \delta(Z_i, Z_j)\left\{ A_{ij}^1(\rho_1^{\theta_i\theta_j} - \rho_0^{\theta_i\theta_j}) + \rho_0^{\theta_i\theta_j} + (A_{ij}^1 - A_{ij}^T)\ell_{10}^{\theta_i\theta_j} \right\}$$

$$+ \frac{1}{2} \sum_{i,j} \delta(Z_i, Z_j) \sum_{t=2}^T \left\{ (\ell_{01}^{\theta_i\theta_j} + \ell_{10}^{\theta_i\theta_j})\left(A_{ij}^t - A_{ij}^{t-1}A_{ij}^t\right) + \ell_{11}^{\theta_i\theta_j} A_{ij}^{t-1}A_{ij}^t - \log \frac{Q_{00}^{\theta_i\theta_j}}{P_{00}^{\theta_i\theta_j}} \right\},$$

where $c(A) = \sum_t c_t(A)$ does not depend on Z. Hence the claim follows.   □

A.2   Approximation of the MLE

Recall the structural assumptions (3)–(4) about the degree correction parameters. Because $P_{01}, Q_{01} = o(1)$, a first-order Taylor expansion yields

$$\log \frac{1 - \theta_i\theta_j Q_{01}}{1 - \theta_i\theta_j P_{01}} = \theta_i\theta_j(P_{01} - Q_{01}) + o\left(P_{01}^2 + Q_{01}^2\right) = \frac{\bar{d}_i\bar{d}_j}{2\bar{m}} + o\left(P_{01}^2 + Q_{01}^2\right),$$

as well as $\ell_{01}^{\theta_i \theta_j} \approx \log \frac{P_{01}}{Q_{01}}$, $\ell_{10}^{\theta_i \theta_j} \approx \log \frac{1-P_{11}}{1-Q_{11}}$ and $\ell_{11}^{\theta_i \theta_j} \approx \log \frac{P_{11}}{Q_{11}}$. Using these approximations in the MLE expression leads to the maximisation of

$$\sum_{t=2}^{T} \sum_{i,j \,:\, z_i = z_j} \left(\tilde{a}_{ij}^t - \theta_i \theta_j \left(P_{01} - Q_{01} \right) \right).$$

where $\tilde{a}_{ij}^t = \alpha \left(A_{\text{new}}^t \right)_{ij} + \beta \left(A_{\text{pers}}^t \right)_{ij}$. Since μ and ν are stationary distributions,

$$\mathbb{E} \left(A_{\text{new}}^t \right)_{ij} = \begin{cases} \theta_i \theta_j \mu_1 (1 - P_{11}) & \text{if } Z_i = Z_j \\ \theta_i \theta_j \nu_1 (1 - Q_{11}) & \text{otherwise,} \end{cases}$$

$$\mathbb{E} \left(A_{\text{pers}}^t \right)_{ij} = \begin{cases} \theta_i \theta_j \mu_1 P_{11} & \text{if } Z_i = Z_j \\ \theta_i \theta_j \nu_1 Q_{11} & \text{otherwise.} \end{cases}$$

Therefore, using $W_{ij} = \sum_{t=2}^{T} \tilde{a}_{ij}$ we have

$$\mathbb{E} W_{ij} = \begin{cases} (T-1) \theta_i \theta_j \mu_1 \left(\alpha (1 - P_{11}) + \beta P_{11} \right) & \text{if } Z_i = Z_j \\ (T-1) \theta_i \theta_j \nu_1 \left(\alpha (1 - Q_{11}) + \beta Q_{11} \right) & \text{otherwise.} \end{cases}$$

Since the community labeling are sampled uniformly at random and using the normalization for the θ_i's, we have

$$\bar{d}_i = (T-1) \theta_i N \frac{\mu_1 \left(\alpha (1 - P_{11}) + \beta P_{11} \right) + (K-1) \nu_1 \left(\alpha (1 - Q_{11}) + \beta Q_{11} \right)}{K}$$

together with $\bar{m} = \frac{N^2}{2} \frac{\mu_1 (\alpha(1 - P_{11}) + \beta P_{11}) + (K-1) \nu_1 (\alpha(1 - Q_{11}) + \beta Q_{11})}{K}$. $\qquad \square$

A.3   Modularity and Normalized Spectral Clustering

The regularized modularity of a partition $Z \in [K]^N$ of the graph A is defined as

$$\mathcal{M} (A, Z, \gamma) = \sum_{i,j} \delta (Z_i, Z_j) \left(A_{ij} - \gamma \frac{d_i d_j}{2m} \right)$$

where $d = A 1_n$ and γ is a resolution parameter. This can be rewritten as

$$\mathcal{M} (A, Z, \gamma) = \text{Tr} \, \tilde{Z}^T \left(A - \gamma \frac{dd^T}{2m} \right) \tilde{Z}$$

where $\tilde{Z} \in \{0, 1\}^{N \times K}$ is the membership matrix associated to the vector Z, that is $\tilde{Z}_{ik} = 1$ for $k = Z_i$, and $\tilde{Z}_{ik} = 0$ otherwise. As maximizing the modularity over $Z \in \mathcal{Z}_{N,K}$ is in general NP-complete [4], it is convenient to perform a continuous relaxation. Following [17], we transform the problem into

$$\hat{X} = \operatorname*{arg\,max}_{\substack{X \in \mathbb{R}^{N \times K} \\ X^T D X = I_K}} \text{Tr} \, X^T \left(A - \gamma \frac{dd^T}{2m} \right) X. \tag{8}$$

The predicted membership matrix $\hat{Z}$ is then recovered by performing an approximated solution to the following k-means problem (see [12])

$$\left(\hat{Z}, \hat{Y}\right) = \underset{Z \in \mathcal{Z}_{N,K}, Y \in \mathbb{R}^{K \times K}}{\arg\min} \left\| ZY - \hat{X} \right\|_F. \tag{9}$$

The Lagrangian associated to the optimization problem (8) is

$$\operatorname{Tr} X^T \left(A - \gamma \frac{dd^T}{2m} \right) X - \operatorname{Tr} \left(\Lambda^T \left(X^T DX - I_K \right) \right)$$

where $\Lambda \in \mathbb{R}^{K \times K}$ is a symmetric matrix of Lagrangian multipliers. Up to a change of basis, we can assume that Λ is diagonal. The solution of (8) verifies

$$\left(A - \gamma \frac{dd^T}{2m} \right) X = DX\Lambda \quad \text{and} \quad X^T DX = I_K,$$

which is a generalized eigenvalue problem: the columns of X are the generalized eigenvectors, and the diagonal elements of Λ are the eigenvalues. In particular, since the constant vector 1_n verifies $(A - \gamma \frac{dd^T}{2m})1_n = (1-\gamma)D1_n$, we conclude that the eigenvalues should be larger than $1 - \gamma$ for the partition to be meaningful.

Multiplying the first equation by 1_n^T leads to $(1-\gamma)d^T X = d^T X\Lambda$, and therefore $d^T X = 0$ (using the previous remark on Λ). The system then simplifies in

$$AX = DX\Lambda \quad \text{and} \quad X^T DX = I_K.$$

Defining a re-scaled vector $U = D^{-1/2}X$ shows that U verifies $D^{-1/2}AD^{-1/2}U = U\Lambda$ and $U^T U = I_K$. Thus, the columns of U are eigenvectors of $D^{-1/2}AD^{-1/2}$ associated to the K largest eigenvalue (or equivalently, the eigenvectors of $\mathcal{L} = I_N - D^{-1/2}AD^{-1/2}$ associated to the K smallest eigenvalues).

References

1. Avrachenkov, K., Dreveton, M., Leskelä, L.: Estimation of static community memberships from temporal network data. arXiv preprint arXiv:2008.04790 (2020)
2. Bhattacharyya, S., Chatterjee, S.: General community detection with optimal recovery conditions for multi-relational sparse networks with dependent layers. arXiv preprint arXiv:2004.03480 (2020)
3. Billingsley, P.: Statistical methods in Markov chains. Ann. Math. Stat. **32**(1), 12–40 (1961)
4. Brandes, U., et al.: On finding graph clusterings with maximum modularity. In: Brandstädt, A., Kratsch, D., Müller, H. (eds.) WG 2007. LNCS, vol. 4769, pp. 121–132. Springer, Heidelberg (2007). https://doi.org/10.1007/978-3-540-74839-7_12
5. Chi, Y., Song, X., Zhou, D., Hino, K., Tseng, B.L.: Evolutionary spectral clustering by incorporating temporal smoothness. In: Proceedings of the 13th ACM SIGKDD International Conference on Knowledge Discovery and Data Mining, KDD 2007, pp. 153–162. Association for Computing Machinery, New York (2007)

6. Fortunato, S.: Community detection in graphs. Phys. Rep. **486**(3), 75–174 (2010)
7. Fournet, J., Barrat, A.: Contact patterns among high school students. PLOS One **9**(9), 1–17 (2014)
8. Ghasemian, A., Zhang, P., Clauset, A., Moore, C., Peel, L.: Detectability thresholds and optimal algorithms for community structure in dynamic networks. Phys. Rev. X **6**(3), 031005 (2016)
9. Holland, P., Laskey, K.B., Leinhardt, S.: Stochastic blockmodels: first steps. Soc. Netw. **5**, 109–137 (1983)
10. Holme, P., Saramäki, J.: Temporal networks. Phys. Rep. **519**(3), 97–125 (2012). Temporal Networks
11. Karrer, B., Newman, M.E.J.: Stochastic blockmodels and community structure in networks. Phys. Rev. E **83**, 016107 (2011)
12. Kumar, A., Kannan, R.: Clustering with spectral norm and the k-means algorithm. In: 2010 IEEE 51st Annual Symposium on Foundations of Computer Science, pp. 299–308. IEEE (2010)
13. Liu, F., Choi, D., Xie, L., Roeder, K.: Global spectral clustering in dynamic networks. Proc. Natl. Acad. Sci. **115**(5), 927–932 (2018)
14. Mastrandrea, R., Fournet, J., Barrat, A.: Contact patterns in a high school: a comparison between data collected using wearable sensors, contact diaries and friendship surveys. PLOS One **10**(9), 1–26 (2015)
15. Matias, C., Miele, V.: Statistical clustering of temporal networks through a dynamic stochastic block model. J. R. Stat. Soc.: Ser. B (Stat. Methodol.) **79**(4), 1119–1141 (2017)
16. Mucha, P.J., Richardson, T., Macon, K., Porter, M.A., Onnela, J.P.: Community structure in time-dependent, multiscale, and multiplex networks. Science **328**(5980), 876–878 (2010)
17. Newman, M.E.J.: Spectral methods for community detection and graph partitioning. Phys. Rev. E **88**, 042822 (2013)
18. Newman, M.E.: Equivalence between modularity optimization and maximum likelihood methods for community detection. Phys. Rev. E **94**(5), 052315 (2016)
19. Newman, M.E., Girvan, M.: Finding and evaluating community structure in networks. Phys. Rev. E **69**(2), 026113 (2004)
20. Pamfil, A.R., Howison, S.D., Lambiotte, R., Porter, M.A.: Relating modularity maximization and stochastic block models in multilayer networks. SIAM J. Math. Data Sci. **1**(4), 667–698 (2019)
21. Reichardt, J., Bornholdt, S.: Statistical mechanics of community detection. Phys. Rev. E **74**(1), 016110 (2006)
22. Rossetti, G., Cazabet, R.: Community discovery in dynamic networks: a survey. ACM Comput. Surv. **51**(2), 1–37 (2018)

Community Detection Using Semilocal Topological Features and Label Propagation Algorithm

Deepanshu Malhotra[1], Ralucca Gera[2], and Akrati Saxena[3(✉)]

[1] University School of Information, Communication and Technology,
Guru Gobind Singh Indraprastha University, New Delhi, India
[2] Naval Postgraduate School, Monterey, CA 93943, USA
rgera@nps.edu
[3] Department of Mathematics and Computer Science,
Eindhoven University of Technology, Eindhoven, The Netherlands
a.saxena@tue.nl

Abstract. The detection of cohesive clusters with similar character-istics in multiple types of networks is of immense informational value to researchers. In this work, we propose a Weighted Semilocal Similarity based Label Propagation Algorithm (WSSLPA) for such community detection. The proposed method detects communities by using the semilocal topological features to overcome the shortcoming of randomness and instability in the existing label propagation algorithms while selecting a community label from multiple maximum labels. We associate user-defined weight parameters with the topological features to help WSSLPA adapt to different networks and enhance the performance metrics scores of detected communities. We compare the performance of the proposed method with other community detection techniques and show that the identified communities of WSSLPA are closer to the ground-truth communities.

Keywords: Complex networks · Community detection · Label propagation

1 Introduction

The emergence of network science has put forth the complex networks that model the intricate relationships among the components of various complex systems [1]. The inception of such complex networks attract researchers' interest to various problems, such as their evolution [25], identify influential nodes [26], link prediction [17], information diffusion [9], and so on. Community detection is one of the fundamental problems in Network Science that aims to find strongly connected clusters of nodes in a network, identified through more intra-cluster edges than inter-cluster edges. A plethora of algorithms exists that are driven by different motivations behind finding the clusters. The well known approaches are based

© Springer Nature Switzerland AG 2021
D. Mohaisen and R. Jin (Eds.): CSoNet 2021, LNCS 13116, pp. 255–266, 2021.
https://doi.org/10.1007/978-3-030-91434-9_23

on modularity optimization [19], information-theoretic techniques [10], genetic algorithms [16], non-negative matrix factorization [3], and label propagation [7].

One algorithm that we build our work on is by Raghavan et al. [22], who proposed the Label Propagation Algorithm (LPA) that utilizes the neighborhood structures of nodes to detect communities in networks. This approach consists of three prominent steps., In the initialization phase, nodes are randomly assigned a unique community label. Following that, the label propagation step begins wherein each node adapts a label that is assigned to the majority of its neighbors. Finally, the algorithm terminates when nodes have a label that belongs to the majority of their neighbors. The nodes having the same label get clustered together, thus producing the community structure of a given network.

The time efficiency and simplicity are advantages of LPA that have encouraged the development of various new approaches in this direction. Poaka et al. [20] developed a new LPA approach wherein they compute link density based measures to avoid ties between multiple maximum labels. They also extended their method by using fuzzy techniques to find overlapping communities in complex networks. Jokar et al. [11] extended the previous approach by developing a new metric that utilized link density to choose the future community of a node in the case of multiple maximum labels; and also presented a balancing parameter that assigned appropriate weights to the similarity measures between the node pairs. Verma et al. [28] developed a semi-supervised learning technique based on LPA to find communities in complex networks; the proposed method initializes the communities using core nodes identified through various centrality measures. Li et al. [14] developed an improved LPA by utilizing the modularity function and node importance, i.e., the normalized degree centrality of the node. The proposed algorithm first initializes the communities using the modularity function and then performs the update step in a specific order by using the node properties.

In light of the context above, we propose a label propagation based community detection method to find better quality communities. We will address two shortcomings [7] of the existing algorithms, (i) the flaw of randomness and (ii) the lack of stability in LPA; that are encountered when the algorithm randomly selects a future community for each node in case of multiple maximum labels. Therefore, LPA is unable to achieve stable community structures that can be observed by its lower metric scores that are computed by taking the mean value on several runs. Our work addresses these issues by utilizing topology based similarity measures, thereby breaking the tie of random assignment from multiple maximum labels in obtaining a community structure. Furthermore, the proposed method produces a stable community structure that is demonstrated by high metric scores obtained on various networks.

In complex networks, the global similarity measures consider the properties of the whole network that lead to high time complexity. On the other hand, local similarity measures examine only the immediate neighborhood information of each node. Henceforth, we use semilocal similarity measures to strike a balance between efficiency and information quantity. Among the semilocal indices, we use the extended Jaccard [2], and 2-hop neighborhood volume [29] to find suitable future communities for the nodes. Our proposed method also integrates user-defined parameters for the above-mentioned measures to better adapt to different networks. We conduct experiments comparing against other existing community detection methods on real-world networks and evaluate the identified communities using different performance metrics.

The rest of the paper is organized as follows. Section 2 describes the algorithm proposed in this study. Section 3 presents the experimental results for the comparison of our method with various existing community detection methods. Lastly, Sect. 4 holds the conclusion of this paper along with future directions.

2 Proposed Method

In this section, we present the details of our proposed method, namely the Weighted Semilocal Similarity based LPA (WSSLPA). As mentioned, the WSSLPA removes the random selection of the future community for a node in case of multiple maximum labels. This is done with the help of topological information that is fine-tuned by parameters.

We first discuss the parameters required for our proposed method. The Extended Jaccard EJ coefficient for two adjacent nodes u and v is defined as,

$$EJ_{(u,v)} = \frac{|\Gamma_2(u) \cap \Gamma_2(v)|}{|\Gamma_2(u) \cup \Gamma_2(v)|}, \tag{1}$$

where $\Gamma_2(u)$ denotes the union of the neighbors of a given node u that are either at one-hop or two-hop distance away from node u.

The 2-Hop Neighborhood volume NV^2 for a node u is defined as,

$$NV_u^2 = \sum_{w \in \Gamma_2(u)} deg(w), \tag{2}$$

where $deg(w)$ denotes the degree of node w and $\Gamma_2(u)$ is defined as in Eq. 1. Additionally, the similarity of a pair of nodes (u, v) is represented by $Sim_{u,v}$ that denotes the weighted sum of $EJ_{(u,v)}$ and NV_v^2 using Eq. 3.

$$Sim_{(u,v)} = k_1 \cdot EJ_{(u,v)} + k_2 \cdot \widetilde{NV_v^2} \tag{3}$$

where $\widetilde{NV_v^2}$ is the normalized value of NV_v^2. The NV^2 values obtained represent the sum of degrees, therefore we normalize the NV^2 values of all the nodes in the network to accommodate them in the interval of $[0, 1]$.

The steps of the WSSLPA algorithm are as follows.

1. WSSLPA takes a network $G(V, E)$ as an input, where V is the set of nodes and E is the set of edges in network G. The initialization phase begins, wherein WSSLPA assigns a unique community label to each node present in the network.

2. The label propagation phase begins. First, the algorithm arranges the nodes in random order. Next, it chooses a maximum community label in the neighborhood of each node and assigns it as the node's future community.

3. In the case of multiple maximum labels being assigned to a single node, say u, we use the Extended Jaccard (EJ) introduced in Eq. 1, and 2-Hop Neighborhood Volume (NV^2) introduced in Eq. 2 as follows.

   (a) Both EJ and $\widetilde{NV^2}$ are pre-computed for the fast execution of the proposed method for all the edges and nodes, respectively.

   (b) Next, $Sim_{(u,v)}$ is computed for every pair of nodes using EJ, $\widetilde{NV^2}$ and the user-defined parameters as explained in Eq. 3.

4. We compute a community wise cumulative sum of the combined similarity measure, that is $Sim_{sum}^{c_i}(u) = \sum_{CommunityLabel(v)=c_i \& (u,v)\in E} Sim_{(u,v)}$, $\forall c_i \in C$, where $C = \{c_1, c_2, \cdots, c_i, \cdots\}$ is the set of community labels. Subsequently, the community label with maximum sum magnitude is selected to be the future community of the given node, therefore, $CommunityLabel(u) = argmax_{c_i}\{Sim_{sum}^{c_i}(u), \forall c_i \in C\}$.

5. The algorithm is terminated if all the nodes have a label that belongs to the majority of their neighbors or the maximum number of iterations (t) is reached.

We now delineate the complete steps for WSSLPA in Algorithm 1.

Algorithm 1: Weighted semilocal similarity based LPA

Input: $G(V, E)$: The Input Network, t: Maximum Iteration Limit

1 For each node u, assign a unique community label $CommunityLabel(u)$
2 $iterations \leftarrow 0$
3 **repeat**
4 | $V' \leftarrow$ Shuffle the list of nodes V to produce a random order
5 | **for** u *in* V' **do**
6 | | **if** *Multiple maximum labels for node* u **then**
7 | | | Compute Sim score using Equation 3
8 | | | Calculate the community wise cumulative sum of Sim score, $Sim_{sum}^{c_i}(u) = \sum_{CommunityLabel(v)=c_i \& (u,v)\in E} Sim_{(u,v)}, \forall c_i \in C$
9 | | | $CommunityLabel(u) = argmax_{c_i}\{Sim_{sum}^{c_i}(u), \forall c_i \in C\}$
10 | | **else**
11 | | | $CommunityLabel(u) \leftarrow$ Maximum label among the neighbors of node u
12 | **end**
13 **until** *All nodes have a label equal to the majority of their neighbors or iterations* $> t$
14 **return** $CommunityLabel$

2.1   Time Complexity

In this section, we present the time complexity of the proposed method. Let n denotes the number of nodes in the network, k_{avg} is the average degree for the nodes, t is the maximum number of iterations if the termination criteria is not satisfied (namely that of all nodes have a label matching most of their neighbors' label).

The first step of the proposed algorithm includes the calculation of semilocal measures. The time complexity for calculating these structural measures is equal to $O(nk_{avg}^2)$. Next, the algorithm begins with the initialization phase, which takes $O(n)$ time. Subsequently, the label propagation step of the proposed method is executed that has $O(tnk_{avg})$ time complexity. Finally, in the termination step, every node's neighborhood is utilized to check the termination criteria. This step has $O(nk_{avg})$ complexity. Henceforth, the overall time complexity of the proposed technique is $O(nk_{avg}^2 + tnk_{avg})$.

3   Experimental Analysis

In this section, we introduce the real-world and synthetic datasets used in the experiments. We then follow it with the performance analysis of the WSSLPA algorithm as compared to the baseline community detection algorithms.

3.1   Datasets

To evaluate the performance of the proposed algorithm, we use various real-world datasets, including Karate, Dolphins, Polbooks, Football, Cora, Citeseer, and AS internet network. The availability of ground-truth community structure is a critical requirement in our experiments, and thus we consider datasets having predefined ground-truth structures. Furthermore, we also test the algorithms on *LFR* benchmark network. In LFR, the minimum degree and minimum community size was set to 20, and the maximum degree and maximum community size was set to 50. Table 1 summarizes metrics of these networks.

3.2   Experimental Settings

For the analysis, we run each algorithm (WSSLPA and baselines) 10 times and report their averages when comparing against WSSLPA. The performance metrics we use to measure the overall quality of the community structures are the Normalized mutual information (NMI) [5], and modularity [19]. We use the termination criteria to set up a maximum number of iterations our proposed method can execute. This is applied if some nodes do not have a label that belongs to the majority of their neighbors, and the maximum number of iterations is set to 1000 for the majority of networks (the exception is *AS Internet* network, where we set this number to 100 as it is a large network).

The weight parameters, namely k_1 and k_2, constitute an important part of experimental settings for WSSLPA. They help the proposed method adapt to

Table 1. Description of datasets used in this study.

Dataset	Acronym	Nodes	Edges	#Ground-truth communities	Ref
Karate	Kar	34	78	2	[30]
Dolphins	Dol	62	159	2	[15]
Polbooks	Pol	105	441	3	[12]
Football	Foot	115	613	12	[8]
Cora	Cora	2708	5278	7	[27]
Citeseer	Cite	3327	4676	7	[27]
AS Internet	AS	23752	58416	176	[4]
LFR	LFR	500	$\mu(0.1 - 0.9)$	20–50	[13]

different networks efficiently and maintain a fine balance between the extended Jaccard and 2-hop neighborhood volume. The default value of weight parameters is set as $k_1 = 0.8$ and $k_2 = 0.2$ for WSSLPA as experimentation says that these settings provide better results compared to baselines for most of the datasets. Table 2 presents parameter values for all datasets that provides best results (shown in Table 3) based on the experimental observation.

Table 2. Parameters value (k_1, k_2) for the best results of WSSLPA.

Datasets	Karate	Dolphins	Polbooks	Football	Cora	Citeseer
k_1, k_2	0.9, 0.2	0.2, 0.8	0.9, 0.1	0.9, 0.2	0.1, 1.0	1.0, 0.1
Datasets	AS	LFR($\mu = 0.1$)	LFR($\mu = 0.3$)	LFR($\mu = 0.5$)	LFR($\mu = 0.7$)	LFR($\mu = 0.9$)
k_1, k_2	0.9, 0.2	1.0, 0.1	0.2, 0.1	0.7, 0.5	0.5, 0.5	0.5, 0.5

3.3   Performance Analysis

Table 3 presents the performance comparison of WSSLPA with four community detection techniques. We compare against four established methods: leading eigenvector algorithm (Lead) [18], LPA [22], walktrap (Walk) [21], and infomap (Info) [23,24] algorithms. For the WSSLPA method we show both the best results achieved, as well as the results for the default parameter setting.

We observe that WSSLPA achieves the best NMI scores on the majority of networks ($0.8209, 0.8483, \& 0.3227$ are the best NMI scores achieved by WSSLPA on *Kar, Dol, Cite*, respectively), and competitive scores on others ($0.5619, 0.9150, \& 0.3342$ are the second best NMI scores obtained on *Pol, Foot, AS*, respectively). WSSLPA with default parameters also obtains better NMI results (on *Kar, Cite, AS*) in comparison with other community detection methods. Infomap and LPA achieve the best NMI scores on some networks; however, WSSLPA gives competitive results on them. Additionally, the Lead and Walk

Table 3. Performance comparison using the NMI and modularity metrics

Data Set	Algorithm	Lead	LPA	Walk	Info	WSSLPA (best)	WSSLPA (0.8/0.2)
Kar	NMI	0.6771	0.6815	0.6110	0.6994	0.8209	0.7901
	Modularity	0.3934	0.3604	0.3431	0.4020	0.3970	0.3925
Dol	NMI	0.4489	0.6377	0.5372	0.5844	0.8483	0.5761
	Modularity	0.4911	0.4795	0.4888	0.5269	0.4331	0.4955
Pol	NMI	0.5201	0.5655	0.5081	0.4934	0.5619	0.5483
	Modularity	0.4671	0.4989	0.4961	0.5228	0.4904	0.4893
Foot	NMI	0.6986	0.8679	0.7451	0.9241	0.9150	0.9006
	Modularity	0.4926	0.5871	0.5883	0.6005	0.5807	0.5739
Cora	NMI	0.3820	0.4233	0.4011	0.4128	0.4111	0.4002
	Modularity	0.7318	0.7401	0.5888	0.7178	0.7339	0.6237
Cite	NMI	0.3011	0.3114	0.3181	0.3119	0.3227	0.3223
	Modularity	0.8541	0.8221	0.8089	0.8207	0.7531	0.7534
AS	NMI	0.0000	0.2302	0.2553	0.4412	0.3342	0.3092
	Modularity	0.0000	0.2010	0.1649	0.5195	0.3540	0.3139

algorithms obtain low NMI scores. For modularity analysis, WSSLPA performs competitively in most cases while it gets low modularity scores on some networks.

The main aim of this study was to develop a community detection technique that could produce near ground-truth community structures. The proposed WSSLPA method performs excellently on the NMI measure as it produces high quality community structures. The modularity community score suffers from the resolution limit problem wherein it rewards the large size communities while ignoring the small communities [6]. This is one of the main reasons for the average performance of WSSLPA on the modularity metric. Overall, WSSLPA produces better performances across various networks as observed in Table 3 summarizing the results of our experiments.

Additionally, we perform an experiment to evaluate different community detection methods utilized in this study on the LFR benchmark datasets, running each experiment 10 times and showing the average values. We create the LFR network with 500 nodes by varying the mixing parameter $\mu \in [0.1, 0.9]$. Figure 1 presents the results for this experiment, where the x-axis denotes the different mixing parameter values, while the y-axis represents the NMI scores obtained by different algorithms.

We observe that the infomap method obtains the highest NMI score for smaller μ (namely $\mu \in \{0.1, 0.3\}$), but its NMI values fall sharply after that to their lowest point. Similar results are observed for LPA wherein the NMI values fall sharply after the mixing parameter value of 0.1. WSSLPA achieves the highest score at $\mu = 0.1$, and it drops as observed for other methods. However, the WSSLPA achieves the highest NMI values in comparison to other algorithms

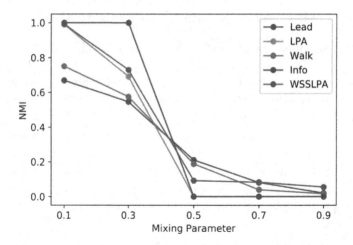

Fig. 1. Comparison on the LFR network using the NMI performance metric.

for $\mu \in \{0.7, 0.9\}$. This observation shows the robust nature of the proposed community detection technique for different mixing parameter values. Lead and Walktrap methods achieve low NMI scores in the majority of the cases except at $\mu = 0.5$, where they obtain higher NMI values compared to other algorithms.

The above experiments help in providing a deep insight into the performances of various community detection techniques. They also help in exhibiting the consistent performances of WSSLPA across a variety of datasets with the help of different evaluation metrics. WSSLPA obtains higher NMI scores on real-world and synthetic networks, signifying the superior quality of identified communities.

3.4 Sensitivity Analysis

We now study the impact of weight parameters, i.e., k_1, k_2, on the performance of WSSLPA for different networks. For this analysis, the one parameter (k_1 or k_2) will be set to 0.5 and the other will be varied in the range of $[0.1, 0.9]$. Figures 2, and 3 show the NMI values of identified communities. In Fig. 2, $k_2 = 0.5$, while $k_1 \in [0.1, 0.9]$, and in Fig. 3, $k_1 = 0.5$, while k_2 varies in the range $[0.1, 0.9]$.

We observe from Fig. 2 that the overall NMI scores of WSSLPA increase as the value of k_1 increases. In Fig. 3, we observe that the overall NMI scores decrease with the increasing value of k_2. The variation in k_2 affects more the performance on small size networks as compared to larger networks.

In Fig. 4, we show the NMI values obtained by WSSLPA on *Kar, Dol, Pol, Foot* networks for all parameter (k_1 and k_2) settings. The results show that a higher value of k_1 and a lower value of k_2 provide good results on most networks, as expected.

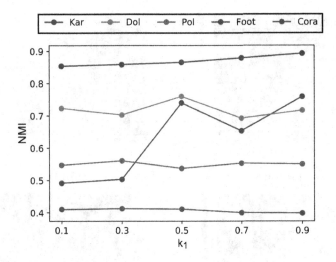

Fig. 2. Analysis of WSSLPA by varying $k_1 \in [0.1 - 0.9]$ while keeping $k_2 = 0.5$.

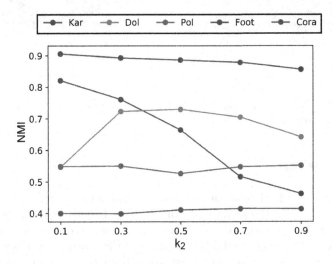

Fig. 3. Analysis of WSSLPA by varying $k_2 \in [0.1 - 0.9]$, and $k_1 = 0.5$.

We, therefore, conclude that on the majority of networks, we obtain better NMI scores through the combination of higher k_1 values and lower k_2 values. An exception is the case of *Cora* network wherein lower k_1 and higher k_2 values give better score. This might be because, in these networks, a higher preference is given to the centrally connected nodes while predicting clusters closer to ground-truth community structures. Henceforth, the parameters for WSSLPA are based on these observations that finally help the proposed algorithm achieve better quality community structures.

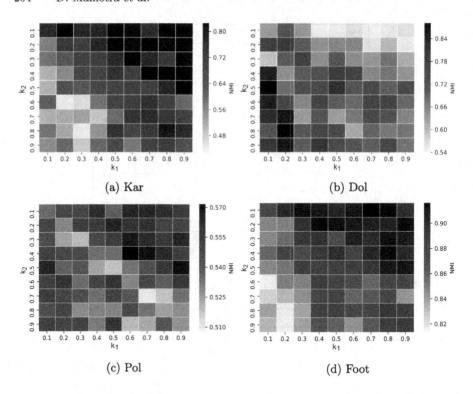

Fig. 4. Analysis of WSSLPA on *Kar, Dol, Pol,* & *Foot* networks for $k_1 \in [0.1 - 0.9]$ and $k_2 \in [0.1 - 0.9]$.

4   Conclusion

The development of efficient and accurate community detection methods is a keen research area in the field of network science. Recent methods propose the utilization of local information to detect the densely connected communities of nodes. Although such methods are efficient, they achieve low performances because of the limited information extracted from the network. Furthermore, global similarity methods consider the topology of the whole network, thereby making them less efficient.

In this study, we presented a Label Propagation Algorithm (LPA), named Weighted Semilocal Similarity based Label Propagation Algorithm (WSSLPA), that detects high quality communities that are similar to the ground-truth communities in complex networks. Our proposed method utilized semilocal similarity measures to counter the shortcoming of randomness in LPA. Consequently, this improved the quality of WSSLPA's detected communities by avoiding the formation of large communities, which is a shortcoming in most modularity-based community detection methods. Additionally, the utilization of semilocal measures helped in retrieving considerable global network information, while the label propagation technique assisted in improving the algorithm run time. The experimental results showed the better performance of WSSLPA on real-world as well as on synthetic networks as compared to baseline methods. The proposed method performs consistently on different networks and achieves competitive NMI scores, thereby signifying the closeness of the identified communities with respect to the ground truth community structure.

One can further extend the proposed method to attributed networks where each node has different properties, and the edges might represent varied relationships. Such diverse information can be harnessed by developing efficient semilocal methods that can be utilized by the community detection algorithms. Furthermore, the existing method could be extended to detect overlapping communities wherein each node can attain multiple community labels. This would help to further develop novel community detection techniques for real-world systems where the objects are heterogeneous, and a single object might be linked with multiple communities.

References

1. Barabási, A.L.: Network science. Philos. Trans. R. Soc. Math. Phys. Eng. Sci. **371**(1987), 20120375 (2013)
2. Berahmand, K., Bouyer, A., Vasighi, M.: Community detection in complex networks by detecting and expanding core nodes through extended local similarity of nodes. IEEE Tran. Computat. Soc. Syst. **5**(4), 1021–1033 (2018)
3. Binesh, N., Rezghi, M.: Fuzzy clustering in community detection based on nonnegative matrix factorization with two novel evaluation criteria. Appl. Soft Comput. **69**, 689–703 (2018)
4. Boguná, M., Papadopoulos, F., Krioukov, D.: Sustaining the internet with hyperbolic mapping. Nat. Commun. **1**(1), 1–8 (2010)
5. Danon, L., Diaz-Guilera, A., Duch, J., Arenas, A.: Comparing community structure identification. J. Stat. Mech: Theor. Exp. **2005**(09), P09008 (2005)
6. Fortunato, S., Barthelemy, M.: Resolution limit in community detection. Proc. Natl. Acad. Sci. **104**(1), 36–41 (2007)
7. Garza, S.E., Schaeffer, S.E.: Community detection with the label propagation algorithm: a survey. Physica A **534**, 122058 (2019)
8. Girvan, M., Newman, M.E.: Community structure in social and biological networks. Proc. Natl. Acad. Sci. **99**(12), 7821–7826 (2002)
9. Guille, A., Hacid, H., Favre, C., Zighed, D.A.: Information diffusion in online social networks: a survey. ACM SIGMOD Rec. **42**(2), 17–28 (2013)
10. Hajek, B., Wu, Y., Xu, J.: Information limits for recovering a hidden community. IEEE Trans. Inf. Theor. **63**(8), 4729–4745 (2017)
11. Jokar, E., Mosleh, M.: Community detection in social networks based on improved label propagation algorithm and balanced link density. Phys. Lett. A **383**(8), 718–727 (2019)
12. Krebs, V.: Proxy networks. analyzing one network to reveal another. Bull. de Méthodol. Sociol. Bull. Sociol. Methodol. (79), 61–70 (2003)
13. Lancichinetti, A., Fortunato, S., Radicchi, F.: Benchmark graphs for testing community detection algorithms. Phys. Rev. E **78**(4), 046110 (2008)
14. Li, H., Zhang, R., Zhao, Z., Liu, X.: LPA-MNI: an improved label propagation algorithm based on modularity and node importance for community detection. Entropy **23**(5), 497 (2021)

15. Lusseau, D., Schneider, K., Boisseau, O.J., Haase, P., Slooten, E., Dawson, S.M.: The bottlenose dolphin community of doubtful sound features a large proportion of long-lasting associations. Behav. Ecol. Sociobiol. **54**(4), 396–405 (2003)

16. Malhotra, D.: Community detection in complex networks using link strength-based hybrid genetic algorithm. SN Comput. Sci. **2**(1), 1–16 (2021)

17. Martínez, V., Berzal, F., Cubero, J.C.: A survey of link prediction in complex networks. ACM Comput. Surv. (CSUR) **49**(4), 1–33 (2016)

18. Newman, M.E.: Finding community structure in networks using the eigenvectors of matrices. Phys. Rev. E **74**(3), 036104 (2006)

19. Newman, M.E., Girvan, M.: Finding and evaluating community structure in networks. Phys. Rev. E **69**(2), 026113 (2004)

20. Poaka, V., Hartmann, S., Ma, H., Steinmetz, D.: A link-density-based algorithm for finding communities in social networks. In: Link, S., Trujillo, J.C. (eds.) Advances in Conceptual Modeling, ER 2016. LNCS, vol. 9975, pp. 76–85. Springer, Cham (2016). https://doi.org/10.1007/978-3-319-47717-6_7

21. Pons, P., Latapy, M.: Computing communities in large networks using random walks. In: Yolum, p, Güngör, T., Gürgen, F., Özturan, C. (eds.) Computer and Information Sciences - ISCIS 2005, ISCIS 2005. LNCS, vol. 3733, pp. 284–293. Springer, Heidelberg (2005). https://doi.org/10.1007/11569596_31

22. Raghavan, U.N., Albert, R., Kumara, S.: Near linear time algorithm to detect community structures in large-scale networks. Phys. Rev. E **76**(3), 036106 (2007)

23. Rosvall, M., Axelsson, D., Bergstrom, C.T.: The map equation. Eur. Phys. J. Spec. Top. **178**(1), 13–23 (2009)

24. Rosvall, M., Bergstrom, C.T.: An information-theoretic framework for resolving community structure in complex networks. Proc. Natl. Acad. Sci. **104**(18), 7327–7331 (2007)

25. Saxena, A.: A survey of evolving models for weighted complex networks based on their dynamics and evolution. arXiv preprint arXiv:2012.08166 (2020)

26. Saxena, A., Iyengar, S.: Centrality measures in complex networks: A survey. arXiv preprint arXiv:2011.07190 (2020)

27. Sen, P., Namata, G., Bilgic, M., Getoor, L., Galligher, B., Eliassi-Rad, T.: Collective classification in network data. AI Mag. **29**(3), 93–93 (2008)

28. Verma, P., Goyal, R.: Influence propagation based community detection in complex networks. Machine Learning with Applications **3**, 100019 (2021)

29. Wehmuth, K., Ziviani, A.: Daccer: distributed assessment of the closeness centrality ranking in complex networks. Comput. Netw. **57**(13), 2536–2548 (2013)

30. Zachary, W.W.: An information flow model for conflict and fission in small groups. J. Anthropol. Res. **33**(4), 452–473 (1977)

Twitter Analysis of Covid-19 Misinformation in Spain

Diego Saby[1(✉)], Olivier Philippe[1], Nataly Buslón[1], Javier del Valle[1], Oriol Puig[1], Ramón Salaverría[2], and María José Rementeria[1]

[1] Barcelona Supercomputing Center, Barcelona, Spain
[2] Universidad de Navarra, Pamplona, Spain

Abstract. A graph analysis on the tweets and users networks from a set of curated news was done to study the existing difference in communication patterns between legitimate and misinformation news. Our findings suggest there is no difference in the influence of misinformation and legitimate news but misinformation news tend to be more shared and present than legitimate news, meaning that while misinformation tweets do not have more influence, their authors are more prolific. Misinformation reach wider audience even if the tweets, individually, are not more influential. A subsequent qualitative analysis on the users reveal that there is also influence of misinformation spreading in Spain from other Spanish speaking countries.

Keywords: Misinformation diffusion · Twitter influence · Network analysis

1 Introduction

The diffusion of misinformation and disinformation through modern communication and social networking sites is one of today's most urgent problems. The situation has worsened in recent years since its circulation has reached unforeseen scales, severely affecting domains that range from politics and economy to public health. In healthcare, the proliferation of manipulated medical information has been perceived as especially harmful due to the impact that this content might have on people's lives [15].

We understand disinformation as verifiably false or misleading information created, presented and disseminated for economic gain or to intentionally deceive the public, while misinformation is verifiably false information that is spread without the intention to mislead[1]. With the rapid spread of misinformation in social networks during the COVID-19 pandemic and the extensive use of Twitter as information platform [8], a large number of studies have been carried out since then. For example, Singh et al. analyse the most prevalent myths about the COVID-19 and the number of conversations about those myths over time [14]. Authors in [4] classify misinformation in some categories and try to find

---

[1] https://digital-strategy.ec.europa.eu/en/policies/online-disinformation.

© Springer Nature Switzerland AG 2021
D. Mohaisen and R. Jin (Eds.): CSoNet 2021, LNCS 13116, pp. 267–278, 2021.
https://doi.org/10.1007/978-3-030-91434-9_24

who is interacting with influential tweets containing misinformation and where those tweets are propagated in the global network. In [7], authors develop a twitter misinformation dataset and analyses the communities inside the users interactions network. Mcquillan et al. present a network community analysis and topic analysis about COVID-19 topics in Twitter [6].

As part of the RRSSalud project *Dynamics of dissemination in social networks of false news about health, 2020–2022*[2], we leveraged data from Twitter to study the dynamics of dissemination of health, politics and technology misinformation and disinformation.

Additionally, we examined the dynamics of tweets written in Spanish and their profiles in the context of the COVID-19 pandemic.

2   Dataset and Methodology

Our study has been carried out with a dataset composed by misinformation and legitimate news. The subset was built during the RRSSalud project [13] selecting 10 misinformation news on health, technology and politics published between March and April 2020[3]. We completed our dataset with 10 verified news, established by MyNews repository, a platform that records all the information published in the main Spanish media on the same dates, and on similar topics. Tables 1 and 2 show the 20 selected claims, 10 misinformation news and 10 legitimate news.

Table 1. Misinformation news

New id	Claim
11	Alimentos que más inmunizan contra el coronavirus
12	Hantavirus. La OMS advierte al mundo del nuevo virus que viene de China
13	Madrid denuncia que el Gobierno paralizó en Zaragoza 5.000 kilos de mascarillas para Madrid porque «Aduanas cierra a las 15h»
14	Bill Gates anuncia que implantará microchips para combatir Covid-19 y rastrear las vacunas
15	Demuestran científicamente la relación causal entre la tecnología 5G y el COVID-19
16	Todo apunta a que el COVID-19 "se escapó" del Instituto de Virología de Wuhan: ¿Desarrollaban los chinos un arma de guerra biológica?
17	El uso prolongado del tapaboca produce hipoxia
18	Stefano Montario: las mascarillas incuban el cáncer
19	El Ministerio del Interior alemán define al coronavirus como «falsa alarma global» en un informe filtrado a la prensa
20	La OMS alerta sobre el Virus Nipah que puede ser peor que el Covid-19

[2] This project is financed by the BBVA Foundation in its 2019 call for *Ayudas a Equipos de Investigación Científica en el área de Economía y Sociedad Digital.* https://www.rrssalud.org.
[3] Sources: EFE Verifica, Maldita and Newtral.

Table 2. Legitimate news

New id	Claim
1	La Junta iniciará el lunes el reparto de menús para menores en riesgo de exclusión en Córdoba
2	El CSIC busca una vacuna contra el Covid-19 a partir del virus que erradicó la viruela
3	Coronavirus: llega a Madrid el material sanitario al que Huawei se comprometió con el rey Felipe VI
4	No hay evidencia de que los perros transmitan el virus
5	Moscú aprovecha el coronavirus para imponer el control a la población con códigos QR
6	El coronavirus solo se desactiva por completo a más de 90 grados
7	El uso de la mascarilla agrava el aislamiento de las personas sordas
8	Utilizar guantes no tiene ningún sentido
9	Evitar el coronavirus con la vacuna de la polio
10	Sanidad estudia adelantar este año la vacuna de la gripe

In addition, we have collected tweets using the official Twitter API related to the news from our dataset. For each of them, we searched a first set of tweets containing the URL of these news or their claim, starting on March 2020 until May 2021. From this initial set of 2.5k tweets we then captured the whole conversation around those tweets, the retweets and the quotes and collect a total of 20.5k tweets. Finally we collected the users set composed of the tweets' authors and the users who liked or retweeted any tweet from the initial set (N = 40.1k).

With this dataset we built the tweet network T and user network U, as well as the resulting two one-mode projection of the bipartite network B, where the nodes $N_t \cup N_u$. N_t are the nodes representing the tweets and N_u are the nodes representing users. An edge e_{tu} represents the interaction (like or retweet) from the user u to the tweet t. Since we are interested in the influence and spreading of the news, we based our graph on retweets and likes. This is an important aspect to retain as all the metrics under this graph can only be interpreted based on its foundation. An edge between node t_1 and node t_2 is created if they have a common user that has interacted with both tweets. Finally, graph U represents the users network and its nodes represent users. Again, an edge between user u_1 and user u_2 is created if they interacted with a common tweet.

Since the tweets are collected from different sources of selected misinformation news, tweets that are related to a misinformation news are labelled as *misinformation tweets*, and tweets that are related to a legitimate new are labelled as *legitimate tweets*.

3   Tweets Analysis

3.1   Graphs Analysis

In terms of the type of news gathered, our graph T presents an unbalance in favour for the misinformation tweets as depicted in Fig. 1. We have identified

16601 tweets related to the misinformation news, while only 3905 are about legitimate news. Thus, misinformation news were tweeted four times more than legitimate news.

Fig. 1. Graph T with pink coloured misinformation news and blue coloured legitimate news (Color figure online)

We also found unbalance in regards to the news proportion. As shown in Table 3 most of the tweets are about the news #14, and we have barely some tweets from news #15 and #12 and nothing for news #11 and #20.

For the legitimate news tweets, as shown in Table 4, most of the tweets also comes from a single item, the news #3. Moreover, we did not find any tweet coming from news #1, #2, #5, and #7. Results show that legitimate news have more potential to go unnoticed, compared to misinformation news.

Table 3. Misinformation news distribution

New id	Nodes percentage	Mean degree	Mean retweet influence coefficient
11	0%	–	–
12	0.63%	5.80	0.37
13	4.38%	18.13	0.24
14	54.59%	21.31	0.22
15	0.55%	9.15	0.51
16	2.03%	4.74	0.30
17	1.80%	11.04	0.04
18	8.03%	20.00	0.19
19	16.45%	22.07	0.22
20	0%	–	–

To find if different tweets are related by the same group of people we built the sub-graph $T_{initial}$ of the graph T where the nodes are only the tweets collected in the initial phase of the data collection, and discarded the nodes that are part of

Table 4. Legitimate news distribution

New id	Nodes percentage	Mean degree	Mean retweet influence coefficient
1	0%	–	–
2	0%	–	–
3	12.45%	34.67	0.79
4	0.03%	1	0.88
5	0%	–	–
6	2.19%	17.68	0.79
7	0%	–	–
8	2.19%	14.27	0.76
9	0.03%	3.50	0.99
10	0.09%	4	0.93

the conversation of those tweets (see Sect. 2). The goal was to remove the tweets that were related to each other as they were part of the same conversation in order to analyse the connections between different and independent tweets. The resulting graph $T_{initial}$ is a graph where most of the nodes have no connection at all. However, those nodes who have connections belong only to misinformation news.

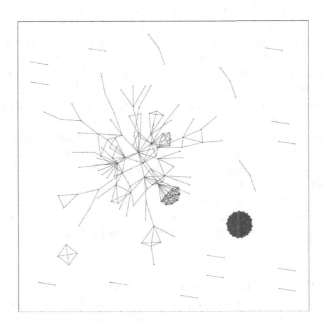

Fig. 2. Subgraph of initial tweets

The Fig. 2 represents the graph of the nodes that had at least one connection from the mentioned graph. Tweets mentioning misinformation news are depicted with red dots, while tweets mentioning legitimate news are in blue. As observed, only red nodes are represented[4]. We interpret this as there is only a connection between the users interacting with misinformation news. Thus, a user who interacts with a tweet containing misinformation is more probable for it to interact with another tweet containing misinformation. This reinforces the findings from [7]. To compare how nodes from the same type of news tend to connect to nodes from their similar type we calculated the discrete assortativity coefficient [9] for the graph T described in Sect. 2. We get a discrete assortativity of 0.898. A high assortativity coefficient means that nodes do not tend to be connected if they are different, so there are few links between misinformation news nodes and legitimate news nodes which means there are few users that interact with both misinformation tweets and legitimate tweets. This reinforce the idea that users tend to interact with the same type of news.

3.2 Tweets Influence

On a general perspectives, influence has been extensively studied using users on Twitter [12], as well as the measure of the tweet influence itself [4]. These metrics measure the influence of tweets by counting their number of retweets following the logic that if a tweet is retweeted it means the message in the tweet was strong enough for the user to share it with others and being presented to a new set of users, contributing to the spread of the news. Previous studies have shown that the number of retweets follows a power law distribution [5]. Hence, our following analysis aims to verify whether the number of tweet count per number of retweets also followed a power law. One way to measure the influence of a tweet could by the degree centrality measure [10]. However, this method is not adequate in our case, due to the limitations for the graph construction It does not represent a global measure of influence but rather in the context of the graph, and the way it was built. Another way to measure the influence of a tweet is by counting how many times the tweets has been retweeted.

We fitted the tweet counts by number of retweets to a power law using the tool developed in [1] and observed that they follow a power law with an $\alpha = 1.80$ as seen in Fig. 3. This means that tweets that have a lot of retweets, and by our definition are more influential, are less frequent. We can have a coefficient of how influential a tweet is by the retweets and the Probability Density Function (PDF) of the power law.

$$c_i(t) = 1 - r_t^{-\alpha} \tag{1}$$

Equation 1 shows us this Retweet Influence Coefficient (RIC) c_i for a tweet t which has r_t retweets and in our case $\alpha = 1.80$. Since we want the tweets with more influence to have a higher RIC we took the inverse of the PDF by subtracting it to 1.

---

[4] The figure shows a big clique of 45 nodes, those tweets are the same tweets being posted by the same user and this user liking its own tweets.

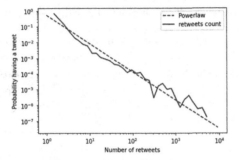

Fig. 3. Powerlaw fitting to the tweet count by number of retweets

In Table 3 we can see that the tweets that are the most connected are not necessarily the one with the most influence. One of the causes of this, might be to the graph construction.

Doing a qualitative investigation about the difference between the tweets with high degree centrality and the tweets with high RIC, we found out that tweets that have high degree centrality do not have any particularity to be the most influential. Some were comments, some were original tweets, some had lots of replies and others barely where liked.

Therefore we cannot consider the degree centrality of a tweet as reliable measure of influence for this type of graphs.

On contrary, using our measure of RIC, the tweets with the highest RIC were tweets that carried an important message and/or where published by someone well known like an important politician.

When comparing the misinformation news RIC score (RIC = 0.164) and the legitimate news' one (RIC = 0.156) no significant difference were found (one-way ANOVA: $F(2, 4751) = 1.05$, p-value = 0.41). Legitimate news and misinformation news have the same probability to be retweeted.

4   User Analysis

4.1   User Graph

For the user graph the nodes are the users, and a link between two users if they interacted with the same tweet. The resulting graph contains 32500 nodes and 1006998 edges. The mean degree in the graph is 63.61, the maximum 821, and the minimum is 1. Having a higher degree than in the tweets graphs is probably due to the fact that users tend to interact with the tweets in the same context and even more in the same conversation.

Table 5 shows that the most connected users are also the ones who are the most active (produce more tweets), have more followers and follow more people.

In case a user interacted or authored more tweets related to misinformation news it was marked as misinformation user, otherwise it was marked as legitimate

Table 5. Highest degree nodes compared to the average node

Measure	General mean	Mean 100 highest degree	Mean 10 highest degree
Tweets posted	30141.27	119146.60	219308.70
Followers	1466.63	4427.49	2405.50
Following	1204.13	2055.07	2852.60

user. We found that for a total of 31629 users, 25703 users were misinformation users and 5926 were legitimate users. This is coherent with the nature of our dataset.

If we calculate the discrete assortativity coefficient [9] in the users graph U to measure how nodes connect to similar nodes when comparing about misinformation users and legitimate users, we get a 0.88 coefficient which is high. This is expected since legitimate users will interact with legitimate news and misinformation users interact with misinformation news according to our definition.

In addition, we clustered the users based on their graph U using the Louvain algorithm [3]. Then we identified those clusters where there were a mix of legitimate and misinformation users. There are 70 groups, with each group having mean size of 295.3 users. With the smallest group being of 3 users and the biggest group being of 1877 users.

Table 6. 10 biggest communities that contain misinformation and legitimate users

Community	Legitimate users	Misinformation users	Proportion
62	469	8	98.32% - 1.68%
14	5	508	0.97% - 99.03%
49	106	557	**15.99% - 84.01%**
11	602	67	**89.99% - 10.01%**
47	1	709	0.14% - 99.86%
57	710	55	**92.81% - 7.19%**
12	6	813	0.73% - 99.27%
54	2	1010	0.19% - 99.81%
6	55	1000	5.21% - 94.79%
10	23	1854	1.23% - 98.77%

Table 6 shows that most of the communities are very unbalanced in favour of a type of user. Nonetheless, we created the community sub-graph of the *most balanced* communities to see how users from other groups interacted with each other, which is represented in Fig. 4. We can observe that even by creating mixed communities sub-graphs, the communities remains homogeneous. For example

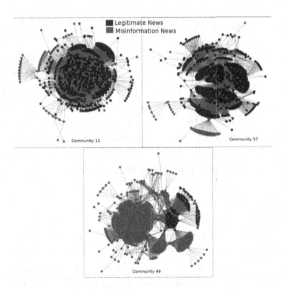

Fig. 4. Mixed user communities

in community 49, mostly composed of misinformation users, there are 2 clusters of legitimate users. In communities 11 and 57, mostly composed by legitimate users, most of the misinformation users are in the peripheries of the cluster or linked by bridge users to the legitimate users. In conclusion legitimate users tend to connect to legitimate users, and misinformation users tend to connect to misinformation even in the case of mixed communities and even when connection between the two communities exist, they are only connected through a few users.

4.2   User Influence

Our analysis to measure the influence of users in Twitter is based on a metric proposed by [11].

They defined the *Retweet Impact* as shown in Eq. 2. Where OG is the number of original tweets, and RT is the number of users who have retweeted the author's tweets.

$$RI = OG \cdot log(RT) \tag{2}$$

For our case we used the total count of retweets instead of the total of different users that have retweeted at least once one of the author's tweets to account from the Twitter's API limits of 100 users retweets data collection.

A qualitative analysis of the 100 most influential users, we found that 78 were misinformed (38 from Spain) and 22 were legitimate users (18 from Spain). Despite having 81% of the total users misinformed, this proportion is not maintained with the most influential users. Even though there users interacting with misinformation, those users are not necessarily more influential than legitimate users.

We also found out that most of the misinformed users from this list of 100 most influential post content about politics, and religion. Where 13 of those users post content related to extreme right ideas, 4 of those users post content related to extreme left ideas, and finally 3 of those users post religious content. Most of the legitimate users from the list of 100 most influential users post content about scientific diffusion and fight against misinformation.

Finally we found that from this list of 100 most influential users, 44 users are not from Spain or their region could not be identified.

4.3   Users Location

We developed a tool to identify the location of the users based on their location input and their description. Even though we selected news from Spanish sources we could not avoid getting users from other countries specially from Latin America. As shown in Table 7, for Spanish users, our user sample reflect the proportion of the different provinces.

Table 7. Users, authors of initial dataset, per Spanish localisation

Location	Number of users
Unknown	2513
Comunidad de Madrid	168
Cataluña	106
España	65
Andalucía	46
Galicia	44
Comunidad Valenciana	43
Castilla y León	10
Principado de Asturias	19
Canarias	16
Región de Murcia	15
Others Spanish regions	41

A qualitative review from a sample from users that have an unknown location showed that around one third of those users still are from Spain. The rest of the users where in minority from an unknown location or from a Spanish speaking country. Nevertheless we cannot ignore the rest of the users coming from another country. We could say that the level of misinformation also depends from the level of misinformation in other Spanish speaking countries.

5   Discussion and Future Work

In this study we analysed the tweet influence among conversations on Twitter about misinformation and legitimate news. Our goal was to develop a better understanding of the different news spread patterns, as well as tackling some limitations inherent to influence metrics. A common issue raised by metrics, such as degree centrality measure, its their dependency upon the graph construction process itself. This led us to develop the RIC measure which has the advantages to be independent to the graph structure and capturing the influence of the tweets appropriately.

While being more adequate to unveil influential tweets, our influence metric did not show any significant difference between misinformation and verified news, arguing for a similar impact of both type of news reveal a difference in the users behaviour. Users that spread and/or interact with misinformation tend to post more about it than users who post about verified news. This effect was also found in [2], where authors observed that misinformation users are more focused on spreading their ideas than legitimate users.

But since there is not a significant difference in the influence of misinformation tweets compared to legitimate tweets, both news are shared the same amount of times. Misinformation is being posted more often, resulting in a wider spread thus a different pattern of communication where users exhibit different behaviours if they tend to post about misinformation rather than legitimate news. No only does the users from different groups behave differently, they also do not interact with each others. They tend to interact with other users that share the same type of news than them and thus are prone to the echo chamber effect, confirming finding from previous studies [2,7].

Another important aspect in term of combating misinformation is the language. Users are connected with people from other countries that speak the same language. This raises the concern of how do spread patterns of misinformation behave. Our findings indicate that language prevails over geographic factors, exacerbating the difficulties for governments to control the dissemination of misleading information.

References

1. Alstott, J., Bullmore, E., Plenz, D.: PowerLaw: a Python package for analysis of heavy-tailed distributions. PLoS One 9(1), e85777 (2014). https://doi.org/10. 1371/journal.pone.0085777. http://arxiv.org/abs/1305.0215. arXiv: 1305.0215
2. Bessi, A., Coletto, M., Davidescu, G.A., Scala, A., Caldarelli, G., Quattrociocchi, W.: Science vs conspiracy: collective narratives in the age of misinformation. PLOS One 10(2), e0118093 (2015). https://doi.org/10.1371/journal.pone.0118093. https://journals.plos.org/plosone/article?id=10.1371/journal.pone.0118093. Publisher: Public Library of Science
3. Blondel, V.D., Guillaume, J.L., Lambiotte, R., Lefebvre, E.: Fast unfolding of communities in large networks. J. Stat. Mech.: Theory Exp. 2008(10), P10008 (2008). https://doi.org/10.1088/1742-5468/2008/10/P10008. http://arxiv. org/abs/0803.0476. arXiv: 0803.0476

4. Huang, B., Carley, K.M.: Disinformation and misinformation on Twitter during the novel coronavirus outbreak. arXiv:2006.04278 [cs], June 2020. http://arxiv.org/abs/2006.04278

5. Mathews, P., Mitchell, L., Nguyen, G., Bean, N.: The nature and origin of heavy tails in retweet activity. In: Proceedings of the 26th International Conference on World Wide Web Companion - WWW 2017 Companion, pp. 1493–1498. ACM Press, Perth (2017). https://doi.org/10.1145/3041021.3053903. http://dl.acm.org/citation.cfm?doid=3041021.3053903

6. McQuillan, L., McAweeney, E., Bargar, A., Ruch, A.: Cultural convergence: insights into the behavior of misinformation networks on Twitter. arXiv:2007.03443 [physics], July 2020. http://arxiv.org/abs/2007.03443. arXiv: 2007.03443

7. Memon, S.A., Carley, K.M.: Characterizing COVID-19 misinformation communities using a novel twitter dataset. arXiv:2008.00791 [cs], September 2020. http://arxiv.org/abs/2008.00791

8. Myers, S.A., Sharma, A., Gupta, P., Lin, J.: Information network or social network?: the structure of the twitter follow graph. In: Proceedings of the 23rd International Conference on World Wide Web, pp. 493–498. ACM, Seoul Korea, April 2014. https://doi.org/10.1145/2567948.2576939. https://dl.acm.org/doi/10.1145/2567948.2576939

9. Newman, M.E.J.: Mixing patterns in networks. Phys. Rev. E **67**(2), 026126 (2003). https://doi.org/10.1103/PhysRevE.67.026126. http://arxiv.org/abs/cond-mat/0209450. arXiv: cond-mat/0209450

10. Nieminen, J.: On the centrality in a graph. Scand. J. Psychol. **15**(1), 332–336 (1974). https://doi.org/10.1111/j.1467-9450.1974.tb00598.x. https://onlinelibrary.wiley.com/doi/abs/10.1111/j.1467-9450.1974.tb00598.x

11. Pal, A., Counts, S.: Identifying topical authorities in microblogs. In: Proceedings of the fourth ACM international conference on Web search and data mining - WSDM 2011, p. 45. ACM Press, Hong Kong (2011). https://doi.org/10.1145/1935826.1935843. http://portal.acm.org/citation.cfm?doid=1935826.1935843

12. Riquelme, F., González-Cantergiani, P.: Measuring user influence on Twitter: a survey. Inform. Process. Manage. **52**(5), 949–975 (2016). https://doi.org/10.1016/j.ipm.2016.04.003. http://arxiv.org/abs/1508.07951. arXiv: 1508.07951

13. Salaverría, R., Buslón, N., López-Pan, F., León, B., López-Goñi, I., Erviti, M.C.: Desinformación en tiempos de pandemia: tipología de los bulos sobre la Covid-19. El Prof. Inform. **29**(3) (2020). https://doi.org/10.3145/epi.2020.may.15. https://revista.profesionaldelainformacion.com/index.php/EPI/article/view/epi.2020.may.15

14. Singh, L., et al.: A first look at COVID-19 information and misinformation sharing on Twitter. arXiv:2003.13907 [cs], March 2020. http://arxiv.org/abs/2003.13907

15. Viviani, M., Pasi, G.: Credibility in social media: opinions, news, and health information—a survey. WIREs Data Min. Knowl. Discov. **7**(e01209), 1–25 (2017). https://doi.org/10.1002/widm.1209. https://onlinelibrary.wiley.com/doi/abs/10.1002/widm.1209

Comparing Community-Aware Centrality Measures in Online Social Networks

Stephany Rajeh$^{(\boxtimes)}$ ⓘ, Marinette Savonnet ⓘ, Eric Leclercq ⓘ,
and Hocine Cherifi ⓘ

Laboratoire d'Informatique de Bourgogne, University of Burgundy, Dijon, France
stephany.rajeh@u-bourgogne.fr

Abstract. Identifying key nodes is crucial for accelerating or impeding dynamic spreading in a network. Community-aware centrality measures tackle this problem by exploiting the community structure of a network. Although there is a growing trend to design new community-aware centrality measures, there is no systematic investigation of the proposed measures' effectiveness. This study performs an extensive comparative evaluation of prominent community-aware centrality measures using the Susceptible-Infected-Recovered (SIR) model on real-world online social networks. Overall, results show that K-shell with Community and Community-based Centrality measures are the most accurate in identifying influential nodes under a single-spreader problem. Additionally, the epidemic transmission rate doesn't significantly affect the behavior of the community-aware centrality measures.

Keywords: Complex networks · Centrality · Influential nodes · Community structure · SIR model

1 Introduction

With the plethora of data flowing into online social networks, representing the main entities and their interactions is essential. Networks offer an ideal representation of such complex systems to investigate their structure and dynamics. Identifying influential nodes is crucial for many applications such as designing lucrative marketing campaigns, targeting terrorist attacks, controlling epidemic spreading, detecting financial risks, and extracting salient features from visual content [3,9,14,16,20,23,24]. Centrality is one of the main approaches employed to do so. Classically, centrality measures exploit the topology and dynamics of networks [16]. They can be classified into two main groups, namely local and global. The former uses the node's neighborhood, while the latter incorporates all of the network's information to quantify a node's influence. They can also be combined [10,28].

Many real-world networks contain densely connected zones that are loosely linked to each other. This so-called community structure is a ubiquitous feature in natural and artificial systems [6]. The network's structure and dynamics are

D. Mohaisen and R. Jin (Eds.): CSoNet 2021, LNCS 13116, pp. 279–290, 2021.
https://doi.org/10.1007/978-3-030-91434-9_25

significantly affected by communities [19]. Recently developed centrality measures exploit this information to identify influential nodes [5, 7, 8, 17, 18, 30, 32]. We refer to them as "community-aware" centrality measures. Unlike classical centrality measures, community-aware centrality measures differentiate between the node's intra-community links (links between nodes in the same community) and inter-community links (links between nodes in different communities). Intra-community links exert influence at the community level, while inter-community links exert influence at the network level [22]. The difference between community-aware measures is mainly based on how intra-community links and inter-community links are associated together. For example, Comm centrality [8] preferentially selects bridges over hubs by prioritizing inter-community links over intra-community links. Community-based Mediator [30] favors nodes with unbalanced intra-community and inter-community links.

With limited resources, it is essential to identify top influential nodes either for maximizing or for minimizing the diffusion in online social networks. The Susceptible-Infected-Recovered (SIR) model is commonly used to model disease and rumor spreading [1]. Starting with a small set of initial spreaders defined by a specific centrality measure, the goal is to evaluate its ability to reach the maximum outbreak size.

The SIR model has been widely used to investigate the behavior of various classical centrality measures [2, 11, 15]. Studies on community-aware centrality measures examine either a small number of the proposed solutions in the literature or experiments are performed on a small sample of networks [8, 17, 18, 30, 32]. Therefore, there is no consensus about the effectiveness of the most popular measures on online social networks, where communities are naturally prevalent [4, 13, 29]. This paper aims to fill this gap. An extensive investigation of seven community-aware centrality measures is performed on ten real-world online social networks using the SIR diffusion model under a single-spreader scheme.

The paper is organized as follows. Section 2 introduces the community-aware centrality measures. Section 3 presents the networks, the tools, and the methodology applied. Experimental results are provided in Sect. 4. The main findings are discussed in Sect. 5. Finally, in Sect. 6, the conclusion is given.

2 Community-Aware Centrality Measures

In this section, we briefly recall the definitions of the seven community-aware centrality measures under test. Let $G(V, E)$ be an undirected and unweighted graph where V is the set of nodes, E is the set of edges, and $N = |V|$ is the size of the network. It is partitioned into N_c non-overlapping communities where c_k is k-th community. A node i possess k_i^{intra} intra-community links and k_i^{inter} inter-community links such that $k_i^{tot} = k_i^{intra} + k_i^{inter}$ represents its degree. Note that if the community structure is unknown, a community detection algorithm is needed to uncover it.

    1. Community Hub-Bridge [5] weights the intra-community links of a node by its community size. The inter-community links are weighted by the number of communities reached by the node. It is defined as follows:

$$\alpha_{CHB}(i) = |c_k| \times k_i^{intra} + NNC_i \times k_i^{inter} \tag{1}$$

where $|c_k|$ is the size of the community of node i and NNC_i is the number of communities linked to node i.

2. Participation Coefficient [7] gives more importance on the heterogeneity of the inter-community links of a node. If the node's links are uniformly distributed across the communities, its centrality value is one. It is defined as follows:

$$\alpha_{PC}(i) = 1 - \sum_{c=1}^{N_c} \left(\frac{k_{i,c}}{k_i^{tot}} \right)^2 \tag{2}$$

where $k_{i,c}$ is the number of links node i has in a given community c.

3. Community-based Mediator [30] uses entropy to quantify the node's importance through its intra-community and inter-community links. It is defined as follows:

$$\alpha_{CBM}(i) = H_i \times \frac{k_i^{tot}}{\sum_{i=1}^{N} k_i^{tot}} \tag{3}$$

where $H_i = [-\sum \rho_i^{intra} log(\rho_i^{intra})] + [-\sum \rho_i^{inter} log(\rho_i^{inter})]$ is the entropy of node i based on its ρ^{intra} and ρ^{inter} which represent the node's ratio of intra-community and inter-community links.

4. Comm Centrality [8] weights the intra-community links and inter-community links by the ratio of external links. It also prioritizes bridges over hubs. It is defined as follows:

$$\alpha_{Comm}(i) = (1 + \mu_{c_k}) \times \chi + (1 - \mu_{c_k}) \times \varphi^2 \tag{4}$$

where μ_{c_k} is the proportion of inter-community links over the total community links in community c_k, $\chi = \frac{k_i^{intra}}{max_{(j \in c)} k_j^{intra}} \times R$, $\varphi = \frac{k_i^{inter}}{max_{(j \in c)} k_j^{inter}} \times R$, and R is a constant to scale intra-community and inter-community values to the same range.

5. Modularity Vitality [18] is based on the modularity variation due to the node removal from the network. Removal of a bridge node increases the modularity, while removal of internal a hub decreases the modularity. It is defined as follows:

$$\alpha_{MV}(i) = M(G) - M(G_i) \tag{5}$$

where $M(G)$ is the modularity of a network and $M(G_i)$ is the network's modularity after the removal of node i. Note that Modularity Vitality is a signed centrality. In this study, we use its absolute value to rank the nodes.

6. Community-based Centrality [32] is based on weighting the node's intra-community and inter-community links by the subsequent sizes of their belonging communities. It is defined as follows:

$$\alpha_{CBC}(i) = \sum_{c=1}^{N_c} k_{i,c} \left(\frac{n_c}{N} \right) \tag{6}$$

where n_c is the number of nodes in community c and $k_{i,c}$ is the number of links node i has in a given community c.

7. K-shell with Community [17] is based on the k-shell (also called k-core) hierarchical decomposition of the network composed of intra-community links and the network composed of inter-community links, separately. A weighting parameter then combines the two values to prioritize the selection of hubs or bridges. It is defined as follows:

$$\alpha_{ks}(i) = \delta \times \alpha^{intra}(i) + (1 - \delta) \times \alpha^{inter}(i) \tag{7}$$

where $\alpha^{intra}(i)$ and $\alpha^{inter}(i)$ stand for the k-shell value of node i by only considering intra-community links and inter-community links, respectively. δ is set to 0.5 in this study.

3   Data, Tools, And Methods

3.1   Data

This study uses ten unweighted and undirected online social networks publicly available. They originate from various online platforms (Facebook, Twitter, Deezer, Hamsterster, and Pretty Good Privacy). Table 1 reports their basic topological characteristics. As their community structure is unknown, it is uncovered by Infomap [26].

1. Facebook Friends [21]: Nodes are users from a Facebook ego network extracted in April 2014. Edges between two users mean they are "friends" on Facebook.

2. Retweets Copenhagen [25]: Nodes are Twitter users tweeting while the United Nations conference in Copenhagen about climate change was taking place. Edges represent retweets.

3. Caltech [25]: Nodes are users on Facebook enrolled at Caltech University. Edges between two users mean they are "friends" on Facebook.

4. Ego Facebook [25]: Nodes are users on Facebook participating in a survey conducted on Facebook. Edges between two users mean they are "friends" on Facebook.

5. Hamsterster [12]: Nodes represent users from an online social pet network hamsterster.com. Edges represent friendships between the users.

6. Facebook Organizations [21]: Nodes are users on Facebook who work in the same company. Edges between two users mean they are "friends" on Facebook.

7. Facebook Politician Pages [25]: Nodes are Facebook pages of politicians from different countries. Edges represent mutual likes of Facebook users among the given pages.

Table 1. Topological features of the networks. N is the number of nodes. $|E|$ is the number of edges. $< k >$ is the average degree. ζ is the transitivity. μ is the mixing parameter. λth is the epidemic threshold. * means the largest connected component of the network is taken if it is disconnected.

Network	N	$|E|$	$< k >$	ζ	μ	λth
Facebook Fri.*	329	1,954	11.88	0.512	0.112	0.048
Retweets Co.	761	1,029	2.70	0.060	0.287	0.139
Caltech*	762	16,651	43.70	0.291	0.410	0.048
Ego Facebook	4,039	88,234	43.69	0.519	0.077	0.009
Hamsterster*	1,788	12,476	13.49	0.090	0.298	0.022
Facebook Org.	5,524	94,219	34.11	0.222	0.366	0.016
Facebook Pol	5,908	41,729	14.12	0.301	0.111	0.024
Princeton*	6,575	293,307	89.21	0.163	0.365	0.006
PGP	10,680	24,316	4.55	0.378	0.172	0.056
DeezerEU	28,281	92,752	6.55	0.095	0.429	0.066

8. Princeton [25]: Nodes are users on Facebook enrolled at Princeton University. Edges between two users mean they are "friends" on Facebook.

9. PGP [12]: Nodes are users from the web of trust, utilizing Pretty Good Privacy (PGP) encryption for sharing information online. Edges between users represent sharing data under secure connections.

10. DeezerEU [27]: Nodes represent users from Deezer, a European platform for music streaming. Edges represent online friendships between users.

3.2 Susceptible-Infected-Removed Model

The Susceptible-Infected-Removed (SIR) model is one of the widely used diffusion models in networks. Initially, a single node or a set of nodes (f_o) is in the infectious state (I) while the remaining nodes are in the susceptible state (S). At each iteration, an infectious node infects its susceptible neighbors at a rate λ. Previously infected nodes recover and are removed from the network at a rate γ. The spreading continues until there are no infectious nodes. At this point, the number of nodes in the "Recovered" state indicates the spreading power of the single node or the initial set of nodes (f_o). Each network has an epidemic threshold (λ_{th}) controlling the epidemic spreading. It is defined as [31]:

$$\lambda_{th} = \frac{< k >}{< k^2 > - < k >} \tag{8}$$

where $< k >$ and $< k^2 >$ are the first and second moments of the network's degree distribution. The epidemic threshold values are reported in Table 1.

3.3   Imprecision Function

The imprecision function [11] measures the performance of a centrality measure in predicting influential spreaders. It is based on the average number of infections due to an infected seed node. It is defined as follows:

$$\epsilon_c(p) = 1 - \frac{M_c(p)}{M_{eff}(p)} \qquad (9)$$

where p is a value between $[0,1]$, $M_c(p)$ is the average spreading power of top pN nodes ranked according to a specific centrality measure c, and $M_{eff}(p)$ is the average spreading power of top pN nodes ranked according to their influence in the SIR model (N is the number of nodes). The smaller the value of $\epsilon_c(p)$, the better the performance of the centrality measure c.

3.4   Methods

The SIR model runs on each network using different transmission rates around the epidemic threshold ($\frac{\lambda_{th}}{2}$, $\frac{\lambda_{th}}{1.5}$, $1.5 \times \lambda_{th}$, $2 \times \lambda_{th}$). The recovery rate γ is set to 1 to measure the spreading ability of the seed node initiating the spreading only. For each transmission rate, 1000 independent simulations of the SIR model are performed in networks with less than 6000 nodes and 100 independent simulations otherwise. The SIR spread size of each node in the network is computed after setting it as the seed of diffusion. The set ordered from highest to smallest SIR spread size is called the reference set. The community-aware centrality measures are computed, and nodes are ranked from higher to lower centrality value. For each transmission rate (λ), we calculate the imprecision function over the top fraction pN nodes.

4   Experimental Results

4.1   Performance of the community-Aware Centrality Measures Within Networks

Figure 1 illustrates the performance of the community-aware centrality measures for the ten networks under study. The transmission rate is set equal to the epidemic threshold (λ_{th}) for each network. Each figure reports the evolution of the seven community-aware centrality measures' imprecision function when the top spreading nodes' size ranges from $p = 0.02$ to $p = 0.2$ of the network size. Remember that the lower the value of the imprecision function, the more effective the centrality measure. One can observe that the performance generally increases with the proportion of top spreading nodes. Furthermore, no community-aware centrality measure outperforms the others in all the situations. Overall, there is a high variability of community-aware centrality measures performances within and across networks. For example, In Ego Facebook, at $p = 0.02$, the imprecision value of K-shell with Community is 0.38, followed by Modularity Vitality at 0.6.

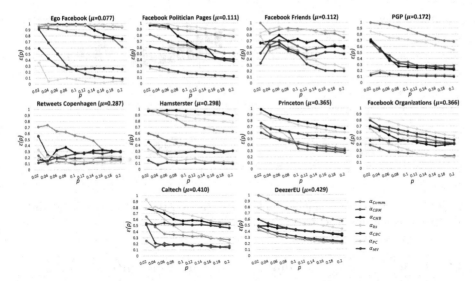

Fig. 1. The imprecision function $\epsilon(p)$ for the 7 community-aware centrality measures on each network. The transmission rate is set to λ_{th} and the recovery rate is set to 1. μ is the mixing parameter, the ratio of inter-community over total community links in a network. The community-aware centrality measures are: Comm Centrality = α_{Comm}, Community-based Mediator = α_{CBM}, Community Hub-Bridge = α_{CHB}, K-shell with Community = α_{ks}, Community-based Centrality = α_{CBC}, Participation Coefficient = α_{PC}, and Modularity Vitality = α_{MV}.

Then all others have an imprecision value between 0.9 and 1. The variability among the community-aware centralities persists till $p = 0.2$. K-shell with Community now has a value of 0.05, indicating its high accuracy at higher p. In the same vein comes Community-based Centrality, which has an imprecision value of 0.1. Its accuracy improves by almost 90% compared to its value at $p = 0.02$. Modularity Vitality follows with an imprecision value of 0.25, improving in almost half value of $\epsilon(p)$. Community-based Mediator improves from 0.92 ($p = 0.02$) to a value of 0.61 ($p = 0.20$). Community Hub-Bridge also improves, but in a lower proportion. Finally, Participation Coefficient and Comm Centrality show a negligible improvement. There is also a high variability for the same community-aware centrality measures across networks. For example, in Facebook Politician Pages, the imprecision value of Community-based Mediator at $p = 0.02$ is 0.81, while in Caltech, it amounts to 0.25. Another example is Community-based Centrality in Ego Facebook amounting to 0.9 at $p = 0.02$ while it amounts to 0.11 in PGP.

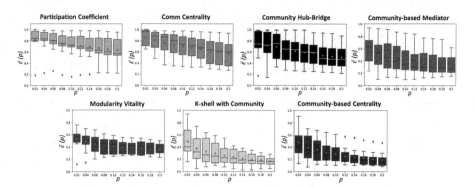

Fig. 2. The average of the imprecision function $\bar{\epsilon}(p)$ over the 10 online social networks. The transmission rate is set to λ_{th} and the recovery rate is set to 1.

4.2 Performance of the community-Aware Centrality Measures Across Networks

Each community-aware centrality measure's imprecision function is averaged over the ten networks for all p values. The goal is to better understand the performance consistency. Figure 2 illustrates these results. The most stable (low variability) community-aware centrality measure is Modularity Vitality. Despite the change in p, the imprecision function values remain stable and condensed. Then comes K-shell with Community and Community-based Centrality. Even though they show high variability when $p \leq 0.08$, both are very consistent afterward. On the opposite, the remaining community-aware centrality measures show higher variability as p increases. The average imprecision function $\bar{\epsilon}(p)$ illustrates the high accuracy of K-shell with Community and Community-based Centrality for all p values. It ranges from 0.5 for the lowest p value to 0.1 at the highest p value. Then comes Modularity Vitality, with $\bar{\epsilon}(p) = 0.55$ at $p = 0.02$ and $\bar{\epsilon}(p) = 0.40$ at $p = 0.20$. Community-based Mediator has similar $\bar{\epsilon}(p)$ values as Modularity Vitality, yet it has high variability. Community Hub-Bridge shows $\bar{\epsilon}(p)$ between 0.75 and 0.5 at $p = 0.02$ and $p = 0.20$, respectively. Participation Coefficient and Comm Centrality perform poorly. Their minimum for $\bar{\epsilon}(p)$ is around 0.6, and their maximum is around 0.8. These results confirm the results of Fig. 1.

4.3 Influence of the transmission Rate

In this experiment, we study the effect of varying the transmission rate (λ) in the SIR model around the epidemic threshold (λ_{th}). Figure 3 shows the average imprecision function $\bar{\epsilon}(p)$ of the seven community-aware centrality measures at five different transmission rates. The average imprecision function $\bar{\epsilon}(p)$ is calculated considering a low portion of top nodes ($p = 0.02$), a medium portion of top nodes ($p = 0.10$), and a high portion of top nodes ($p = 0.20$).

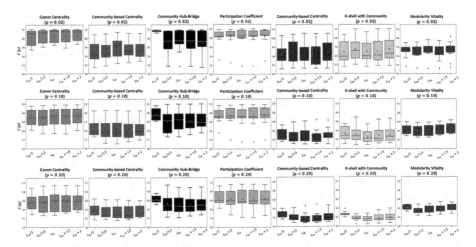

Fig. 3. The average of the imprecision function $\bar{\epsilon}(p)$ over the 10 online social networks as a function of five different transmission rates $(\frac{\lambda_{th}}{2}, \frac{\lambda_{th}}{1.5}, \lambda_{th}, 1.5 \times \lambda_{th}, 2 \times \lambda_{th})$. The recovery rate is set to 1. The upper, middle, and bottom figures show the results at $p = 0.02$, $p = 0.10$, and $p = 0.20$, respectively.

At low p values ($p = 0.02$), results are generally comparable. For example, the mean of $\bar{\epsilon}(p)$ for Comm Centrality at the five different transmission rates is in the vicinity of 0.8. Also, the boxplots' interquartile range is quite similar, indicating that the values are consistent across λ. Participation Coefficient, Community-based centrality, K-shell with Community, and Modularity Vitality also show consistent results. In contrast, Community Hub-Bridge is the most sensitive to the variation of the transmission rate. When $\lambda = \frac{\lambda_{th}}{2}$, Community Hub-Bridge cannot differentiate the nodes. Indeed, the mean $\bar{\epsilon}(p)$ is 0.98, and the interquartile range's height is very narrow. When the transmission rate is set to $\frac{\lambda_{th}}{1.5}$, λ_{th}, $1.5 \times \lambda_{th}$, and $2 \times \lambda_{th}$, $\bar{\epsilon}(p)$ becomes quite comparable. The consistency of the mean and the interquartile range of $\bar{\epsilon}(p)$ persist at $p = 0.10$ and at $p = 0.20$. Indeed, they share approximately the same values of $\bar{\epsilon}(p)$ for all community-aware centrality measures except for Community Hub-Bridge. Although now its interquartile range is wider compared to that of $p = 0.02$ when $\lambda = \frac{\lambda_{th}}{2}$, the mean and interquartile range are much different than the others.

5   Discussion

This study aims to investigate the behavior of popular community-aware centrality measures in online social networks. Community-aware centrality measures quantify a node's importance based on its local influence (inside its community using intra-community links) and its global impact (outside of its community using inter-community links). Yet, each community-aware centrality measure processes these two types of links distinctively.

A comparative evaluation of seven community-aware centrality measures is performed using the SIR diffusion model under a single-spreader scheme. The imprecision function quantifies the centrality measure's accuracy by comparing the spreading power of top nodes according to a centrality measure compared to their ground truth spreading efficiency. Results indicate that K-shell with Community and Community-based Centrality outperform the alternative community-aware centrality measures. K-shell with Community exploits the hierarchical structure of the networks while taking into consideration its community structure. This result corroborates the study reported in [11].

Indeed, under a single-spreader setting, nodes identified using k-shell are the most accurate in predicting spreading outbreaks in networks. The performance of Community-based Centrality is also on the same line as the findings of the authors who proposed this measure [32]. This study shows that this community-aware centrality measure is accurate in online social networks with communities of different sizes. Results also show that Community-based Mediator is somewhat sensitive to the community structure strength. Indeed, as shown in Fig. 1, when the network has a strong community structure ($\mu \leq 0.172$), it performs poorly. Yet, as the network has a weaker community structure, it becomes as accurate as K-shell with Community and Community-based Centrality. This centrality exploits the heterogeneity of links to assess the node's importance. Indeed, in a weak community structure, a node possesses a higher number of inter-community links than intra-community links. It explains why it performs better in a weak community structure. Modularity Vitality is the most consistent across networks, regardless of the strength of the community structure. The low accuracy of Participation Coefficient, Comm Centrality, and Community Hub-Bridge may be linked to the fact that they give a lot of importance to bridges. Besides bridges, online social networks also include hubs inside their communities that play a major role in information dissemination.

6   Conclusion

Identifying influential nodes in online social networks is fundamental for maximizing information diffusion and inhibiting fake news spreading. The community structure of a network plays a crucial role in the dynamics of these spreading processes. This work investigates the effectiveness of prominent community-aware centrality measures to target influential nodes using the SIR diffusion process under a single-spreader scheme. Results show that K-shell with Community and Community-based Centrality are the most accurate community-aware centrality measures. Additionally, performances are pretty insensitive to variation of the transmission rate. Therefore, this work gives clear indications about which community-aware centrality measure to use when resources are restrained to target single nodes. Nevertheless, practitioners need to be conscious that the community aware-centrality measures accuracy depends on the seed node size. As in numerous situations, the community structure is unknown. Future work will investigate the results consistency using alternative community detection algorithms.

Another direction of research is to study the influence of the community-aware centrality measures using different propagation processes. Finally, we are planning to link the performances to the network topological properties.

References

1. Anderson, R.M., May, R.M.: Population biology of infectious diseases: part i. Nature **280**(5721), 361–367 (1979)
2. Bucur, D.: Top influencers can be identified universally by combining classical centralities. Sci. Rep. **10**(1), 1–14 (2020)
3. Demirkesen, C., Cherifi, H.: A comparison of multiclass SVM methods for real world natural scenes. In: Blanc-Talon, J., Bourennane, S., Philips, W., Popescu, D., Scheunders, P. (eds.) Advanced Concepts for Intelligent Vision Systems, ACIVS 2008. LNCS, vol. 5259, pp. 752–763. Springer, Heidelberg (2008). https://doi.org/10.1007/978-3-540-88458-3_68
4. Gaisbauer, F., Pournaki, A., Banisch, S., Olbrich, E.: Ideological differences in engagement in public debate on twitter. PLoS ONE **16**(3), e0249241 (2021)
5. Ghalmane, Z., Hassouni, M.E., Cherifi, H.: Immunization of networks with non-overlapping community structure. Soc. Netw. Anal. Min. **9**(1), 1–22 (2019). https://doi.org/10.1007/s13278-019-0591-9
6. Girvan, M., Newman, M.E.: Community structure in social and biological networks. PNAS **99**(12), 7821–7826 (2002)
7. Guimera, R., Amaral, L.A.N.: Functional cartography of complex metabolic networks. Nature **433**(7028), 895–900 (2005)
8. Gupta, N., Singh, A., Cherifi, H.: Centrality measures for networks with community structure. Physica A **452**, 46–59 (2016)
9. Hassouni, M., Cherifi, H., Aboutajdine, D.: Hos-based image sequence noise removal. IEEE Trans. Image Process. **15**(3), 572–581 (2006)
10. Ibnoulouafi, A., El Haziti, M., Cherifi, H.: M-centrality: identifying key nodes based on global position and local degree variation. J. Stat. Mech: Theor. Exp. **2018**(7), 073407 (2018)
11. Kitsak, M., Gallos, L.K., Havlin, S., Liljeros, F., Muchnik, L., Stanley, H.E., Makse, H.A.: Identification of influential spreaders in complex networks. Nat. Phys. **6**(11), 888–893 (2010)
12. Kunegis, J.: Handbook of network analysis [konect-the koblenz network collection]. arXiv preprint arXiv:1402.5500 (2014)
13. Labatut, V., Dugué, N., Perez, A.: Identifying the community roles of social capitalists in the twitter network. In: 2014 IEEE/ACM International Conference on Advances in Social Networks Analysis and Mining (ASONAM 2014), pp. 371–374. IEEE (2014)
14. Lasfar, A., Mouline, S., Aboutajdine, D., Cherifi, H.: Content-based retrieval in fractal coded image databases. In: Proceedings 15th International Conference on Pattern Recognition. ICPR-2000, vol. 1, pp. 1031–1034. IEEE (2000)
15. Liu, Y., Tang, M., Zhou, T., Do, Y.: Core-like groups result in invalidation of identifying super-spreader by k-shell decomposition. Sci. Rep. **5**(1), 1–8 (2015)
16. Lü, L., Chen, D., Ren, X.L., Zhang, Q.M., Zhang, Y.C., Zhou, T.: Vital nodes identification in complex networks. Phys. Rep. **650**, 1–63 (2016)
17. Luo, S.L., Gong, K., Kang, L.: Identifying influential spreaders of epidemics on community networks. arXiv:1601.07700 (2016)

18. Magelinski, T., Bartulovic, M., Carley, K.M.: Measuring node contribution to community structure with modularity vitality. IEEE Trans. Netw. Sci. Eng. **8**(1), 707–723 (2021)

19. Nematzadeh, A., Ferrara, E., Flammini, A., Ahn, Y.Y.: Optimal network modularity for information diffusion. Phys. Rev. Lett. **113**(8), 088701 (2014)

20. Pastrana-Vidal, R.R., Gicquel, J.C., Blin, J.L., Cherifi, H.: Predicting subjective video quality from separated spatial and temporal assessment. In: Human Vision and Electronic Imaging XI, vol. 6057, p. 60570S. International Society for Optics and Photonics (2006)

21. Peixoto, T.P.: The netzschleuder network catalogue and repository (2020). https://networks.skewed.de/

22. Rajeh, S., Savonnet, M., Leclercq, E., Cherifi, H.: Characterizing the interactions between classical and community-aware centrality measures in complex networks. Sci. Rep. **11**(1), 1–15 (2021)

23. Rital, S., Bretto, A., Cherifi, H., Aboutajdine, D.: A combinatorial edge detection algorithm on noisy images. In: International Symposium on VIPromCom Video/Image Processing and Multimedia Communications, pp. 351–355. IEEE (2002)

24. Rital, S., Cherifi, H., Miguet, S.: Weighted adaptive neighborhood hypergraph partitioning for image segmentation. In: Singh, S., Singh, M., Apte, C., Perner, P. (eds.) Pattern Recognition and Image Analysis, ICAPR 2005. LNCS, vol. 3687, pp. 522–531. Springer, Heidelberg (2005). https://doi.org/10.1007/11552499_58

25. Rossi, R.A., Ahmed, N.K.: The network data repository with interactive graph analytics and visualization. In: AAAI (2015)

26. Rosvall, M., Bergstrom, C.T.: Maps of random walks on complex networks reveal community structure. PNAS **105**(4), 1118–1123 (2008)

27. Rozemberczki, B., Sarkar, R.: Characteristic functions on graphs: Birds of a feather, from statistical descriptors to parametric models (2020)

28. Sciarra, C., Chiarotti, G., Laio, F., Ridolfi, L.: A change of perspective in network centrality. Sci. Rep. **8**(1), 1–9 (2018)

29. Traud, A.L., Kelsic, E.D., Mucha, P.J., Porter, M.A.: Comparing community structure to characteristics in online collegiate social networks. SIAM Rev. **53**(3), 526–543 (2011)

30. Tulu, M.M., Hou, R., Younas, T.: Identifying influential nodes based on community structure to speed up the dissemination of information in complex network. IEEE Access **6**, 7390–7401 (2018)

31. Wang, W., Liu, Q.H., Zhong, L.F., Tang, M., Gao, H., Stanley, H.E.: Predicting the epidemic threshold of the susceptible-infected-recovered model. Sci. Rep. **6**(1), 1–12 (2016)

32. Zhao, Z., Wang, X., Zhang, W., Zhu, Z.: A community-based approach to identifying influential spreaders. Entropy **17**(4), 2228–2252 (2015)

Two-Tier Cache-Aided Full-Duplex Content Delivery in Satellite–Terrestrial Networks

Quynh T. Ngo[1]($\boxtimes$), Khoa T. Phan[1], Wei Xiang[1], Abdun Mahmood[1], and Jill Slay[2]

[1] School of Engineering and Mathematical Sciences, Department of Computer Science and Information Technology, La Trobe University, Melbourne, Australia
{T.Ngo,K.Phan,W.Xiang,A.Mahmood}@latrobe.edu.au
[2] School of Computer and Information Science, University of South Australia, Adelaide, Australia
Jill.Slay@unisa.edu.au

Abstract. Enabling global Internet access is challenging for the Internet of Things due to limited range of terrestrial network services. One viable solution is to deploy satellites into terrestrial systems for coverage extension. However, operating a hybrid satellite-terrestrial system incurs potentially high satellite bandwidth consumption and excessive service latency. This work aims to reduce the content delivery delay and bandwidth consumption from the Internet-connected gateway to remote users in satellite terrestrial networks, using a two-tier cache-enabled full-duplex system model where content caches are placed at the satellite and the ground station. A closed-form solution for successful delivery probability of content files within allocated time slot under general caching policy is derived considering the requested content distributions and channel statistics. Illustrative results demonstrate superior performance of the proposed system over those of single-tier cache-aided. The trade-off between successful delivery probability and satellite bandwidth consumption, in addition to insights on the network support ability are also investigated.

Keywords: Satellite-terrestrial networks · Edge caching · Full-duplex · Successful delivery probability · Cache placement design

1 Introduction

Although being around for many years, cellular-based technologies for the Internet of Things (IoT) have not fulfilled the demand of globally connecting everyone and everything. To offer service across all geographic regions, particularly, in rugged or dispersed terrain, integrating satellites into IoT networks has been proposed. Providing better service reliability and coverage, satellite is a more effective solution for IoT than cellular network. Embedding satellite into IoT

© Springer Nature Switzerland AG 2021
D. Mohaisen and R. Jin (Eds.): CSoNet 2021, LNCS 13116, pp. 291–302, 2021.
https://doi.org/10.1007/978-3-030-91434-9_26

networks, however, poses challenges in service delays and satellite bandwidth cost. A content delivered to end-users from the Internet-connected gateway is relayed through satellite(s) and ground station(s), which extends the serving time in addition to very pricey and often limited satellite bandwidth. Excessive content delivery delay can hinder potential use of satellite IoT systems to support delay-sensitive applications. To address with these challenges, caching has been proposed to move content storage to the edge devices closer to the users. Caching popular contents at the network edge can reduce network congestion and decrease content delivery latency [12].

Employing edge caching technique in hybrid satellite-terrestrial relay networks, [2] proposes an amplify and forward relaying protocols where the cache is enabled at relays on the ground (a.k.a. ground stations in our current work). The work in [2] considers the most popular and uniform content-based cache placement schemes, that shows substantial improvement over the traditional approach without caching in terms of outage probability. To off-load the backhaul of terrestrial network, [9] proposes using the hybrid satellite terrestrial network in combination with an off-line edge caching algorithm. The performance of the proposed off-line caching algorithm is measured through cache hit ratio. Using the same network model as in [9,13] investigates the system performance in two use cases: in dense urban areas and in sparsely populated regions. The effectiveness of the system is studied through cache hit ratio and cost per bit for satellite transmission. It should be emphasized that [2,9,13] consider dual-hop downlink satellite–relay(s)–users transmissions with single-tier cache placed at the relays. The satellite uplink communications from the Internet-connected gateway to the satellite is not considered, which is usually the bottleneck in satellite IoT systems due to large bandwidth consumption.

Aiming at reducing the satellite uplink bandwidth consumption, additional caching at the satellite has been proposed. The work [15] proposes a two-tier caching model where the first and second cache tiers are placed in the ground stations, and in the satellite, respectively. Caches at each ground station are used for the popular contents in its local area, while the satellite's cache is used for the most popular contents in its coverage (containing multiple ground stations) to take advantage of the satellite's broadcast nature to the ground stations. Non-cached contents can be retrieved from the gateway if needed. While [15] aims to minimize the satellite bandwidth consumption, the file delivery time, a critical concern in satellite IoT operation, is not investigated that requires more elaborated analysis.

In this paper, we consider a two-tier cache-aided model in hybrid satellite-terrestrial systems [15] considering end-to-end gateway to end-users data transmissions over realistic channel models. Our work focuses on the content delivery time analysis in terms of successful delivery probability (SDP) considering full-duplex (FD) transmissions at the satellite and ground station, albeit with imperfect self-interference cancellation (SIC). Deploying FD communications will potentially shorten the service delivery time comparing to half-duplex mode providing sufficiently effective SIC, and hence, increasing the SDP. A closed-form expression for the SDP under FD mode is derived taking into account the requested content characteristics, channel statistics, and network configurations. The results enable convenient evaluations of the trade-off between SDP and satellite bandwidth consumption. Numerical results are presented for the network performance behavior under two caching policies: uniform caching and popular caching. They also show the proposed system's merits over single-tier cache systems [2,9,13] in both SDP and satellite bandwidth consumption.

2   System Model

The satellite assisted IoT system composes of a satellite S, an Internet connected satellite gateway G, a ground station G_s and a set of K end-users $U_i, i = 1, \ldots, K$ as depicted in Fig. 1. In this system, S is a geosynchronous equatorial orbit (GEO) satellite; G_s is a low power IoT base station equipped with a satellite receiver. Both S and G_s are cache enabled. End-users U_i are IoT devices on the ground. Assuming there is no direct link from users U_i to satellite S and from the ground station to gateway due to weather factors, long distance, and/or heavy shadowing.

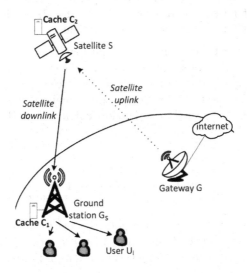

Fig. 1. Satellite IoT networks architecture.

2.1   Two-Tier Caching Model

The caching model consists of two tiers: the first tier is at the ground station with storage capacity of C_1 (bits) and the second tier is at the satellite with storage capacity of C_2 (bits). Gateway is connected to the Internet and hosts N files $W_1, \ldots, W_N$, which are assumed to be equal size of F (bits). This assumption is for scenario of heterogeneous IoT applications. For other IoT applications, the analysis in Sects. 3, 4 and 5 can be easily extended to unequal file size. A typical caching protocol consists of two phases: the placement of files into caches and the delivery of files to users. The focus of this work is on the performance analysis of the content delivery phase. Cache placement design for content placement phase will be discussed later in the text.

The probability for a file W_n being requested follows Zipf distribution, which is $q_n = \frac{n^{-\alpha}}{\sum_{m=1}^{N} m^{-\alpha}}$ where $0 < \alpha < 1$ denotes the Zipf skewness factor [4]. A large α means the requests on the high popularity files, whereas a small α is related to the requests with heavy-tailed popularity. Without loss of generality, we have assumed that files $W_1, \ldots, W_N$ have decreasing popularity.

2.2   Channel Model

Consider block-based communications where a transmission section is accomplished within a coherence time T (seconds). Both large-scale fading and small-scale fading are considered. The large-scale fading is modeled by the distance-dependent power-law path-loss attenuation $d_{m-n}^{-\alpha_i}$ where d_{m-n} denotes the distance between nodes m and n, and α_i represents the path-loss exponent. For small-scale fading, the channel model proposed in [1] has been commonly used for satellite terrestrial communications [2,3,7], and Rayleigh fading channel is commonly used for terrestrial wireless communications.

The Satellite-Terrestrial Links have multipath fading and shadow fading. The multipath fading composes of one line-of-sight (LOS) and many weak scatter components. The shadow fading composes of the LOS shadow fading and multiplicative shadow fading. The channel power gain of the satellite-terrestrial links $h_{S-G_s}(t)$ has the probability distribution function (PDF) of $f_{h_{S-G_s}}(x)$ given by [1,3] as:

$$
f_{h_{S-G_s}}(x) = \left(\frac{b_{1,0} m_1}{b_{1,0} m_1 + \Omega_1} \right)^{m_1} \frac{1}{b_{1,0}} \exp\left(-\frac{x}{b_{1,0}} \right) {}_1F_1\left(m_1, 1, \frac{\Omega_1 x}{b_{1,0}(b_{1,0} m_1 + \Omega_1)} \right), x \geq 0
\tag{1}
$$

where $b_{1,0}$ represents the average power of the scatter components; Ω_1 represents the average power of the LOS component; m_1 is the Nakagami parameter; ${}_1F_1(.,.,.)$ is the confluent hypergeometric function [6, Eq.(9.210.1)].

The Terrestrial Links are modeled as Rayleigh fading channels with channel power gain $h_{G_s-U_i}(t)$ having the PDF given by

$$
f_{h_{G_s-U_i}}(x) = \frac{1}{\bar{h}_i} \exp\left(-\frac{x}{\bar{h}_i} \right), x \geq 0
\tag{2}
$$

with $\bar{h}_i$ is the average channel power gains taking into effects of small-scale fading.

Note that we have omitted the time-dependent index due to stationarity assumption.

2.3   Transmission Scheme

When caching IoT data, it is essential to maintain the data freshness [14]. This model assumes that the users will be served in a time-division multiple access (TDMA) manner which allows each user to be consecutively active in T/K (seconds). The time slot T/K will be large enough to ensure the data freshness in practical IoT applications. Under TDMA, inter-user interference does not exist. Both satellite and ground station are operating in FD mode. When channel state information is known at the transmitter, rate adaption is employed, the achievable rates (bps) on the links are

$$R_{G_s-U_i} = B_{G_s-U_i} \log_2 \left(1 + \frac{P_{G_s} |h_{G_s-U_i}|^2 d_{G_s-U_i}^{-\alpha_g}}{\sigma^2} \right),$$

$$R_{S-G_s} = B_{S-G_s} \log_2 \left(1 + \frac{P_S |h_{S-G_s}|^2 d_{S-G_s}^{-\alpha_s}}{I_{G_s} + \sigma^2} \right), \tag{3}$$

$$R_{G-S} = B_{G-S} \log_2 \left(1 + \frac{P_G |h_{G-S}|^2 d_{G-S}^{-\alpha_s}}{I_S + \sigma^2} \right),$$

where B_{G-S}, B_{S-G_s} and $B_{G_s-U_i}$ are the bandwidths (Hz) of the satellite uplink, downlink and the terrestrial downlink, respectively; P_S, P_G, and P_{G_s} denote the transmit powers of satellite, gateway and ground station, respectively; σ^2 is the additive white Gaussian noise power; I_S and I_{G_s} represent the residual self-interference power at the satellite and ground station, respectively, which are assumed to be proportional to the transmit power P_S and P_{G_s} with coefficient $\beta \geq 0$ being the SIC quality parameter.

When a user requests file W_n, the gateway needs to transfer the non-cached $(1 - \mu_1^n - \mu_2^n)$ portion to the satellite, which also needs to send accumulated $(1 - \mu_1^n)$ portion to the ground station. By employing the FastForward protocol [5] with FD mode at the satellite, the completion time is

$$\max \left\{ (1 - \mu_1^n - \mu_2^n)F/R_{G-S}, (1 - \mu_1^n)F/R_{S-G_s} \right\}.$$

Similarly, by deploying FastForward protocol with FD mode at the ground station, the (end-to-end) delivery time for transferring the whole file to the user is determined by the time for sending the $(1 - \mu_1^n)$ portion over the satellite downlink to the ground station and the time for sending the whole file over the terrestrial link to the user, that is:

$$t_{n,i} = \max \left(\frac{F}{R_{G_s-U_i}}, \frac{(1 - \mu_1^n)F}{R_{S-G_s}}, \frac{(1 - \mu_1^n - \mu_2^n)F}{R_{G-S}} \right). \tag{4}$$

Note that we have neglected the propagation delays for simplicity. Otherwise, they can be straightforwardly included in the delivery time expression as approximate constants and the mathematical derivations in the next Section remain unchanged.

3   Successful Delivery Probability

The successful delivery probability of serving a requested file to a user is defined as the probability that user receives the file within the user's active time slot. Assume that $K \ll N$, i.e., the library size is much larger than the number of users. It implies in this work that users request different files. In general, when users request similar files, more efficient delivery mechanism can be developed utilizing the broadcasting. However, it is out of the scope of this work. The following analysis is for user U_i requesting file W_n, and is true for all other users. The SDP of file W_n requested by user U_i is:

$$\psi_{n,i} = \mathbf{Pr}\left(t_{n,i} \leq \frac{T}{K}\right) \tag{5}$$

The probability operator is taken with respect to the channel power gain variables. The closed-form expression for $\psi_{n,i}$ is

$$
\psi_{n,i}(\mu_1^n, \mu_2^n) = \exp\left(-\frac{2^{\frac{FK}{TB_{G_s-U_i}}} - 1}{\overline{\gamma_1}}\right)
$$
$$
\times \left(1 - \sum_{k=0}^{m_1-1}\sum_{q=0}^{\infty} \frac{(-1)^k}{(k!)^2}\cdot\frac{\Gamma(1-m_1+k)}{\Gamma(1-m_1)}\cdot\frac{\alpha_1\delta_1^k(\delta_1-\chi_1)^q}{l_1^{(q+k+1)}}\cdot\frac{(2^{\frac{FK}{TB_{S-G_s}}(1-\mu_1^n)}-1)^{q+k+1}}{q!(q+k+1)}\right)
$$
$$
\times \left(1 - \sum_{k=0}^{m_2-1}\sum_{q=0}^{\infty} \frac{(-1)^k}{(k!)^2}\cdot\frac{\Gamma(1-m_2+k)}{\Gamma(1-m_2)}\cdot\frac{\alpha_2\delta_2^k(\delta_2-\chi_2)^q}{l_2^{(q+k+1)}}\cdot\frac{(2^{\frac{FK}{TB_{G-S}}(1-\mu_1^n-\mu_2^n)}-1)^{q+k+1}}{q!(q+k+1)}\right). \tag{6}
$$

Note that the distances between nodes in (3) are constants. Since S is a GEO satellite, all the nodes in the system have fixed locations.

Proof. Let denote:

$$\gamma_1 = \frac{P_{G_s}|h_{G_s-U_i}|^2 d_{G_s-U_i}^{-\alpha_g}}{\sigma^2}, \gamma_2 = \frac{P_S|h_{S-G_s}|^2 d_{S-G_s}^{-\alpha_s}}{I_{G_s}+\sigma^2}, \text{ and } \gamma_3 = \frac{P_G|h_{G-s}|^2 d_{G-S}^{-\alpha_s}}{I_s+\sigma^2}.$$

Following (2), the PDF of γ_1 is $f_{\gamma_1}(x) = \frac{e^{-\frac{x}{\overline{\gamma_1}}}}{\overline{\gamma_1}}, \overline{\gamma_1} = \frac{P_{G_s}d_{G_s-U_i}^{-\alpha_g}\bar{h}_i}{\sigma^2}$ for $x \geq 0$.

Let denote $\alpha_1 = \left(\frac{b_{1,0}m_1}{b_{1,0}m_1+\Omega_1}\right)^{m_1}\frac{1}{b_{1,0}}, \chi_1 = \frac{1}{b_{1,0}}, \delta_1 = \frac{\Omega_1}{b_{1,0}(b_{1,0}m_1+\Omega_1)}$ (refer to (1)), the PDF of γ_2 is $f_{\gamma_2}(x) = \frac{\alpha_1}{l_1}e^{-\frac{\delta_1-\chi_1}{l_1}x}\cdot\sum_{k=0}^{m_1-1}\frac{(-1)^k(1-m_1)_k(\frac{\delta_1}{l_1}x)^k}{(k!)^2}$, where $l_1 = \frac{P_S d_{S-G_s}^{-\alpha_s}}{I_{G_s}+\sigma^2}$, $(.)_k$ denotes the Pochhammer symbol with $(1-m_1)_k = \frac{\Gamma(1-m_1+k)}{\Gamma(1-m_1)}$. Similarly, the PDF of γ_3 is $f_{\gamma_3}(x) = \frac{\alpha_2}{l_2}e^{-\frac{\chi_2}{l_2}x}\cdot{}_1F_1(m_2;1;\frac{\delta_2}{l_2}x)$ where $m_2, \alpha_2, \chi_2, \delta_2$ are corresponding denotations in the PDF $f_{\gamma_2}(x)$; and $l_2 = \frac{P_G d_{G-S}^{-\alpha_s}}{I_s+\sigma^2}$.

Substituting (4) into (5) we have

$$\psi_n = \mathbf{Pr}\left(\max\left(\frac{F}{R_{G_s-U_i}}, \frac{(1-\mu_1^n)F}{R_{S-G_s}}, \frac{(1-\mu_1^n-\mu_2^n)F}{R_{G-S}}\right) \le \frac{T}{K}\right)$$

$$= \underbrace{\mathbf{Pr}\left(\frac{F}{B_{G_s-U_i}\log_2(1+\gamma_1)} \le \frac{T}{K}\right)}_{A_1} \cdot \underbrace{\mathbf{Pr}\left(\frac{(1-\mu_1^n)F}{B_{S-G_s}\log_2(1+\gamma_2)} \le \frac{T}{K}\right)}_{A_2}$$

$$\cdot \underbrace{\mathbf{Pr}\left(\frac{(1-\mu_1^n-\mu_2^n)F}{B_{G-S}\log_2(1+\gamma_3)} \le \frac{T}{K}\right)}_{A_3}.$$

Since γ_1 follows exponential distribution, A_1 can be directly obtained as $A_1 = \exp\left(-\frac{2^{\frac{FK}{TB_{G_s-U_i}}}-1}{\bar{\gamma}_1}\right)$. A_2 and A_3 can be obtained in the same approach as follows $A_2 = \mathbf{Pr}\left(\frac{(1-\mu_1^n)F}{B_{S-G_s}\log_2(1+\gamma_2)} \le \frac{T}{K}\right) = 1 - F_{\gamma_2}(2^{\frac{FK}{TB_{S-G_s}}(1-\mu_1^n)} - 1)$, where $F_{\gamma_2}(x)$ denotes the CDF of γ_2,

$$F_{\gamma_2}(x) = \sum_{k=0}^{m_1-1} \frac{\alpha_1}{l_1} \frac{(-1)^k(1-m_1)_k(\frac{\delta_1}{l_1})^k}{(k!)^2} \int_0^x e^{\frac{\delta_1-\chi_1}{l_1}y} y^k dy$$

$$= \sum_{k=0}^{m_1-1} \sum_{q=0}^{\infty} \frac{(-1)^k(1-m_1)_k}{(k!)^2} \cdot \frac{\alpha_1\delta_1^k(\delta_1-\chi_1)^q}{l_1^{q+k+1}} \cdot \frac{x^{q+k+1}}{q!(q+k+1)}.$$

($e^{\frac{\delta_1-\chi_1}{l_1}y} = \sum_{q=0}^{\infty} \frac{(\frac{\delta_1-\chi_1}{l_1})^q y^q}{q!}$ is expanded using Maclaurin series.)

The average SDP of the system is defined as the weighted sum of users' SDPs:

$$\psi(\boldsymbol{\mu}_1, \boldsymbol{\mu}_2) = \sum_{i=1}^{K} \omega_i \psi_i(\boldsymbol{\mu}_1, \boldsymbol{\mu}_2) = \sum_{i=1}^{K} \omega_i \sum_{n=1}^{N} q_n \psi_{n,i}(\mu_1^n, \mu_2^n) \tag{7}$$

with weighting coefficients $\omega_i \in (0,1)$ and $\sum_i \omega_i = 1$. In the following, w.l.o.g., assume homogeneous users with $\omega_i = 1/K$, and average channel gains $\bar{h}_i = \bar{h}_k, \forall i, k$. Thus, $\psi = \psi_i$.

4   Satellite Bandwidth Consumption

Since satellite service is more expensive, we focus on the satellite uplink and downlink bandwidth consumption during content delivery phase. Assume that the total number of requests follow Poisson distribution with an average arrival rate of $\bar{\lambda}$. Because each of all N files has some portions cached at both the satellite and the ground station, the average request arrival rate for file W_n at the two caches is the same, which is $\lambda_n = q_n\bar{\lambda}$. Let p_n denote the probability that the request for file W_n is received during time T with $p_n = \lambda_n e^{-\lambda_n T}$. The satellite bandwidth consumption to deliver file W_n includes the downlink consumption when transmitting the cached portion of the file $p_n \mu_2^n F$, the relay portion of the file $p_n(1 - \mu_1^n - \mu_2^n)F$ and the uplink consumption when receiving the portion $p_n(1 - \mu_1^n - \mu_2^n)F$ from gateway G. Hence, the satellite bandwidth consumption is

$$B(\mu_1, \mu_2) = \sum_{n=1}^{N} p_n(2 - 2\mu_1^n - \mu_2^n)F. \tag{8}$$

Increasing the caching portions at the ground station and/or satellite can reduce the bandwidth consumption.

5   Cache Placement Design

The cached contents are placed into caches C_1 and C_2 via satellite uplink and downlink transmissions during off-peak hours. In this section, the cached portion μ_1^n of all N files at ground station G_s and μ_2^n portion at satellite S are optimized in order to balance the average SDP and bandwidth consumption (maximizing the SPD and minimizing the bandwidth consumption). Hence, this optimization problem is formed as the following a bi-objective optimization problem:

$$\text{minimize} \left(-w \cdot \psi(\mu_1, \mu_2) + (1 - w) \cdot B(\mu_1, \mu_2)\right) \tag{9}$$

$$\text{subject to} \sum_{n=1}^{N} \mu_1^n \cdot F \leq C_1, \quad \sum_{n=1}^{N} \mu_2^n \cdot F \leq C_2 \tag{10a}$$

$$\sum_{n=1}^{N} p_n(2 - 2\mu_1^n - \mu_2^n)F \leq B_{S-max} \tag{10b}$$

$$0 \leq \mu_1^n + \mu_2^n \leq 1 \text{ with } n = 1, 2, ..., N \tag{10c}$$

where $w \in (0, 1)$ is a non-negative weight representing the performance important between SDP and bandwidth consumption. A larger value of w means the amount of cached content is decided in favour of having a higher SDP and vice versa. The set of constraints (10a) ensures that the cached contents will not exceed the cache size of C_1 and C_2. Constraint (10b) makes sure that the bandwidth consumption stays within the satellite bandwidth capacity B_{S-max}. Constraint (10c) shows the nature of partial content of all files caching scheme.

Eq. (9) can be solved using the genetic algorithm described in [11]. We omit the details of the genetic algorithm here for brevity.

6   Numerical Results

In this section, numerical results are presented for two different caching policies in terms of system average SDP and satellite bandwidth consumption. The results are compared between the proposed two-tier cache system and the single-tier cache system adopted from [2,9,13]. The caching policy for single-tier cache system is corresponding to the policy cached at ground station in the two-tier case.

Uniform Caching Policy: All of N files are cached with the same portion $\hat{\mu}_1$ at the satellite and $\hat{\mu}_2$ at the ground station. In uniform caching, we have $\mu_1 = [\hat{\mu}_1, \hat{\mu}_1, ..., \hat{\mu}_1]$ and $\mu_2 = [\hat{\mu}_2, \hat{\mu}_2, ..., \hat{\mu}_2]$ where $0 < \hat{\mu}_1 + \hat{\mu}_2 < 1$.

Popular Caching Policy: The most popular files are cached with a larger portion at the ground station and a smaller portion at the satellite. In popular caching, we have $\mu_1 = [\eta\hat{\mu}_1^1, \eta\hat{\mu}_1^2, .., \eta\hat{\mu}_1^k, 0, ..., 0]$ and $\mu_2 = [\hat{\mu}_1^1, \hat{\mu}_1^2, .., \hat{\mu}_1^k, 0, ..., 0]$ where $\eta \in \mathbb{Z}$, $0 < \eta\hat{\mu}_1^i + \hat{\mu}_1^i < 1$ $(i = 1, 2, ..., k)$ and $\hat{\mu}_1^1, ..., \hat{\mu}_1^k$ follows popular file distribution.

The values for key parameters [8] used in this section are presented in Table 1. Because it is required to have more than $100\,dB$ of SIC for FD system to achieve the same signal-to-noise-ratio-plus-interference as that of the HD system [10], we choose the SIC quality parameter $\beta = 0.0001$ for most of our system configuration under FD transmission. With this value of β, the SIC is achieved from $110\,dB$ to $120\,dB$. The fading states for the satellite-terrestrial links are defined by $\{m_1, b_1, \Omega_1\} = \{5, 0.251, 0.279\}$ and $\{m_2, b_2, \Omega_2\} = \{4, 0.126, 0.835\}$ [7] when approaching average shadowing. End-users are randomly distributed inside a circle with radius $10\,km$ and center at G_s.

Table 1. Parameters used for numerical results

Parameter	Value	Parameter	Value
N	500 files	F	100 Mb
α	0.8	β	0.0001
B_{m-n}/B_{S-G_s}	2/10 Gb	$P_S/P_G/P_{G_s}$	10/30/2 W
$B_{G_s-U_i}$	2 Gb	σ^2	-120 dBm/Hz
d_{S-G_s}/d_{G-S}	35,786 km	α_s/α_g	2/3

We first investigate the effect of percentage of file cached at ground station μ_1 under uniform caching policy. With 10 end-users, the numerical results are shown in Fig. 2. The total caching capacity in Fig. 2 is C_1 for single-tier cache system and $C_1 + C_2$ for two-tier cache system. For both single-tier/two-tier cache systems, the average SDP increases with more percentage of content cached at ground station (Fig. 2(a)). The satellite bandwidth consumption consumes more in single-tier cache system (Fig. 2(b)). It is observed from Fig. 2 that achieving better average SDP with less satellite bandwidth consumption happens in single-tier cache system when μ_1 is no more than 50%. In two-tier cache system, the caching strategy is still effective as long as μ_1 is less than 55%; Otherwise, caching more content will not improve the system performances.

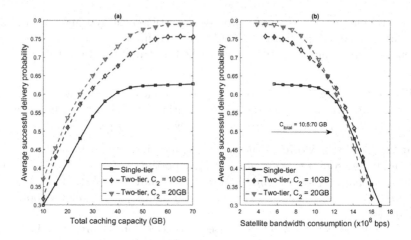

Fig. 2. Two-tier vs. single-tier systems under uniform caching.

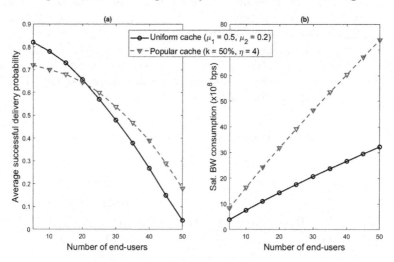

Fig. 3. Two-tier network capacity.

Secondly, we study the two-tier cache system capacity under uniform and popular caching policies. Under uniform caching policy, 50% of each file are prefetched at the ground station and 20% at the satellite. For popular caching policy, the threshold $k = 0.5$ meaning one half of N files are cached and the ratio of percentage cached at the two layers is $\eta = 4$. As shown in Fig. 3(a), the more ground users, the smaller average SDP for both caching policies. For less than 20 users, the uniform caching policy outperforms the popular one. Otherwise, its SDP is surpassed by that of the popular caching policy. Figure 3(b) shows the satellite bandwidth consumption. As serving more users, the popular caching case consumes more bandwidth than the uniform caching case. This result is caused by the amount of data transmitted in satellite uplink/downlink when a requested file is not cached due to its popularity.

7   Conclusion

In this paper, we study the performance of a two-tier cache aided full-duplex satellite-terrestrial network on two metrics: the successful delivery probability and the satellite bandwidth consumption. Both metrics are investigated with their performances presented via numerical results under uniform caching and popular caching policies. Depending on the system use cases, the optimal cache placement scheme can be chosen in order to achieve better average SDP and less satellite bandwidth consumption.

Acknowledgment. This work was supported by the La Trobe University Postgraduate Research Scholarship, the La Trobe University Full-Fee Research Scholarship, and the Net Zero Scholarship sponsored by Sonepar/SLS.

References

1. Abdi, A., Lau, W.C., Alouini, M., Kaveh, M.: A new simple model for land mobile satellite channels: first- and second-order statistics. IEEE Trans. Wireless Commun. **2**(3), 519–528 (2003)
2. An, K., Li, Y., Yan, X., Liang, T.: On the performance of cache-enabled hybrid satellite-terrestrial relay networks. IEEE Wirel. Commun. Lett. **8**(5), 1506–1509 (2019)
3. Bhatnagar, M.R., Arti, M.K.: Performance analysis of AF based hybrid satellite-terrestrial cooperative network over generalized fading channels. IEEE Commun. Lett. **17**(10), 1912–1915 (2013)
4. Breslau, L., Cao, P., Fan, L., Phillips, G., Shenker, S.: Web caching and Zipf-like distributions: Evidence and implications. In: IEEE Conference on Computer Communications Societies (INFOCOM), NY, USA, pp. 126–134 (1999)
5. Dinesh, B., Sachin, K.: Fastforward: fast and constructive full duplex relays. In: Proceedings of the 2014 ACM Conference on SIGCOMM (SIGCOMM 2014), IL, USA, pp. 199–210 (2014)
6. Gradshteyn, I.S., Ryzhik, I.M.: Tables of Integrals, Series and Products, 6th ed. (2000)
7. Guo, K., et al.: On the performance of the uplink satellite multiterrestrial relay networks with hardware impairments and interference. IEEE Syst. J. **13**(3), 2297–2308 (2019)
8. HughesNet: Echostar xix satellite (2020). https://www.hughesnet.com/whats-new/satellite-launch
9. Kalantari, A., Fittipaldi, M., Chatzinotas, S., Vu, T.X., Ottersten, B.: Cache-assisted hybrid satellite-terrestrial backhauling for 5g cellular networks. In: 2017 IEEE Global Communications Conference (GLOBECOM), Singapore, pp. 1–6 (2017)
10. Li, H., et al.: Self-interference cancellation enabling high-throughput short-reach wireless full-duplex communication. IEEE Trans. Wirel. Commun. **17**(10), 6475–6486 (2018)

11. Li, J., Balazs, M.E., Parks, G.T., Clarkson, P.J.: A species conserving genetic algorithm for multimodal function optimization. Evol. Comput. **10**(3), 207–234 (2002)
12. Paschos, G.S., Iosifidis, G., Tao, M., Towsley, D., Caire, G.: The role of caching in future communication systems and networks. IEEE J. Sel. Areas Commun. **36**(6), 1111–1125 (2018)
13. Vu, T.X., Maturo, N., Vuppala, S., Chatzinotas, S., Grotz, J., Alagha, N.: Efficient 5G edge caching over satellite. In: 36th International Communications Satellite Systems Conference (ICSSC 2018), pp. 1–5 (2018)
14. Vural, S., Navaratnam, P., Wang, N., Wang, C., Dong, L., Tafazolli, R.: In-network caching of internet-of-things data. In: 2014 IEEE International Conference on Communications (ICC), Sydney, Australia, pp. 3185–3190 (2014)
15. Wu, H., Li, J., Lu, H., Hong, P.: A two-layer caching model for content delivery services in satellite-terrestrial networks. In: 2016 IEEE Global Communications Conference (GLOBECOM), DC, USA, pp. 1–6 (2016)

Special Track: Fact-Checking, Fake News and Malware Detection in Online Social Networks

Mean User-Text Agglomeration (MUTA): Practical User Representation and Visualization for Detection of Online Influence Operations

Evan Crothers[1]([✉]), Herna Viktor[1], and Nathalie Japkowicz[1,2]

[1] University of Ottawa, Ontario, Canada
ecrot027@uottawa.ca
[2] American University, Washington DC, USA

Abstract. Online influence operations (OIOs) present a serious threat to the integrity of online social spaces and to real-world democratic elections. While many OIO detection approaches have focused on classification algorithms for individual social media posts (often with artificially balanced datasets), we present a novel system centering around a human analyst. This system incorporates a user representation and visualization procedure for unbalanced social media data. Our content-based social media user representation, the Mean User-Text Agglomeration (MUTA), summarizes a user's social media activity with respect to Transformer embeddings of texts authored by the user. We apply MUTA to a real social media dataset in advance of an election event and flag a number of suspicious Reddit users that were later removed by the social media platform. When projected to a 2-dimensional visualizable space, MUTA user representations are shown, via extrinsic cluster quality measures, to outperform BERT representations for analyst identification of OIO accounts.

Keywords: Online influence operations · User representation · Transformer architecture · Natural language processing · Disinformation · Social media analysis

1 Introduction

In the last five years, a significant increase in publicly documented online influence operations (OIOs) has led to concerns regarding the integrity of online social spaces and the potential impact on democratic institutions. To date, public releases by Twitter [34], Facebook [18], and Reddit [20] indicate increased patterns of politically motivated inauthentic behaviour linked to influence campaigns based out of (to date) 20 different countries worldwide. Improvements in OIO detection mechanisms are critical for uncovering the presence of these campaigns, thereby safeguarding authentic expression on social media, and protecting the integrity of democratic elections. There is a particular lack of research

© Springer Nature Switzerland AG 2021
D. Mohaisen and R. Jin (Eds.): CSoNet 2021, LNCS 13116, pp. 305–318, 2021.
https://doi.org/10.1007/978-3-030-91434-9_27

that bridges the gap between recent advances in state-of-the-art natural language processing (NLP), such as Transformer language models, and practical OIO detection workflows on social media data "in the wild".

We introduce an unsupervised approach to detecting online influence operations, which we refer to as Mean User-Text Agglomeration (MUTA), that allows a human analyst to easily leverage Transformer-derived user representations through a visual interface. MUTA is a fully unsupervised approach that computes the average output activations of a Transformer encoder for each user based on all the text submissions available from that user in a collection of social media data. In contrast to other vector representations of social media users, MUTA user representations do not require pretraining of a custom architecture on a data-intensive author identification task. Further, they can be adapted to leverage widely-available Transformer models in a variety of languages and are easily updated as new data becomes available.

Much of the research in OIO detection is focused on automated supervised approaches, where a model is trained to differentiate between activity associated with an influence operation and normal user activity. This is typically accomplished by using data from public datasets of OIO activity released by social media companies, as well as a random sampling of normal user activity to artificially balance the training dataset between influence accounts and normal accounts [2,10,15]. While automated class-balanced supervised approaches are useful for determining the relative predictive power of disparate social media features, these approaches have significant gaps to practical application to detection of OIOs on social media, primarily due to the following causes:

1. OIO accounts account for a small portion of social media users; a supervised classifier trained on artificially class-balanced data does not reflect the problem domain.
2. Supervised models trained on past operations may be skewed by specific subject matter and tradecraft present in prior campaigns, but not in ongoing campaigns.
3. OIO investigations rely heavily on human expertise, and fully-automated OIO detection systems carry ethical risk related to the automated suppression of political speech.

In light of these considerations, pioneering unsupervised detection systems that incorporate a human analyst – though challenging to compare to other work in the field – are a very desirable approach for improving OIO detection capabilities in practice.

The presented methodology takes advantage of common characteristics in user-generated text from online influence operations that manifest in vector sentence representations (embeddings) generated by a high-capacity language model, such as those produced by a pre-trained Transformer encoder architecture. Similarities in linguistic background among cells of OIO operators, as well as the external coordination of subject matter that an orchestrated political OIO necessitates, both impact the resulting embedding from a Transformer encoder [10]. Cosine distances in the embedding space of the Transformer encoder implementation of Bi-directional Encoder Representations from Transformers (BERT), have been shown to be useful for representing relative semantic characteristics among input sentences [27]. In the context of the social media platform Reddit, OIO submission titles will, overall, have embeddings with a lower cosine distance to other OIO title embeddings compared to the embeddings of regular submission titles (even if the distribution of OIO posts is multi-modal). While the embeddings of individual posts vary substantially, by taking the mean embedding of titles submitted by each user, similar OIO accounts are drawn together in the resulting meta-embedding space. In an interactive low-dimensional projection of MUTA user representations, the human analyst's ability to quickly identify patterns in post titles while referencing expert knowledge allows them to rapidly triage what information is relevant to a potential online influence operation, and facilitate OIO detection.

This work includes two major contributions. First, we offer a practical new meta-embedding technique for user representation called MUTA that is computationally cheap to update and easily transferable to more powerful Transformer language models, including those in other languages. Second, we present an end-to-end OIO detection workflow using a case study based on the 2019 Canadian federal election in which MUTA representations are used to identify suspicious accounts on Reddit – several of which were later verified to have been independently removed by the social media platform – demonstrating practical utility in OIO investigations. All code, privacy-protected datasets (see Sect. 2), and associated tools are made available so this process can be easily repeated using other datasets and alternate Transformer architectures [12]. Portions of this work were published in thesis form in partial fulfillment of the requirements for the Master of Computer Science degree at the University of Ottawa [11].

The remainder of this paper is organized as follows. Section 2 addresses the ethical and privacy considerations of OIO detection as they apply to this research. Section 3 discusses related work in user representation and social media analysis. Section 4 contains a complete explanation of the MUTA user analysis methodology. Sections 5 introduces the 2019 Canadian Election Reddit dataset and pre-processing procedure. Section 6 describes the experiments and results. Concluding remarks and future work are presented in Sect. 7.

2   Ethical Considerations and Social Impact

Analysis of social media carries significant ethical and privacy considerations – particularly when moderation of political discourse is concerned.

The inclusion of a human analyst into the analytical process is a significant ethical advantage for a user embedding and visualization approach to OIO detection. Emphatically, even a methodology that incorporates a human analyst (such as MUTA) is capable of manifesting algorithmic bias through how data is presented to the analyst. Additionally, the human analyst's own biases may also influence the results of such a system in practice. Regardless, the integration of a human analyst into the detection methodology is generally regarded as an improvement in the standard of rigour for the deployment of AI systems, as noted by the ethics guidelines for trustworthy AI developed by the High-Level Expert Group on AI [19]. The first of seven key principles of trustworthy AI systems is "human agency and oversight", which is supported by the introduction of a human arbiter. Further, by relying less on data from past operations – an approach that has been demonstrated quantitatively to have potential for discrimination based on shared language backgrounds [10] – it is also an arguable improvement for the fifth key principle: "diversity, non-discrimination and fairness". As such, we believe that our approach aligns very well with current ethical guidance around AI systems for this purpose.

Before large-scale deployment of a MUTA-based OIO detection system, the system should undergo an ethical evaluation. A complete ethical assessment would include a trial applied to a real OIO detection task, including the human oversight component – similar to the proposed EU process for algorithmic impact assessments for automated decision making in policing [21]. Large-scale user trials directly comparing human performance using assorted analyst-driven OIO detection techniques have never been performed publicly to date, and there is limited visibility into the current methods employed by investigative teams at social media companies and security agencies. This represents a challenge for both comparing the efficacy of approaches as well as evaluating the impact of human bias in practice.

All user data utilized within this research is publicly available – however, flagging individual social media accounts may expose certain users to undeserved scrutiny. Furthermore, data legislation such as the General Data Protection Regulation (GDPR), mandates a 'right to be forgotten', and this research includes user accounts that no longer exist due to removal by social media platforms. Respecting these considerations, as well as the interests of reproducible research, we make available all the calculated text and user embeddings, as well as the complete code for creating, processing, visualizing, and evaluating these embeddings. To ensure that others can validate our claims regarding the suspending of the accounts highlighted by this research, we can provide further details on specific banned accounts upon email request.

3   Related Work

While there are limited research works focusing on visualization-based analytical tools for improved OIO detection, past work on OIO detection has demonstrated the broad applicability of content-based features [2]. Extensive work has also been done in the space of social spam detection that overlaps with the area of political OIO detection. Spam operations and OIOs have certain similarities, as both often share a common objective in the dissemination of a particular message to a receptive target audience while evading platform moderators. Many approaches for spam detection exist, such as regex matching on shared spam URLs [16], or modelling friend/follower relationships between users [35]. In political online influence operations, the presence of a malicious website is not required to influence a reader. Furthermore, detection methods based on unusual domains or URLs can be undermined by linking to existing articles that promote the desired narrative, or by hosting bespoke OIO content on popular social media and blogging platforms to avoid registering a custom domain. Modelling friend/follower relationships is also less valuable on a platform such as Reddit, which has a comparatively low emphasis on direct relationships between user accounts.

Transformer-based solutions for detecting malicious behaviour in social media are still in the early stages. Previous research has demonstrated the performance of fine-tuned Transformer models on text classification of malicious comments or falsified reviews [10, 22]. While this research forms useful groundwork for further investigation, the use of balanced training and evaluation datasets means that the resulting systems do not translate well into real-world detection methodologies, where such systems must contend with significant class imbalance. Variation in problem formulation and inconsistent availability of reproducible code mean that quantitative comparison between OIO detection methods is difficult, even when not considering under-researched areas such as analyst-driven approaches.

There are a variety of non-Transformer techniques for creating user representations for detecting hateful or harmful behavior, though many of these rely on platform-specific characteristics (such as "retweet" interactions on Twitter) and do not take a generic content-based approach to user representation [29]. Text-based methods for user representation also exist, but often similarly focus on Twitter users, and involve applying and reducing numerous views, requiring an expensive computational process to update the resulting embedding, rather than a single Transformer model [7].

Previous work has been performed on clustering social media users based on their written text [17, 32], and specifically using BERT embeddings to cluster written arguments between humans [28]. Research on the nature of BERT embeddings has also leveraged visualizations obtained through dimensionality reduction techniques, resulting in insights into the combination of syntax and semantics represented within BERT embeddings [8]. These works do not consider the combination of multiple BERT embeddings to create user representations based on textual content – the approach used by MUTA.

4   Methodology

The proposed OIO detection methodology applies several computational techniques in order to transform user-authored text submissions into a user visualization that assists a human analyst in discovering accounts that warrant further investigation. A central component of this workflow is the Mean User-Text Agglomeration (MUTA), which consists of calculating the mean embedding of multiple user-generated texts, prior to their projection into a low-dimensional space where similar accounts gravitate together. For reference, a flow diagram of the methodology can be found in Fig. 1.

4.1   User Representation

The MUTA user representation approach is broken into two steps: first, the creation of embeddings for each of user-authored texts; and second, the combination of these text embeddings into a user embedding for each author.

In the first step, we generate an embedding for each text using a sentence representation token from a Transformer encoder (specifically, the [CLS] token from the uncased English BERT$_{BASE}$ model [13]). While other Transformer encoders exist, and other methods of using BERT to create sentence representations may offer improved performance, we use the BERT [CLS] method as it is the most easily reproduced, reflects the approach of the original BERT paper, and demonstrates the effectiveness of our approach under the most generic and reproducible settings possible.

The second step groups these text embeddings by the user that authored each text, conceptualizing a "user" as the collection of all texts authored by a particular person. For a user k that has authored n texts, all n text embeddings are averaged to form a single user embedding, u_k. The count n is preserved to allow for weighting the embeddings of future texts before adding to u_k.

This approach has several major practical advantages over alternate approaches to content-based user representation:

1. User representations are easily transferable to Transformer encoders in other languages, or more advanced encoders with greater representational capacity
2. Generating user representations is quick, and does not require a costly and slow pre-training process as is required by custom Transformer architectures for user representation [4]
3. Upon the emergence of new data, new text embeddings can be weighted and added to existing user representations, a process which is computationally cheap and can be executed at the massive scale required for social media analysis

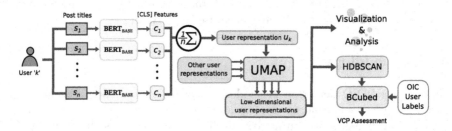

Fig. 1. Flow diagram of user embedding, visualization, and evaluation methodology

Generated user embeddings can be projected into 2 or 3 dimensions for visualization using Uniform Manifold Approximation and Projection (UMAP) [25]. In contrast to other dimensionality reduction methods, such as t-distributed stochastic neighbour embedding (t-SNE) [23], UMAP better preserves the global structure of the underlying data [25], providing a projection where distances between nearby points are meaningful (local structure), as well as the relative position of groups of points within the projection space (global structure).

As the presence of influence accounts is unlabelled in contemporary social media data, we rely on publicly disclosed accounts from past online influence operations as a means of evaluating to what extent influence titles are colocated in projection of the embedding space. It is important to note that this methodology can also be entirely performed without labelled OIO data, making it possible to eliminate topic bias caused by using data from past operations.

5   Datasets and Processing

According to the Alexa rankings, which factor a combination of traffic and engagement, Reddit is currently the seventh most popular website in Canada and the U.S., placing it ahead of Twitter, Instagram, and Wikipedia [1]. As such, Reddit is an important platform for online influence research in the Canadian digital ecosystem. Reddit has been the target of past online influence operations, and has preserved the accounts associated with these OIOs for future research [20]. All submissions and comments from these OIO accounts are available for download on GitHub [9], and (as mentioned in Sect. 4) are used for evaluating the user embedding projection.

To analyze social media activity leading up to the Canadian election, submission titles from 2019 prior to the federal election within Canada-focused subreddits were scraped using the Pushshift API [5] and organized into a dataset. A variety of large Canada-focused subreddits were selected in consultation with OIO subject matter experts based on relevance to the Canadian election and likelihood to influence Canadian voters. While this list is by no means comprehensive, the sampling represents a variety of different Canada-oriented Reddit communities.

As the BERT model used is English-specific, and several Canadian subreddits in the corpus are bilingual, the resulting titles are filtered to remove French submission titles. This is performed by dropping submission titles that are identified as French with a score $p_{fr} \geq 0.99$ by the spaCy language detection library [14]. SpaCy frequently produces high-confidence results, and the high threshold minimizes the amount of posts removed – erring on the side of retaining posts where the title contain both French and English text.

6   Experiments and Evaluation

All experiments were run on a single high-capacity processing machine with 32 GB of RAM, an Intel® i7-6800K CPU @ 3.40 GHz, and a NVIDIA GPU with 6 GB of memory. Submission titles in the entire 2019 Canadian Reddit dataset were grouped by author and the dataset was filtered to only those users who posted at least m submissions over the collected period, where m was set to

10 to remove users with little posting activity. As making new accounts requires additional work – and new accounts are restricted based on age and post history – it is attractive for an attacker to maximize usage of an established account. The BERT embeddings of the [CLS] tokens for these submissions were then averaged together to create the MUTA meta-embedding of each user's pre-election 2019 submission titles. BERT inference on GPU took an average of 29.2 ms per title. This process was repeated for users from past Reddit OIOs who also produced 10 or more submissions.

UMAP was used to reduce these vectors into a visualizable number of dimensions for review by OIO analysts. We perform UMAP dimensionality reduction, using cosine distance as the similarity measure, reflecting other work in comparing sentence representations using BERT embeddings [27]. The number of output components is set to produce a 2-dimensional representation for consumption by embedding visualization tools. The UMAP "number of neighbors" hyperparameter was set to 15 (the default) as adjusting this parameter to 10 or 20 had little impact on the resulting projection, and the usage of the default value better demonstrates the effectiveness of the approach under generic settings. As UMAP projections may have minor variation between runs [25], for reproducibility, the projections in this paper were generated in UMAP's deterministic operating mode, with a random seed of 0. UMAP dimensionality reduction completed in 10.3 s.

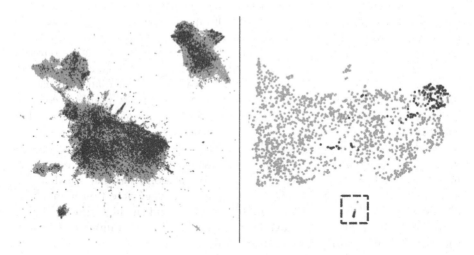

Fig. 2. Projection comparison of raw submission title embeddings (left) and by-user mean title embeddings (right). OIO accounts highlighted in bold red. Averaging multiple title embeddings by user draws them closer together in fewer regions of the projection, making them more likely to be observed by an analyst viewing user submission samples through mouseover events. (Color figure online)

6.1   Assessment of Visualization Quality

A 2D projection is useful for facilitating human OIO detection if the projection places OIO embeddings into visually identifiable structures within the projection, either in a single part of the embedding space, or in several distinct formations. We introduce a Visual Class Perceptibility (VCP) metric for measuring the utility of a point visualization to a human searching for a particular class of points (i.e., the OIO class). This is accomplished by executing an established clustering algorithm (HDBSCAN), followed by an extrinsic cluster quality measurement (BCubed). The usage of a perceptual quality metric to evaluate the quality of a visualization is common practice in information visualization, and relies on selection of criteria that heuristically capture desirable visualization properties. While work in human-computer interaction (HCI) has produced several existing visualization quality metrics for scatter plots [6], these metrics do not address the case where the viewer is searching for a particular class of points. The calculation of the VCP metric is now described, along with a justification as to its appropriateness as a measure of visualization quality for this task.

To calculate VCP, we apply Hierarchical Density-Based Spatial Clustering of Applications with Noise (HDBSCAN) [26] to create a density-based clustering of embedding visualizations from labelled influence operations. Density-based clustering methods, such as DBSCAN and HDBSCAN, are preferable for this problem, due to noise in the data, as well as the emergence of clusters of varying shape in UMAP projections. Between these methods, HDBSCAN has several advantages over DBSCAN that make it more suitable as a quality metric. First, unlike DBSCAN, HDBSCAN is able to identify clusters of variable density, which may emerge following UMAP projection. Second, HDBSCAN automatically sets the epsilon and distance parameters based on stability, eliminating the need for a grid search of DBSCAN hyperparameters and increasing the simplicity of the metric. We fix the minimum cluster size and minimum samples variables to 15 after following the guidance of HDBSCAN documentation for selecting parameters [24].

To obtain the final quantitative VCP metric for point projection evaluation, we leverage BCubed as an extrinsic cluster evaluation measure. BCubed is an extrinsic cluster evaluation measure that can be used to compare the performance of a clustering against a set of labels indicating the ideal cluster assignment for each point [30]. An extended superset of BCubed, known as Extended BCubed [31], also supports overlapping clusterings. BCubed provides precision, recall, and F-score measures for clustering results. The BCubed F-score is a particularly desirable extrinsic evaluation measure as it fulfills "homogeneity", "completeness", "rag bag", and "clusters size versus quantity" conditions surveyed to be important for an intuitive extrinsic cluster evaluation measure [3].

To quantify the utility of title and user projections, we calculate VCP on projections of submission titles and MUTA user representations, using data from past OIO operations as the positive class. The VCP measure reflects the tendency of a point projection to gather OIO accounts into visually distinct clusters.

A projection that places many influence accounts into the same density-based cluster is assessed to have greater value to a human analyst.

For comparison, we perform clustering under two different operating modes of HDBSCAN: excess of mass (EOM) clustering and leaf clustering. These modes determine how HDBSCAN selects clusters from the cluster tree hierarchy. Put simply, EOM combines nearby smaller clusters into larger clusters, while leaf clustering selects only the leaf nodes from the HDBSCAN tree, producing a larger number of smaller homogeneous clusters. Leaf clustering is compared to assess whether there is benefit to further breaking down large clusters considered a single region under EOM.

The VCP results under both settings of HDBSCAN can be found in Table 1. From these results, we find that visualizations of MUTA user embeddings provide a greater perceptibility of OIO content than visualizations of BERT title embeddings. This aligns with qualitative observation of Fig. 2, which are discussed further in the next section. A random baseline is also provided, which reflects an unstructured view of the data (i.e., the results obtained if same number of clusters were assigned randomly, rather than according to proximity in a 2D projection). This is accomplished by assigning labels to each sample from a uniform distribution over $1..k$ where k is the number of clusters—an approach that is regularly used for obtaining a baseline of extrinsic cluster quality [33]. We repeat the random cluster assignment 10 times, and the results are averaged to obtain a robust baseline with a standard deviation $\sigma < 0.0005$ for the random F1 benchmark results in Table 1.

Table 1. Extrinsic BCubed precision, recall, and F1 scores of density-based clusters in 2D projections with ground-truth provided. Cluster counts provided in parentheses.

Proj.	Cluster Selection (&)	Precision	Recall	F-Score
MUTA	HDBSCAN-EOM (4)	0.875	**0.955**	**0.913**
	HDBSCAN-Leaf (22)	**0.893**	0.395	0.547
	Random (4)	0.864	0.251	0.389
	Random (22)	0.875	0.051	0.096
BERT	HDBSCAN-EOM (72)	0.907	0.393	0.549
	HDBSCAN-Leaf (1054)	**0.915**	**0.489**	**0.637**

6.2 Qualitative Analysis and Application

A 2D visualization of the user embeddings can be found on the right of Fig. 2. We find that the MUTA user representation qualitatively colocates users who tend to post similar topics, or structure their submission titles in similar ways. The position of a user account within this projection reflects both the common topics in a user's submission titles, as well as the typical grammatical structure of their sentences. For example, the top left "tip" of the large structure is occupied by users posting news headlines, while the small blob in the upper middle is a group of accounts that had a higher incidence of French and bilingual submission titles. There are three main groups of influence accounts: the majority occupy the upper right area of the projection, a smaller group occupies the less dense area in the middle, and the remaining accounts are in a very dense structure in the distant bottom right. By hovering over accounts, an analyst can identify commonalities in different regions of the projection, and use this to focus their attention according to their domain knowledge.

HDBSCAN-EOM clustering of 2-dimensional MUTA embeddings alongside labelled OIO Reddit users resulted in 3 clusters (plus an additional "noise" cluster containing 4 points). These three clusters are as follows:

1. A large cluster of users containing the majority of the points in the central region of the user representation in Fig. 2.
2. A small cluster of French and bilingual users remaining after the language-filtering step, found in the upper middle of the user representation in Fig. 2.
3. A small dense cluster containing the highest relative proportion of OIO accounts. This cluster was of particular interest and is marked by the red box in Fig. 2.

Cluster 3 was identified as a distinct cluster by both HDBSCAN-Leaf and HDBSCAN-EOM. Content within this cluster is entirely English-language, and characterized by promotional posts, often related to technology. Of the 20 accounts within this cluster, 16 were among those banned and preserved as part of the Reddit transparency report account disclosure. The remaining 4 are users from the 2019 Canadian subreddit dataset. Querying these accounts six months later revealed that each of these users had been suspended, raising the proportion of deactivated accounts in this cluster to 100%.

The removal of these accounts is promising for the usage of MUTA for moderation. Commonality in material and proximity to OIO accounts indicates that these 4 accounts have similar features to a group of similar accounts used during past OIOs on Reddit. This does not necessarily mean that these accounts are owned and operated by the same entity. They may, for example, belong to a spam network that sold accounts to the OIO operator at some point in time. Alternatively, it may be that these accounts were operated using similar software utilities, or even simply that spam strongly resembles other spam much more than regular discourse. Whether or not there is a direct link between the detected spam accounts and OIO accounts, the separation of a cluster that exhibits suspicious characteristics is of benefit to OIO investigation efforts, and will likely be useful in analyzing and characterizing future online influence operations.

7   Conclusion

In this paper we introduced Mean User-Text Agglomeration (MUTA): a practical method for user representation as part of a novel human-centric unsupervised approach to OIO detection. We simply show that this practical meta-embedding technique is effective at creating user visualizations that are quantitatively more useful for identifying OIO users than either an unstructured view of accounts or direct text embeddings, and furthermore demonstrate that this approach has the practical ability to spotlight inauthentic accounts during a genuine election event. Combined, these findings create a compelling case for the value of future research into analyst-driven OIO detection methods – which this work enables through providing reproducible code and privacy-protected datasets.

Many challenges persist as OIOs and detection methodologies continue to evolve. First, the text produced as part of online influence operations can vary significantly over the course of an operation's length, as well as between disparate operations, impacting text representations accordingly (i.e., "concept drift"). Second, there has yet to be any user study in which OIO detection methods are piloted by OIO domain experts to determine real-world effectiveness. Third, alternative methodologies for user representation, such as those that incorporate additional temporal and metadata features, may offer improvements in user representations for OIO detection. Finally, as generative text models for computer-authored comments become more ubiquitous, offensive OIO tradecraft may change dramatically, requiring corresponding adaptation in defensive measures.

References

1. Alexa Internet, I.: Alexa rankings by country (2021). Accessed 06 July 2021
2. Alizadeh, M., Shapiro, J.N., Buntain, C., Tucker, J.A.: Content-based features predict social media infl. operations. Sci. Adv. **6**(30), eabb5824 (2020)
3. Amigó, E., Gonzalo, J., Artiles, J., Verdejo, F.: A comparison of extrinsic clustering evaluation metrics. Inf. Retrieval **12**(4), 461–486 (2009)
4. Andrews, N., Bishop, M.: Learning invariant representations of social media users. In: EMNLP/IJCNLP (2019)
5. Baumgartner, J., Zannettou, S., Keegan, B., Squire, M., Blackburn, J.: The pushshift reddit dataset. ArXiv abs/2001.08435 (2020)
6. Behrisch, M., et al.: Quality metrics for information visualization. In: Computer Graphics Forum. Wiley Online Library, vol. 37, pp. 625–662 (2018)
7. Benton, A., Arora, R., Dredze, M.: Learning multiview embeddings of twitter users. In: 54th Annual Meeting of the ACL (Volume 2: Short Papers), pp. 14–19 (2016)
8. Coenen, A., Reif, E., Yuan, A., Kim, B., Pearce, A., Viégas, F.B., Wattenberg, M.: Visualizing and measuring the geometry of Bert. In: NeurIPS (2019)
9. Coscia, A.: Reddit suspicious accounts dataset (2018). https://github.com/ALCC01/reddit-suspicious-accounts. Accessed 20 Apr 2019
10. Crothers, E., Japkowicz, N., Viktor, H.L.: Towards ethical content-based detection of online influence campaigns. In: IEEE MLSP 2019, pp. 1–6 (2019). https://doi.org/10.1109/MLSP.2019.8918842
11. Crothers, E.: Ethical detection of online influence campaigns using transformer language models. université d'Ottawa/University of Ottawa (2020)
12. Crothers, E.: Muta-2021 (2021). https://github.com/ecrows/MUTA-2021

13. Devlin, J., Chang, M., Lee, K., Toutanova, K.: BERT: pre-training of deep bidirectional transformers. CoRR abs/1810.04805 (2018). http://arxiv.org/abs/1810.04805
14. Explosion: Spacy python library. https://github.com/explosion/spaCy (2019). Version 2.0.16
15. Fornacciari, P., Mordonini, M., Poggi, A., Sani, L., Tomaiuolo, M.: A holistic system for troll detection on twitter. Comput. Hum. Behav. **89**, 258–268 (2018)
16. Gao, H., Hu, J., Wilson, C., Li, Z., Chen, Y., Zhao, B.Y.: Detecting and characterizing social spam campaigns. In: Proceedings of ACM IMC 2010, p. 35–47. ACM, New York, NY, USA (2010). https://doi.org/10.1145/1879141.1879147
17. Gencoglu, O.: Deep representation learning for clustering of health tweets. CoRR abs/1901.00439 (2019). http://arxiv.org/abs/1901.00439
18. Gleicher, N.: Removing coordinated inauthentic behavior (2020). https://about.fb.com/news/2020/07/removing-political-coordinated-inauthentic-behavior/
19. Hleg, E.H.L.E.G.o.A.: Ethics guidelines for trustworthy AI (2019). https://ec.europa.eu/digital-single-market/en/news/ethics-guidelines-trustworthy-ai
20. Huffman, S.: Reddit 2017 transparency report findings (2018). Accessed 23 May 2019
21. Kaminski, M., Malgieri, G.: Algo. impact assessments under the GDPR: Producing multi-layered explanations. SSRN (2019). https://doi.org/10.2139/ssrn.3456224
22. Kennedy, S., Walsh, N., Sloka, K., McCarren, A., Foster, J.: Fact or factitious? contextualized opinion spam detection. In: ACL 57: Student Research Workshop. ACL, Florence, Italy, pp. 344–350 (2019). https://doi.org/10.18653/v1/P19-2048
23. van der Maaten, L., Hinton, G.: Visualizing data using t-SNE. J. Mach. Learn. Res. **9**, 2579–2605 (2008). http://www.jmlr.org/papers/v9/vandermaaten08a.html
24. McInnes, L.: Parameter selection for HDBSCAN (2016). https://hdbscan.readthedocs.io/en/latest/parameter_selection.html
25. McInnes, L., Healy, J.: UMAP: Uniform Manifold Approximation and Projection for dimension reduction. ArXiv abs/1802.03426 (2018)
26. McInnes, L., Healy, J., Astels, S.: HDBSCAN: Hierarchical Density based clustering. JOSS 2(11) (2017). https://doi.org/10.21105/joss.00205, https://doi.org/10.21105
27. Reimers, N., Gurevych, I.: Sentence-BERT: Sentence embeddings using siamese bert-networks. In: EMNLP/IJCNLP (2019)
28. Reimers, N., Schiller, B., Beck, T., Daxenberger, J., Stab, C., Gurevych, I.: Classification and clustering of arguments with contextualized word embeddings. In: ACL 57, pp. 567–578. ACL, Florence, Italy (2019). https://doi.org/10.18653/v1/P19-1054
29. Ribeiro, M., Calais, P., Santos, Y., Almeida, V., Meira Jr, W.: Characterizing and detecting hateful users on twitter. In: ICWSM, vol. 12 (2018)
30. Foundation of evaluation: van Rijsbergen. J. Documentation **30**, 365–373 (1974)
31. Rosales-Méndez, H., Ramírez-Cruz, Y.: CICE-BCubed: a new evaluation measure for overlapping clustering algorithms. In: Ruiz-Shulcloper, J., Sanniti di Baja, G. (eds.) Progress in Pattern Recognition, Image Analysis, Computer Vision, and Applications, CIARP 2013. LNCS, vol. 8258, pp. 157–164. Springer, Heidelberg (2013). https://doi.org/10.1007/978-3-642-41822-8_20
32. Singh, K., Shakya, H., Biswas, B.: Clustering of people in social network based on textual similarity. Perspect. Sci. **8**, 570–573 (2016). https://doi.org/10.1016/j.pisc.2016.06.023

33. Strehl, A., Ghosh, J.: Cluster ensembles - a knowledge reuse framework for combining multiple partitions. JMLR **3**, 583–617 (2003). https://doi.org/10.1162/153244303321897735
34. Twitter: Twitter elections integrity dataset. Internet (2019). Accessed 20 Apr 2019
35. Yang, C., Harkreader, R., Zhang, J., Shin, S., Gu, G.: Analyzing spammers' social networks for fun and profit: A case study of cyber criminal ecosystem on twitter. In: WWW 2012. p. 71–80. ACM, New York, NY, USA (2012). https://doi.org/10.1145/2187836.2187847

The Role of Information Organization and Knowledge Structuring in Combatting Misinformation: A Literary Analysis

Kevin Matthe Caramancion[✉]

University at Albany, State University of New York, Albany, NY 12222, USA
kcaramancion@albany.edu

Abstract. This paper seeks to explore how the three hallmark dimensions of Information Organization—(1) Access, (2) Discovery, and (3) Retrieval—each as a construct paves the way for the rise of misinformation and, consequently, be the essential areas of interest for its control and regulations. Furthermore, the role of social networking platforms, grounded on *folksonomy-designed* environments, is examined on how they function as a seminal contributing factor for the creation and persistence of misinformation. A *taxonomic* approach as an addendum for the remedy of misinformation is assessed, including suggestions for its more robust implementation. This paper concludes with a summative precis of the presented ideas and literature on this subject and stipulates the general limitations of Information Organization and Knowledge Structuring in their applications in the domain of misinformation.

Keywords: Taxonomy · Folksonomy · Misinformation

1 Introduction

Misinformation or *fake news*, albeit its deceptive nature, is still a type of information. Thus, it is still subject to the characteristics and properties of classical correct and truthful information. These include the metadata functioning as its descriptors, ontological clues dictating its domain, and even the semantical provenance that can be used to trace its origins—among other qualities.

Knowledge structure, the interrelated collection of facts or knowledge about a particular topic, is grounded on labels [1] and relations [2]. The Data-Information-Knowledge-Wisdom (DIKW) pyramid of Information Management suggests the processing and transformations that transpire in many models, including information systems and the cognitive behavior of humans. An erroneous value in any of the stages of the DIKW can be a highly probable cause of misinformation. Incorrect data, when not corrected, can lead to inaccurate analysis. Correct data, coupled with incorrect analysis, may yield inaccurate information. Correct data and analysis that leads to verified information may still be rejected by users when their cognitive capabilities can't comprehend or outright reject it due to biases or incompatibility in information-seeking motivation (i.e., impairment of knowledge and wisdom).

© Springer Nature Switzerland AG 2021
D. Mohaisen and R. Jin (Eds.): CSoNet 2021, LNCS 13116, pp. 319–329, 2021.
https://doi.org/10.1007/978-3-030-91434-9_28

The figure [3] below illustrates the connected and interacting elements that, in their trivial ways, when compromised, may contribute to misinformation which is represented by the blue oval as the societal challenge in this paper's context (Fig. 1).

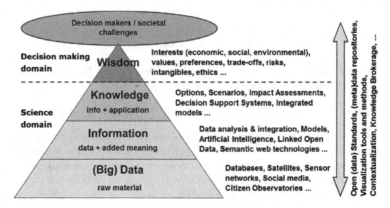

Fig. 1. The DIKW ecosystem & its elements

Given these groundings, falsehoods and misinformation is effectively controlled when prevented in the first place rather than the use of corrective mechanisms. However, the use of the latter should always be available as an essential recourse for its governance in information environments since the infallibility of the integrity in all the content remains elusive.

2   Information Access as a Factor

2.1   Access as a Definitional Construct

In the context of knowledge inquiry, *information access* is the series of actions performed by users to achieve the goals of seeking, organizing, and understanding the phenomena of any particular information [4]. Moreover, information access is the freedom or ability to identify, obtain and make use of any database or information effectively since users typically interact with information technologies for this undertaking. Information access, as an abstraction, takes in the concrete form of search queries, text summarization, and text clustering, among other representations [5]. Information access, as a dimension, covers several key issues prevailing in the current times, including but not limited to copyright issues, open-source, privacy, and even security.

One primal importance of information access is the objective to which it seeks to simplify and make it more effective for human users to access and further process large and unwieldy amounts of data and information. Information consumers engaged in the information-seeking process has one or more goals in their minds and use the search systems as tools to help achieve those goals. The fields of user-centered design and human-computer interaction (HCI) are the prevailing subfields of computing that support bridging the gap between the information-seeking motivation of users and the interface of information technologies. To fulfill the information needs of users, accurate knowledge translations (e.g., from tacitly abstract to more explicit codified forms) and representations between these two entities is imperative.

The figure [6] below shows how this consistent interaction and their respective triggers take place (Fig. 2).

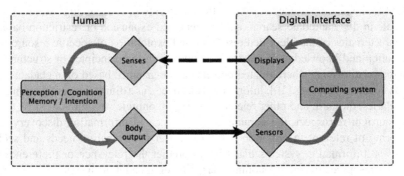

Fig. 2. The human-computer interaction

2.2 Applications on Misinformation

Social Networking Platforms, which have been historically used for entertainment purposes by users [7], have now superseded that affordance. These platforms are now the most used medium for information access, including news consumption which offers more timely, real-time updates from public figures and news channels. These platforms, functioning as containers, now allows webpages to be contained in their respective environments allowing both structured and unstructured data. The very backbone of knowledge inquiry and structuring has been maintained arduously—from queries, keyword searches, even up to reverse audio or image lookup. The flexibility of even the most abstract form of queries to be looked up has been made incrementally possible.

The rise of these platforms mainly altered two crucial features of the knowledge inquiry process. First is the transfer of inceptions in the information-seeking goals of the users to the news feed. Unlike traditional queries where users have an objective prior to their information access, users now typically access the medium first and, from there, decide which information to consume. Next, unlike the traditional information-seeking mediums where resources and artifacts generally are scrutinized and substantiated, social networking websites such as Facebook and Twitter allow almost any user to create and share contents that can be freely accessed, information-wise, by other users [8]. These are the two predominant characteristics of the current platforms that pave the way for the creation and persistence of misinformation. The developments in access technologies, in ways, made it easier for misinformation to reach an even bigger and wider audience.

3   Information Discovery as an Amplifier

3.1   Discovery as a Definitional Construct

In the context of the information search process, *information discovery* is the complex series of tasks that involves locating a particular digital object on the network through iterative research activities, which usually involve the specification of a set of criteria relevant to the resources needs of users, the organization, and ranking resources in the candidate in this candidate search, and the repeated expansion or restriction based on the characteristics of the identified resources and exploration of specific resources [9]. Information and knowledge discovery is grounded on the principles of structures and semantic relationships where related objects can be grouped based on metadata values for classificatory purposes [10]. Information discovery, as a dimension, entails controlled vocabularies, thesauri, and other related forms of taxonomic structures.

In outright retrospect, the paramount importance of information discovery is the uncovering of relevant information based on a user's information needs and seeking objectives. Information systems that seek to predict the relevance or preference to a particular resource are called recommender or recommendation systems.

The figure [11] below outlines the explicit and implicit data points used by recommender systems to attain this objective (Fig. 3).

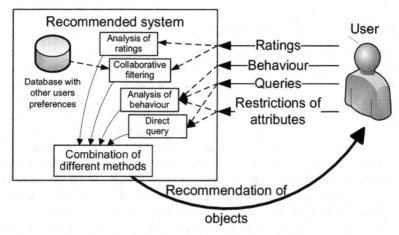

Fig. 3. A summative diagram of modern recommender systems

Information discovery is further attributed as the dimension responsible for filtering irrelevant information in the search process due to information overload [12]. Although it may seem uncomplicated, this task is especially arduous due to the vast number of resources, artifacts, and information that may be accessible to the users, especially in the digital age. Another crucial role of information discovery in the search process is the presentation of related resources that the users may find valuable and relevant for their information-seeking needs. This is due to the fact that the information search process is not static and, as a matter of truth, interactive, where learning and discovery are both unequivocal components of the process.

3.2   Applications on Misinformation

There has never been a failure in the transference of the principles of information discovery from its traditional mediums and formats to the contemporary social networking platforms. As a matter of truth, the components and enforcements of knowledge discovery did not just transfer, but rather, is even more amplified to reach more than its intended potential. This is particularly visible in the emergence of recommender systems in the queries of users, which have been so advanced that issues of privacy have been raised in the data collection methods used in these current technologies. The upgrades in mining and analytic techniques have been ubiquitous [13]. Unlike before, where usually structured data, metadata, and textual information are processed and mined, sophisticated advances now allow unstructured data not just processed but even manipulated. Computer vision as a field, for instance, offers complete analysis and control of images and videos alike [14].

The methods of information discovery transgressed the traditional forms of structured and unstructured data and included an analysis and mining of the behavior of the users. Included herein but is not limited to are mouse clicks, eye movement tracks, and even webpage refresh rates for the purposes of recommending more relevant information and content to the users in the form of extreme personalization, which are originally grounded on the fundamentals of information discovery. Fake news literature predominantly suggests that this model is a chief enabler of misinformation [15–17]. When users tend to acquire content to feed off their confirmation biases, even conspiracies, it creates an environment, typically referred to as a filter bubble, that is dividedly partisan and is usually a potent breeding ground of misinformation.

4   Information Retrieval as the Filter of Fake News

4.1   Retrieval as a Definitional Construct

In the information search process, *retrieval* is the process, methods, and procedures of acquiring particular information from resources relevant to users' information needs expressed through search queries. The technological underpinning that made this possible are two-fold: (a) the indexing system and (b) the query system, functioning as the interface for users. Results displayed are from the indexing system and can either be based on full-text or other specific-text extraction techniques [18]. The widely accepted variables functioning as metrics that measure a retrieval system's effectiveness are *precision and recall*. The former is the fraction of retrieved documents that are relevant to a particular query, while the latter is the fraction of the relevant documents that are successfully retrieved.

For its valuation in knowledge structuring, information retrieval compels search systems (e.g., engines, directories, etc.) to facilitate a rapid and accurate search-indexed structure based on the input(s) of users, usually in the form of keywords, to fulfill their interests and informational needs. Information retrieval, as a science, is in a continuous iterative process to keep improving this indexing design to provide an even more effective and rich search that includes texts from documents, metadata that describe data, and even unstructured data objects such as images, videos, and audio. Fields of computing are born out of information retrieval such as natural language processing (NLP), decision trees (DT), and social search, among others. Due to these emergences, the queries, previously in textual format, can now be in the form of voice [19], reverse image, and even audio footprint—all for recognition purposes.

The figure [20] below displays the differences in indexing both structured and unstructured data objects (Fig. 4).

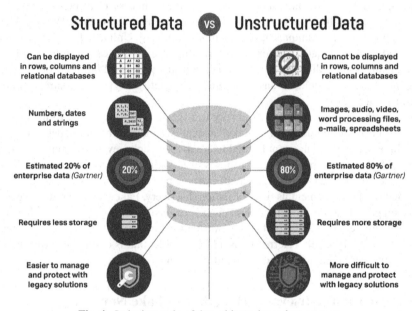

Fig. 4. Indexing style of data objects dependent on type

4.2 Applications on Misinformation

Non-factual information should cease to exist by the very doctrines of information retrieval. One of the hopes of *precision and recall* is to filter out irrelevant content that will not fulfill the information needs of the user performing the query. In the current information environments, however, the actors of misinformation have played the card of relevance all too well. Deceptive contents are indexed as appropriate since the descriptors are designed to truly mislead the users in their quest for information. Even unstructured data objects are not left untouched; manipulated photos and videos that support the claims of fake news are used to mislead the information user further. The paradox is that the quest for speed in retrieving content based on the queries of users will now often result in the inclusion of content that may be unverified.

Although retrieval systems may have been exploited to disseminate misinformation, the remedy still lies in this domain. Ranking strategies in indexing systems should promote verified, truthful content and demote the misleading ones. A strategy already enforced in retrieval systems where the computation of a numeric score on how well each object in the database matches the query and ranks the objects according to this value. The top-ranking objects are then shown to the users. This process may then be iterated if the user wishes to refine the query since the outright removal of contents (i.e., censorship), as of this time, is a highly controversial topic since it intersects with the laws in freedom of speech and press [21]. The ranking strategy, however, is not without its controversies. Issues of discrimination, biases from paid content, privacy intrusion, and even racism are sometimes associated with this area of strategy.

5 The Contribution of Folksonomy in Misinformation Explosion

5.1 Influence in Social Networking Platforms

At the very backbone of social networking platforms lies the democratic model of user-generated content (UGC). These are any forms of content that are contributed by any user in an information environment, be it of any type, including videos and images. The question of whether these environments are anchored from folksonomic or taxonomic structures is visibly evident and has long been settled. The knowledge acquisition obtained through social tagging in almost any content has been long the classical strength of social mediums over the traditional top-down approach to information sharing [22]. The community-based UGCs and tagging lacked the traditional hierarchies of taxonomies, and for a long time, this has been a feature of strength rather than a vulnerability. This bypassed the delay due to the processing time in the indexing of contents, thereby offering anyone of almost instantaneous access to any information, unfortunately including the untruthful ones.

5.2 The Rise of Misinformation

The long flagship feature of these platforms, the democratic folksonomic model, unequivocally paved the way for misinformation. The bottom-up structure offered direction of information flows both vertical and horizontal alike almost without restrictions. One of the hopes of folksonomy is to improve collaboration [23] through the contribution of tags by a broad spectrum of users [24]. But then, the current misinformation environment reveals how impaired conspiratorial and hyper-partisan extremists used this structure to pollute these platforms. Unlike the traditional taxonomic models where the gateways (i.e., information professionals such as subject matter experts, fact-checkers) verify the integrity of information prior to its distribution, the right of the users to free speech overpowered the right to access the truth.

5.3 Crowd-Driven Tags as Enforcers of Misinformation

Tags, as an essential component of this bottom-up structure, have been weaponized to shield misinformation. The characteristics of tags to reflect personal association, categories, and concepts for content representation have been exploited. Debunked beliefs, including conspiracy theories, have been labeled as opinions and personal views by some users [25]. The dangerous contradiction is when an impaired and compromised community cannot accept a scientific truth that goes against their belief. Tags of fake news and lies are contradictorily associated with such truths. For instance, people who refuse to vaccinate spread misinformation and claim how vaccines are the means of government to install a tracker in their bodies or even how vaccines are even more dangerous than the actual diseases. Medical and scientific campaigns that call for vaccine awareness are unfortunately tagged as lies by crowds of misinformation actors. Misinformation draws its strength from the density of users, and this model provides that possibility through the virality of content enhanced by tags.

5.4 Online Communities of Misinformation

A more concrete effect of ultra-personalization in information discovery and retrieval in these social networking platforms is the creation of online communities grounded on false beliefs. Social tags allow users to connect with other users that share common beliefs and interests, including misinformation. Unlike taxonomic classification, the vocabulary in folksonomy directly reflects the user's vocabulary, and people with the same demographic, socioeconomic, culture, and perceptions [26] tend to group themselves in these networks socially. This grouping is made easier by recommender systems. This is a factor of political polarization, a divide of people based on opinions or beliefs. The knowledge structured from communities of misinformation is based on *blind faith*. Scientific and factual truths become relative positions based on the community one belongs.

6 The Classical Functions of Taxonomic Authority as Misinformation Safeguards

6.1 Natural Characteristics of Taxonomy Against Misinformation

Taxonomies, in their own intrinsic authoritative structure, might give the impression of being dictatorial, but it creates an information environment that allows for efficient searching, sorting, and reporting. An information architecture that is grounded on a hierarchical system of classifying information [27] with clear guidelines and categorizations may put misinformation in its rightful place. For instance, the current content integrity descriptors in social networking platforms identifying posts as misleading and verified when enforced efficiently may offer a sense of clarity to users. The authoritative fact-checkers, functioning as third-party subject matter experts, dictate the integrity of the content as per their current scientific state.

This taxonomic system, when communicated regularly, may limit the growth of misinformation through a simple system of categorization. First-person accounts and opinions should be clearly labeled as personal views rather than binding claims or "claimed" personal truths. This is because there are simple truths and facts that are not subject to personal interpretations, such as on the topics of public health and national security.

6.2 The Bias Neutralizer in News and Media

Whereas the truth may be absolute, the way in which it is expressed through linguistics or lexicons may be relatively subjective. For instance, the very keyword misinformation has many associated synonyms, including the colloquialism *fake news* or even deception. In an attempt by most media organizations to keep their subscribers and viewers maximally engaged, the need to cater to the latter's dispositions may be a priority to the former. This is the reason why news headlines may have different framing of their words and lexical structures.

A taxonomic approach to the wordings of news headlines may provide an even more objective approach to their presentation. Current proposals to implement linguistic technologies against misinformation are introduced to detect biases and to grade the political spectrums of vocabularies. These technologies are grounded on semantic relations, including synsets with the more generic, objective terms positioned on top of the hierarchy. Although the synonyms, including hyponyms, hypernyms, meronyms, may have various interpretations that change over time—the authoritative term as the reference index remains the same. This can limit the semantic inconsistencies explicitly used by some news organization that implicitly amplify the social polarization in a nation's citizens and as a natural consequence, promotes the culture of misinformation.

6.3 Achieving Conjunctive Balance with Folksonomy

While the previous sections highlight the shortcomings of the folksonomy and strength of taxonomic classifications in social networks, this paper does not recommend the termination of the former and the complete enforcement of the latter. As always, a harmonious balance between the two philosophies is the ideal model of a functioning platform grounded on efficiency and integrity. An absolute taxonomic approach to control misinformation will result in the content of the environments losing personalized and relatable topics to the users. On the other hand, the current structure in the platforms is on the opposite spectrum, where an almost absolute folksonomic culture of tagging is exactly what amplified (if not directly caused) the societal problems of polarization and misinformation. A social network knowledge model that considers the strength of both approaches where one remedies the weakness of the other creates an information environment where misinformation may be difficult to thrive.

7 Conclusion

Misinformation, in its very existence, is the antithesis of Information Organization. Through the lenses of Knowledge Structures, Representations, and Models, this paper examined how the phenomenon of misinformation emerges and lingers in an information environment. This paper, with regards to the science of misinformation, shifted the analytical focus from the typical domains of such as threat actors, victims, motivations to the system that enables it on the background, the *misinformation environment*. From an ecosystem view, the interacting elements that make it possible to happen are also the key factors for its control and regulation. At the surface, although it may appear that the most trivial of processes may not seem to affect its growth, such as index rankings, labels in content, miswording—however, the current state of the social networking platforms suggests otherwise.

As this paper had revealed, a substantial contributor to the culture of misinformation is the blueprint of the classificatory system in social media. The combination of taxonomy and folksonomy, drawing from the strength of one another, remains the optimal model. As a literary subject, misinformation is highly interdisciplinary [28]—and the place of information science in the core domains investigating its nature will remain unchallenged and imperative.

8 Limitations

As a complex phenomenon, misinformation has many contributory factors outside the scope of knowledge structures and representations. For instance, the determinative effects of sociological design in different countries and societies. Misinformation can be a symptom of an even deeper societal problem, such as the divide in a country's citizens rooted in culture, economic disparity, and politics. In cases like these, misinformation will persist outside online communities and social networking sites. The misinformation flows in these online mediums may be temporarily controlled, but people will simply find new ways of creating and disseminating them.

An even more dangerous subset of misinformation, *dezinformatsiya* or disinformation, is information created and carefully engineered to deceive. Unlike misinformation, disinformation actors smartly play the rules of knowledge structures and representations to completely mislead, usually for the purpose of cyber warfare, computed propaganda, and political manipulations. Their strategies include but are not limited to mass astroturfing and *deepfakes*, among other usually state-sponsored acts. Unlike a misinformation actor that can be corrected through training and awareness programs, disinformation actors act on blind faith. In cases such as these, more aggressive and stringent information policies grounded on legal remedies are necessary for their control. This subset of misinformation is outside the scope of this paper and falls under the domains of cybersecurity and digital forensics.

References

1. Fox, M.J., Wilkerson, P.L.: Introduction to Archival Organization and Description. Getty Publications (1999)
2. Hedden, H.: Chapter 4 creating relationships. In: The Accidental Taxonomist, pp.97–127. Information Today, New Jersey (2010)
3. Lokers, R., Knapen, R., Janssen, S., van Randen, Y., Jansen, J.: Analysis of big data technologies for use in agro-environmental science. Environ. Model. Softw. **84**, 494–504 (2016)
4. Dezső, Z., Almaas, E., Lukács, A., Rácz, B., Szakadát, I., Barabási, A.L.: Dynamics of information access on the web. Phys. Rev. E **73**(6), 066132 (2006)
5. Tombros, A., Jose, J.M., Ruthven, I.: Clustering top-ranking sentences for information access. In: Koch, T., Sølvberg, I.T. (eds.) ECDL 2003. LNCS, vol. 2769, pp. 523–528. Springer, Heidelberg (2003). https://doi.org/10.1007/978-3-540-45175-4_47
6. Papetti, S.: Design and perceptual investigations of audio-tactile interactions. In: Proceedings of AIA DAGA (2013)

7. Mastley, C.P.: Social media and information behavior: a citation analysis of current research from 2008–2015. Ser. Libr. **73**(3–4), 339–351 (2017)
8. Intezari, A., Taskin, N., Pauleen, D.J.: Looking beyond knowledge sharing: an integrative approach to knowledge management culture. J. Knowl. Manage. (2017)
9. Proper, H.A., Bruza, P.D.: What is information discovery about? J. Am. Soc. Inf. Sci. **50**(9), 737–750 (1999)
10. Riley, J.: Understanding Metadata: What is Metadata, and What is it For? NISO Press, Washington (2017)
11. Eckhardt, A.: Various aspects of user preference learning and recommender systems. In: DATESO, pp. 56–67 (2009)
12. Koltay, T.: Information overload, information architecture and digital literacy. Bull. Am. Soc. Inf. Sci. Technol. **38**(1), 33–35 (2011)
13. Ristoski, P., Paulheim, H.: Semantic Web in data mining and knowledge discovery: a comprehensive survey. J. Web Semant. **36**, 1–22 (2016)
14. Smiraglia, R.P.: Domain analysis of domain analysis for knowledge organization: observations on an emergent methodological cluster. KO Knowl. Organ. **42**(8), 602–614 (2015)
15. Caramancion, K.M.: An exploration of disinformation as a cybersecurity threat. In: 2020 3rd International Conference on Information and Computer Technologies (ICICT), pp. 440–444. IEEE, March 2020
16. Caramancion, K.M.: Understanding the impact of contextual clues in misinformation detection. In: 2020 IEEE International IOT, Electronics and Mechatronics Conference (IEMTRONICS), pp. 1–6. IEEE, September 2020
17. Caramancion, K.M.: The demographic profile most at risk of being disinformed. In: 2021 IEEE International IOT, Electronics and Mechatronics Conference (IEMTRONICS), pp. 1–7. IEEE, April 2021
18. Denny, M.J., Spirling, A.: Text preprocessing for unsupervised learning: why it matters, when it misleads, and what to do about it. Polit. Anal. **26**(2), 168–189 (2018)
19. Sa, N., Yuan, X.: Examining users' partial query modification patterns in voice search. J. Am. Soc. Inf. Sci. **71**(3), 251–263 (2020)
20. Lawtomated: Structured Data vs. Unstructured Data: what are they and why care? (2019). lawtomated.com
21. Caramancion, K.M.: Understanding the association of personal outlook in free speech regulation and the risk of being mis/disinformed. In: 2021 IEEE World AI IoT Congress (AIIoT), pp. 0092–0097. IEEE, May 2021
22. Deng, S., Lin, Y., Liu, Y., Chen, X., Li, H.: How do personality traits shape information-sharing behaviour in social media? Exploring the mediating effect of generalized trust. Inf. Res. Int. Electron. J. **22**(3), n3 (2017)
23. Karunakaran, A., Reddy, M.C., Spence, P.R.: Toward a model of collaborative information behavior in organizations. J. Am. Soc. Inform. Sci. Technol. **64**(12), 2437–2451 (2013)
24. Mai, J.E.: Folksonomies and the new order: authority in the digital disorder. KO Knowl. Organ. **38**(2), 114–122 (2011)
25. Cooke, N.A.: Posttruth, truthiness, and alternative facts: information behavior and critical information consumption for a new age. Libr. Q. **87**(3), 211–221 (2017)
26. Begany, G.M., Sa, N., Yuan, X.: Factors affecting user perception of a spoken language vs. textual search interface: a content analysis. Interact. Comput. **28**(2), 170–180 (2016)
27. Rosenfeld, L., Morville, P.: Information Architecture for the World Wide Web. O'Reilly Media, Inc. (2002)
28. López-Huertas, M.J.: Domain analysis for interdisciplinary knowledge domains. KO Knowl. Organ. **42**(8), 570–580 (2015)

Fake News Detection Using LDA Topic Modelling and K-Nearest Neighbor Classifier

Mario Casillo[1], Francesco Colace[1], Brij B. Gupta[2] ⓘ, Domenico Santaniello[1](✉) ⓘ, and Carmine Valentino[1] ⓘ

[1] University of Salerno, Fisciano, SA, Italy
{mcasillo,fcolace,dsantaniello,cvalentino}@unisa.it
[2] National Institute of Technology Kurukshetra, Kurukshetra, India
gupta.brij@ieee.org

Abstract. The spread of the COVID 19 virus has dramatically impacted global society by modifying its lifestyle. Social networks, video streaming tools, virtual collaborative environments have been the primary source of communication through the Internet. This suspension of the "real" has led all activities to be declined through new places and contexts of virtual discussion, increasing new problems, including the most important related to the spread of so-called Fake News. The spread of such news can be devastating: consider what is happening during the critical vaccination phase for COVID 19. In this scenario, systems able to recognize, in a practical way, the truthfulness of news are becoming more and more valuable.

This paper aims to present an approach that combines probabilistic and machine learning techniques such as Latent Dirichlet Allocation and K-NN in combination with Context-Awareness techniques to identify the veracity of the news. Adopting Context-Awareness techniques within the proposed system allows a better definition of the operational context Fake News refers to, reducing the problems of semantic polysemy. The first results obtained through standard datasets or using data from real contexts are very interesting and promising.

Keywords: Fake news · Fake news detection · Latent Dirichlet allocation · Machine learning · Text classification

1 Introduction

The last year has seen a series of events that have had a devastating impact on the whole of world society and, more generally, on its way of life [1]. For example, the epidemic linked to the spread of the COVID 19 virus, which effectively suspended the main activities linked to our daily lives, transferring the stage for social interactions to those communication systems that use the Internet as the technological tool needed to spread them. Video-conferencing environments, social networks, chats, collaborative environments, and e-commerce sites have for a long time constituted the technological backbone enabling social interaction between people. This suspension of the "real" has meant that all activities have been declined through this new dynamic: think, for example, of the presidential elections in the United States held in 2020. The whole process has

© Springer Nature Switzerland AG 2021
D. Mohaisen and R. Jin (Eds.): CSoNet 2021, LNCS 13116, pp. 330–339, 2021.
https://doi.org/10.1007/978-3-030-91434-9_29

seen social networks as the places where it has been possible to share information and discuss. However, these new dynamics have led to new problems, the most important of which is linked to the spread of so-called Fake News. The expression Fake News indicates articles or publications on social networks written with invented, misleading, or distorted information, made public with the deliberate intent to misinform or create scandal through the media or attract clicks on the Internet. The effect of Fake News can be devastating: think of what is happening during the delicate phase of vaccination for COVID 19 [2, 3]. Everyday posts question the vaccine's usefulness or describe its potentially harmful effects: because of these posts in some areas, the vaccination process has slowed down significantly. Therefore, it is increasingly essential to develop methodologies that can automatically recognize and label posts or news items Fake News. The process of classifying Fake News is not a simple one, typically consisting of numerous steps, and its complete automation has not yet led to particularly effective software. In this paper, an automatic approach for identifying Fake News will be presented through the combined use of methodologies based on probabilistic approaches such as Latent Dirichlet Allocation and machine learning such as K-NN. The proposed approach aims to identify topics and characteristic words to build a Fake News model for specific topics and operational contexts. Adopting Context-Aware techniques within the proposed system allows a better definition of the operational context to which the Fake News refers and reduces the problems of semantic polysemy. After this first phase, it is possible to describe the Fake News through the main characteristic components and their clustering. To classify a news item, the system uses the same techniques adopted previously to have its modelling by components and then use a K-NN-based approach for the actual classification. The first results obtained through standard datasets or data coming from real contexts are fascinating and promising.

This paper is organized as follows:

- Section 2 presents the background and related works;
- Section 3 contains the description of the proposed approach;
- Section 4 shows the dataset used to test the proposed approach and the experimental results;
- Section 5 contains the conclusion and future works.

2   Background

Fake News can be defined as intentionally and verifiably false [4, 5]. The definition given is not the only one provided by state of the art, but it is adequate for this paper. Fake News Detection can be achieved through various strategies.

Knowledge-Based Methods [6, 7] are mainly based on Fact-Checking, which aims to determine the truthfulness of news by verifying the content of the news itself through other already verified news. Style-Based Methods [6, 8] aim to determine the truthfulness of the news by studying the language used and the lexical style. This approach is widely used to exploit machine learning techniques [9, 10].

Network-Based Detection Methods [11] exploit representation using graphs to describe links on the network. Based on the type of links described by the graph, it is possible to have:

- The Friendship network, exploited in the context of social networks and which aims to describe the connection between the users of the social;
- The diffusion network, which describes the directions of information propagation;
- The Propagation Network, which aims to describe the propagation of news through macro-level and micro-level. The macro-level describes the propagation of news and tweets and retweets related to the news itself. The micro-level is represented through a tree that describes the propagation through shares.

Finally, the Source-Based Methods [6] aim to verify the truthfulness of the news through the analysis of the reliability of the source. The proposed approach aims to determine the truthfulness of the news through syntactic and semantic analysis. In particular, Latent Dirichlet Allocation (LDA) [12, 13] will be exploited to determine the main topics of the news and the K-Nearest Neighbor (K-NN) algorithm for the final classification of the news. In addition to the coefficients associated with the topics, other news features related to the semantic and syntactic environment will be exploited. These will be described in detail in the section dedicated to the proposed approach.

The Latent Dirichlet Allocation (LDA) is a generative probabilistic model of a documents collection (Blei et al., 2003) that allows determining the topics of a text [14]. This topic classification assumes that there are a fixed number of latent topics within the document under analysis that can be determined through a probabilistic study of the problem under consideration.

The K-NN [15–17] is, instead, a technique that allows the clustering of data. A distance metric [18] is exploited for classification. In particular, the Euclidean distance will be exploited in the proposed approach.

$$d(x, y) = \sqrt{\sum_{i=1}^{p} (x_i - y_i)^2} \quad x, y \in \mathcal{R}^p \tag{1}$$

K-NN can be used as both a classification and regression technique [19] and allows for label prediction by analyzing the k elements closest to the analyzed data. The K-NN technique is used by [19] for Fake News Detection.

A further tool exploited by the proposed approach is the Context Dimension Tree (CDT) [20], an undirected acyclic graph $G = <r, N, E>$ that allows context management (Fig. 1) [21, 22]. Such a graph G is composed of the root node r, the set of edges E and the set of nodes N. In particular, the nodes can be divided into:

- Dimension nodes, which describe the dimension domain;
- Concept nodes, which describe the values that the dimension domain can take;
- Parameters, which allow specifying the information associated with a concept node.

The following section will show how the tools described work in synergy for Fake News Detection.

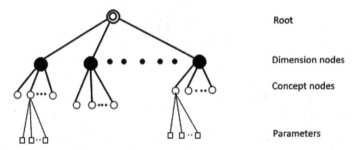

Fig. 1. The context dimension tree

3   The Proposed Approach

This section aims at describing the Fake News Detection approach that is designed by exploiting the LDA, CDT and K-NN methodologies. Figure 2 shows the architecture of the proposed system, which graphically represents the workflow of the proposed methodology. The system analyzes news through a Feature Extraction phase. Three modules are involved in this phase:

- The topic analyzer module, which extracts, through the LDA approach, the topic to which the news belongs;
- The syntactic analyzer module, which analyzes the news parameters;
- The Sentiment analyzer module, which is able to extract the sentiment of the news.

The proposed architecture presents a Context analyzer module, which is a particular module that can store contextual information to improve the Feature Extraction Phase. In particular, this module exploits the Context Dimension Tree (CDT), acquiring temporal information regarding the origins and topic of belonging.

The Context analyzer module is fed by internal modules such as the Topic analyzer and some Semantic analyzer data and external Open data API services that can further enrich the description of the news context. All the modules described above give life to the second phase of the proposed approach, which concerns the evaluation of the truthfulness of the news. In particular, in this phase, the inference engine is able to exploit the K-NN approach to analyze the news by comparing them and expressing a reliability index.

In summary, the proposed approach can be divided into two main phases. The first phase is the identification of the news features. This phase concerns the probability that the news falls within the reference topics, the semantic and syntactic characteristics of the news. The second phase involves the use of the K-NN method for news classification. This method is trained using a dataset containing both real and fake news. In addition, the news in the dataset is profiled using the features determined in the first phase. These phases will be presented in detail below. The first phase aims at news profiling and data discretization. In fact, otherwise, it would not be possible to use the K-NN method, which works by calculating distances between points in a Euclidean space.

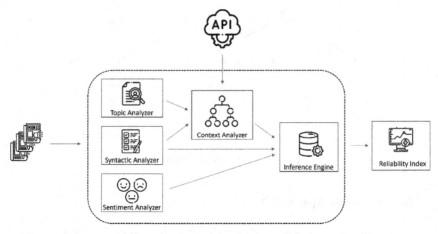

Fig. 2. System architecture

3.1 Feature Extraction

In order to analyze the news through the K-NN method is necessary to generate profiles that represent the news itself. The profiling is performed by exploiting the topic to which the news belongs and the semantic and syntactic characteristics.

In order to use the Latent Dirichlet Allocation, stop-words [23], which are common words in the language and do not allow to associate the news to a specific topic, are initially eliminated from the news. It should be noted that there are methods to avoid the elimination of stopwords [24]. At this point, it is possible to proceed with the method that will allow determining the belonging to the reference topics.

Syntactic and semantic features are added to the topic analysis. The syntactic features are the following:

- The number of characters in the text. This parameter allows the description of the length of the news;
- Flesch Index F, an index that allows measuring the readability of the news [25]. This index is derived through the following formula:

$$F = 206.834 - (84.6 \times S) - (1.015 \times P) \tag{2}$$

where S is the average number of syllables per word and P represents the average number of words per sentence. A low Flesch index indicates text that is difficult to read, while a high Flesch index indicates easy to read.

- Gunning Fog Index GI, an index used to determine the comprehensibility of text [25]. The following preliminary steps must be taken to calculate this index:

  - Fix a part of the text of about 100 words that does not truncate any sentence.
  - In the selected section of the text, count the words that have three or more syllables (complex words), eliminating from the count the common suffixes related to the plural of words or verb declensions. Compound words and proper nouns should not be counted.

The index is calculated using the following formula:

$$GI = \frac{4}{10}\left(m_s + 100\frac{w_c}{w_t}\right) \tag{3}$$

Where m_s indicates the average number of words per sentence, w_c indicates the number of complex words, and w_t represents the number of total words. A high Gunning Fog index indicates high news comprehensibility, a low Gunning Fog index indicates low news comprehensibility.

Through the Context Dimension Tree (CDT), the information associated with the features just determined is further filtered. In particular, the temporal context and the news sources are exploited and integrated with the syntactic features to make an appropriate selection of topics determined by the LDA method. This allows limiting the study to the most popular topics in the selected contextual area, bringing improvements to the performance of the K-NN method. The semantic analysis, in addition to being linked to topics, as mentioned above, is integrated through Sentiment Analysis [26, 27]. In particular, two different features are integrated:

- The probability of negative news sentiment;
- The probability of positive news sentiment.

Let the news be $n \in N$, where N is the set of all news, $f_1^{(n)}$ denotes the index related to the number of news characters, $F^{(n)}$ denotes the Flesch index of news n and $GI^{(n)}$ denotes the Gunning Fog index of n. Let the vector $t^{(n)}$ relative to news topic and let $f_2^{(n)}$ and $f_3^{(n)}$ the coefficients related to negative and positive sentiment of the news n respectively. Because of the coefficients f_1, F, GI are more significant than others, a normalization is needed through the maximum of respective features.

Data la news $n \in N$, con N insieme di tutte le news, denotiamo con $f_1^{(n)}$ l'indice legato al numero di caratteri della news n, con $F^{(n)}$ il flesch index della news e con $GI^{(n)}$ il Gunning Fog index di n. Sia $t^{(n)}$ il vettore legato ai topic della news e siano $f_2^{(n)}$ e $f_3^{(n)}$ le stime legate alla probabilità che il sentiment sia negativo e positivo rispettivamente. Si sottolinea che, poiché i f_1, F, GI sono solitamente molto più grandi rispetto agli dati, verranno normalizzati attraverso il massimo delle rispettive caratteristiche. Therefore, the following operations are done.

$$\overline{f_1^{(n)}} = \frac{f_1^{(n)}}{\max_{m \in N} f_1^{(m)}} \quad n \in N \tag{4}$$

$$\overline{F^{(n)}} = \frac{F^{(n)}}{\max_{m \in N} F^{(m)}} \quad n \in N \tag{5}$$

$$\overline{GI^{(n)}} = \frac{GI^{(n)}}{\max_{m \in N} GI^{(m)}} \quad n \in N \tag{6}$$

The profile of each news $n \in N$ is obtained as the vector $f^{(n)}$ described in (7).

$$f^{(n)} = \left(\overline{f_1^{(n)}}, \overline{F^{(n)}}, \overline{GI^{(n)}}, t^{(n)}, f_2^{(n)}, f_3^{(n)} \right) \tag{7}$$

At this point the data is ready to be analyzed through the K-NN method.

3.2 News Classification

The K-Nearest Neighbor method includes a training phase. In this phase, a dataset containing both real and false news are used to train the algorithm. This news has been profiled as just shown. After the training phase, the classifier is able to label the news introduced. Once the news profiling is given as input by LDA, syntactic analysis and Sentiment Analysis is done, it is labelled by calculating the closest K news. This calculation is performed based on the Euclidean distance described in (1). The classification will be based on how much news with a given label will be identified as close to the input news.

4 Experimental Results

This section will describe the experimental phase exploited to test the proposed approach. Three different datasets were exploited for this purpose:

- ISOT[1]: a dataset containing 21417 real news taken from Reuters.com and 23481 fake news taken through different sources labelled through polifacts.com. Each news item is described by title, text, category and date of publication.
- CREDBANK dataset [28] composed of more than 60 million tweets collected between October 2014 and December 2015;
- UNICO: dataset containing 14237 posts collected via Facebook and Twitter and related to the ongoing coronavirus and pandemic issue;

The evaluation metrics consist of identifying true positives (T_p), false positive (F_p), true negative (T_n) e false negative (F_n) to calculate the precision P, recall R and F1-score F_1 described below.

$$P = \frac{T_p}{T_p + F_p} \tag{8}$$

$$R = \frac{T_p}{T_p + F_n} \tag{9}$$

$$F_1 = 2\frac{PR}{P + R} \tag{10}$$

---

[1] https://www.uvic.ca/ecs/ece/isot/datasets/fake-news/index.php.

Table 1 shows the results obtained on the respective datasets.

Table 1. Numerical results of the proposed approach on the datasets ISOT, CREDBANK and UNICO.

Dataset	Precision	Recall	F_1 score
ISOT	0,81	0,84	0,82
CREDBANK	0,77	0,80	0,78
UNICO	0,82	0,86	0,84

Table 1 shows that the proposed system achieves acceptable results using the ISOT dataset. This result is still interesting even though the contextual parameters used were only temporal, and therefore the full potential of the proposed system was not exploited.

However, on the CREDBANK dataset, the proposed approach offers worse results. These results are probably due to the dataset's limitations concerning both syntactic and contextual feature extraction.

Finally, the proposed approach has been applied with satisfactory results using the UNICO dataset. This result is due to the ability of the system to exploit the dataset elements, which allow the extraction of all the syntactic, semantic and contextual features described in Sect. 3.

Although the proposed approach encountered difficulties in using conventional datasets found in the literature, the results obtained are satisfactory and can be improved by adding more data or features to the system.

5 Conclusions

In the present paper, after an overview of models for Fake News Detection, an approach that exploits the syntactic and semantic properties of news for the classification of real and fake news was presented. In particular, the Syntactic Analyzer, the Sentiment Analyzer and the Topic Analyzer allow news profiling. Moreover, through the contextual analysis that exploits the Topic Analyzer and the Syntactic Analyzer data, it is possible to adjust the data according to the specific context. The processed data are exploited by the K-NN classifier that has been previously trained through a dataset made available.

In Sect. 5, the experimental data allow us to state that the results obtained are encouraging. In particular, good results are obtained on the UNICO and ISOT datasets.

In the future, to improve the proposed approach, the news analysis from a social point of view will also be integrated.

References

1. di Renzo, L., et al.: Eating habits and lifestyle changes during COVID-19 lockdown: an Italian survey. J. Transl. Med. **18**(1) (2020). https://doi.org/10.1186/s12967-020-02399-5

2. Herrera-Peco, I., et al.: Antivaccine movement and COVID-19 negationism: a content analysis of Spanish-written messages on Twitter. Vaccines **9**(6) (2021). https://doi.org/10.3390/vaccin es9060656

3. York, C., Ponder, J.D., Humphries, Z., Goodall, C., Beam, M., Winters, C.: Effects of fact-checking political misinformation on perceptual accuracy and epistemic political efficacy. J. Mass Commun. Q. **97**(4) (2020). https://doi.org/10.1177/1077699019890119

4. Shu, K., Sliva, A., Wang, S., Tang, J., Liu, H.: Fake news detection on social media. ACM SIGKDD Explor. Newsl. **19**(1) (2017). https://doi.org/10.1145/3137597.3137600

5. Allcott, H., Gentzkow, M.: Social media and fake news in the 2016 election. J. Econ. Perspect. **31**(2) (2017). https://doi.org/10.1257/jep.31.2.211

6. Zhou, X., Zafarani, R.: A survey of fake news: fundamental theories, detection methods, and opportunities. ACM Comput. Surv. **53**(5) (2020). https://doi.org/10.1145/3395046

7. Sahoo, S.R., Gupta, B.B.: Real-time detection of fake account in Twitter using machine-learning approach. In: Gao, X.-Z., Tiwari, S., Trivedi, M.C., Mishra, K.K. (eds.) Advances in Computational Intelligence and Communication Technology. AISC, vol. 1086, pp. 149–159. Springer, Singapore (2021). https://doi.org/10.1007/978-981-15-1275-9_13

8. Przybyła, P.: Capturing the style of fake news (2020). https://doi.org/10.1609/aaai.v34i01. 5386

9. Nagaraja, A., Soumya, K.N., Naik, P., Sinha, A., Rajendrakumar, J.V.: Fake news detection using machine learning methods (2021). https://doi.org/10.1145/3460620.3460753

10. Ozbay, F.A., Alatas, B.: Fake news detection within online social media using supervised artificial intelligence algorithms. Physica A: Stat. Mech. Appl. **540** (2020). https://doi.org/ 10.1016/j.physa.2019.123174

11. Shu, K., Liu, H.: Detecting fake news on social media. Synthesis Lectures Data Min. Knowl. Discov. **11**(3) (2019). https://doi.org/10.2200/s00926ed1v01y201906dmk018

12. Blei, D.M., Ng, A.Y., Jordan, M.I.: Latent Dirichlet allocation. J. Mach. Learn. Res. **3**(4–5) (2003). https://doi.org/10.1016/b978-0-12-411519-4.00006-9

13. Clarizia, F., Colace, F., Lombardi, M., Pascale, F., Santaniello, D.: Sentiment analysis in social networks: a methodology based on the latent Dirichlet allocation approach, August 2019. https://doi.org/10.2991/eusflat-19.2019.36

14. Colace, F., Casaburi, L., de Santo, M., Greco, L.: Sentiment detection in social networks and in collaborative learning environments. Comput. Hum. Behav. **51** (2015). https://doi.org/10. 1016/j.chb.2014.11.090

15. Hu, Q., Yu, D., Xie, Z.: Neighborhood classifiers. Expert Syst. Appl. **34**(2) (2008). https:// doi.org/10.1016/j.eswa.2006.10.043

16. Dong, W., Charikar, M., Li, K.: Efficient K-nearest neighbor graph construction for generic similarity measures (2011). https://doi.org/10.1145/1963405.1963487

17. Jiang, L., Cai, Z., Wang, D., Jiang, S.: Survey of improving K-nearest-neighbor for classification. In: Proceedings - Fourth International Conference on Fuzzy Systems and Knowledge Discovery, FSKD 2007, vol. 1 (2007). https://doi.org/10.1109/FSKD.2007.552

18. Weinberger, K.Q., Saul, L.K.: Distance metric learning for large margin nearest neighbor classification. J. Mach. Learn. Res. **10** (2009). https://doi.org/10.1145/1577069.1577078

19. Kesarwani, A., Chauhan, S.S., Nair, A.R.: Fake news detection on social media using k-nearest neighbor classifier (2020). https://doi.org/10.1109/ICACCE49060.2020.9154997

20. Casillo, M., Conte, D., Lombardi, M., Santaniello, D., Valentino, C.: Recommender system for digital storytelling: a novel approach to enhance cultural heritage. In: Del Bimbo, A., et al. (eds.) ICPR 2021. LNCS, vol. 12667, pp. 304–317. Springer, Cham (2021). https://doi.org/ 10.1007/978-3-030-68787-8_22

21. Colace, F., Lombardi, M., Pascale, F., Santaniello, D.: A multi-level approach for forecasting critical events in smart cities (2018). https://doi.org/10.18293/DMSVIVA2018-002

22. Clarizia, F., Colace, F., de Santo, M., Lombardi, M., Pascale, F., Santaniello, D.: A context-aware chatbot for tourist destinations. In: 2019 15th International Conference on Signal-Image Technology & Internet-Based Systems (SITIS), pp. 348–354, November 2019. https://doi.org/10.1109/SITIS.2019.00063

23. Fariña, A., Brisaboa, N.R., Navarro, G., Claude, F., Places, Á.S., Rodríguez, E.: Word-based self-indexes for natural language text. ACM Trans. Inf. Syst. **30**(1) (2012). https://doi.org/10.1145/2094072.2094073

24. Wilson, A.T., Chew, P.A.: Term weighting schemes for latent Dirichlet allocation (2010)

25. Casillo, M., et al.: A multi-feature Bayesian approach for fake news detection. In: Chellappan, S., Choo, K.-K., Phan, NhatHai (eds.) CSoNet 2020. LNCS, vol. 12575, pp. 333–344. Springer, Cham (2020). https://doi.org/10.1007/978-3-030-66046-8_27

26. Kula, S., Choraś, M., Kozik, R., Ksieniewicz, P., Woźniak, M.: Sentiment analysis for fake news detection by means of neural networks. In: Krzhizhanovskaya, V.V., et al. (eds.) ICCS 2020. LNCS, vol. 12140, pp. 653–666. Springer, Cham (2020). https://doi.org/10.1007/978-3-030-50423-6_49

27. Bhutani, B., Rastogi, N., Sehgal, P., Purwar, A.: Fake news detection using sentiment analysis (2019). https://doi.org/10.1109/IC3.2019.8844880

28. Mitra, T., Gilbert, E.: CREDBANK: a large-scale social media corpus with associated credibility annotations (2015)

Machine Learning Technique for Fake News Detection Using Text-Based Word Vector Representation

Akshat Gaurav[1], B. B. Gupta[2(✉)], Ching-Hsien Hsu[3,4],
Arcangelo Castiglione[5], and Kwok Tai Chui[6(✉)]

[1] Ronin Institute, Montclair, NJ 07043, USA
akshat.gaurav@ronininstitute.org

[2] National Institute of Technology Kurukshetra, Kurukshetra 136119, Haryana, India

[3] Department of Computer Science and Information Engineering,
Asia University, Taichung, Taiwan

[4] Department of Computer Science and Information Engineering,
National Chung Cheng University, Minxiong, Taiwan

[5] University of Salerno, Fisciano, Salerno, Italy
arcastiglione@unisa.it

[6] Department of Technology, School of Science and Technology,
Hong Kong Metropolitan University, Hong Kong, China
jktchui@ouhk.edu.hk

Abstract. In the modern era, social media has taken off, and more individuals may now utilise it to communicate and learn about current events. Although people get much of their information online, some of the Internet news is questionable and even deceptively presented. It is harder to distinguish fake news from the real news as it is sent about in order to trick readers into believing fabricated information, making it increasingly difficult for detection algorithms to identify fake news based on the material that is shared. As a result, an urgent demand for machine learning (ML), deep learning, and artificial intelligence models that can recognize fake news arises. The linguistic characteristics of the news provide a simple method for detecting false news, which the reader does not need to have any additional knowledge to make use of. We discovered that NLP techniques and text-based word vector representation may successfully predict fabricated news using a machine learning approach. In this paper, on datasets containing false and genuine news, we assessed the performance of six machine learning models. We evaluated model performance using accuracy, precision, recall, and F1-score.

Keywords: Machine learning · NLP · LR · SVM · Linear regression · Fake news

1 Introduction

Due to the fast growth of the Internet, social networks have become a significant vehicle for the dissemination of false news, distorted information, fraudulent

© Springer Nature Switzerland AG 2021
D. Mohaisen and R. Jin (Eds.): CSoNet 2021, LNCS 13116, pp. 340–348, 2021.
https://doi.org/10.1007/978-3-030-91434-9_33

reviews, rumours, and satires [12, 19]. Many people believe that false news played a role in the United States' 2016 presidential election campaign; as a result of this election, the phrase has entered the popular lexicon [10, 24]. Therefore, academia and industry are collaborating to study and create methods for analysing and identifying false news. In addition, the battle against fake news is closely linked to social networks and data consumption issues. By distributing harmful information, a user wastes the network and processing resources, while also jeopardising the service's reputation. Deceptive news leads to an increase in distrust, which is reflected in the Quality of Trust metric [13, 20, 21, 27].

Fake news is disseminated on social media to trick readers or start rumours. The proliferation of social media platforms has accelerated the transmission of rumors and incorrect information, resulting in an increase in the distribution of fake news [9, 13]. Due to the widespread mistrust of conventional media, social network users often depend on false news, which is frequently shared by friends or confirms previous information. Additionally, when consumers are constantly bombarded with false information, it becomes difficult to tell the difference between real and fake news. In terms of 3 V [16], as shown in Fig. 1, it also posed a danger to many communities and had a profoundly detrimental effect on people through widespread advertising, online purchasing, and social messaging.

Several academics have been working on developing effective and automated frameworks for detecting online fake news in recent years in order to distinguish spurious news from legitimate news [1, 3, 10, 23]. Numerous researchers presented their models through the use of machine learning and deep learning methods [7, 25]. However, finding false news on social networks is difficult. To begin with, collecting statistics on false news is challenging. Additionally, manually identifying false news is a challenge. Due to their intent to mislead readers, they are difficult to identify just on the basis of the news substance [2, 18]. It is difficult to evaluate the validity of newly released and time-bound news, since they provide an insufficient training dataset for the application. Significant methods for recognizing trustworthy individuals, extracting valuable news characteristics, and developing an authentic information distribution system are only a few of the critical study areas that need further exploration [18]. However, the suggested techniques have significant limits in terms of accuracy. A new technique is required to address these problems and efficiently identify false news.

We used many classification algorithms, including the Multinomial NB Algorithm, Logical Regression, Gradient Boosting classifier, Random forest, and support vector machine to see if they might be used to detect fake news in this research. To increase the overall performance of each model, the stacking approach was used. Kaggle dataset was used in our study. To tokenize the text and title features of these two datasets, we employed counter-vectorization methods. Performance is frequently measured using the following four categories: accuracy, recall, f1 score, and precision. New experimental findings were compared with results from previous literature to assess the effectiveness of the suggested stacking approach. All experimental data have been gathered in separate tables

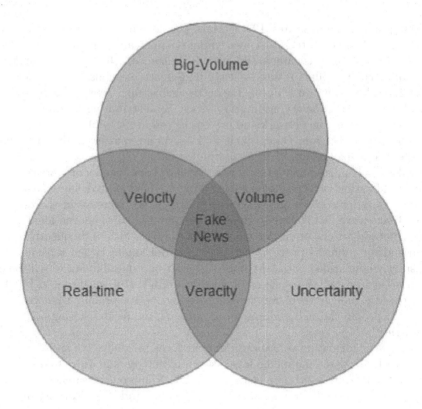

Fig. 1. 3 V's of fake news [16]

and are shown graphically in various figures, with the intention of facilitating comprehension.

This paper is split into the following sections: The second part covers similar research in the field of false news detection. Section 3 describes the approach, while Sect. 4 summarizes the findings. Finally, Sect. 5 brings the paper to a conclusion.

2   Related Work

Numerous researchers have proposed a variety of approaches for detecting different cyber attacks [6,28,29] and fake news [5,26]. We presented some of the most widely used false news detection algorithms in this section.

The authors in [10] evaluated the performance of ML models and DL models on two different-sized fake and real news datasets. In order to build ML and DL models for text representation, the authors employed term frequency, term frequency-inverse document frequency techniques. Similarly, the authors in [14] tested twenty-three supervised AI systems on three datasets to determine the most effective approach for detecting false news. According to their findings, the decision tree technique outperformed all other algorithms in all assessment metrics except recall.

The authors in [16] provide an automated technique for detecting false news on Facebook using the Chrome browser. The authors utilize deep learning to evaluate a Facebook account's activity based on a range of account characteristics and some components of news content. The experimental examination of real-world data demonstrates that the proposed work intent was to distribute false information. The authors in [22] developed a novel approach termed 'Traceminor' to identify fake news broadcasts across a different network using deep learning classifiers. Additionally, it investigates how the material reacts when certain bits of information are omitted. In [11] authors proposes a novel hybrid deep learning model for classifying fake news that combines CNN and RNN. The model was successfully validated on two fake news datasets, providing detection results that were significantly superior to non-hybrid baseline approaches. To determine the trustworthiness of sources and, by extension, the news they publish or disseminate, the hybrid model of the author in [8] combines graph embeddings from the Twitter user follower network with individual attributes. The authors in [15] proposed a method involves creating more characteristics for NLP by using data storage mining; as a result, the efficiency of identifying fake news increases. Additionally, the authors compare the proposed approach's outcomes with those of existing machine learning methods, such as LSTM.

The authors in [17] proposed a model that illustrates how diverse methods of debunking falsehoods influence the distribution of misinformation among populations. We have confidence in this method since it is capable of locating and removing false news from OSNs. This suggested method establishes a key parameter known as the basic reproduction number (Rn) in the investigation of message propagation in OSNs. When Rn is smaller than one, it is possible to limit the distribution of fraudulent messages inside the OSN. Otherwise, the rumour will be unstoppable.

3   Methodology

Six fundamental machine learning algorithms are used in the proposed technique to detect false news. The suggested strategy begins with the elimination of superfluous characters, tokenization, and stop wording. Each ML technique's performance is evaluated using accuracy, precision, recall, and F-Measure. Figure 2 represents the flow diagram of the proposed technique.

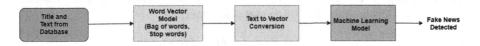

Fig. 2. Proposed approach

3.1 Dataset

Classifying a piece of news as "fake news" may be a time-consuming and complex task. As a consequence, a previously collected and recognized dataset of fake news was utilized. This project used data from the Kaggle dataset [4]. The dataset has a header and columns for title and content, as well as a flag indicating if the news item is fake or real.

3.2 Pre-processing of Dataset

Prior to incorporating text data into machine learning models, it must be pre-processed using techniques such as stop word removal, phrase segmentation, and punctuation removal. These approaches have the potential to significantly assist us in identifying the most pertinent keywords and optimizing model performance. Because our datasets are taken from real-world news stories, they contain a large amount of meaningless text and unusual characters. As a result, we eliminated duplication in our data collection by removing these unwanted characters. The next stage of preprocessing is to eliminate stop words. Stop words are frequently used in English sentences to complete the phrase structure, even if they serve no use in expressing specific concepts. As a result, we excluded them from all of our tests because of the possibility that they caused excessive noise.

3.3 Machine Learning Models

We preprocessed the data set using the machine learning models in the preceding part. However, because the utilized machine learning models can not accept text input, we transform the text information to machine-readable form using a counter vectorized. After converting the text to a vector format, we supplied the news headlines and body content to the machine learning model for training and testing purposes. Finally, statistical approaches are used to compare the performance of several machine learning models.

4 Result and Discussions

We began by removing stopwords and special characters from our dataset in this research. Then, we utilised tokenization and counter vectorization algorithms to extract and transform the undesired text from the dataset. Following that, we trained individual models using five different machine learning techniques, including LR, DT, KNN, RF, and SVM. The performance of these machine learning models is evaluated using statistical metrics.

$$Accuracy = \frac{T_P + T_N}{T_P + T_N + F_P + F_N} \tag{1}$$

$$Racall = \frac{T_P}{T_P + F_N} \tag{2}$$

$$Precision = \frac{T_P}{T_P + F_P} \tag{3}$$

$$F1\text{-}Score = 2 \times \frac{Precision \times Recall}{Precision + Recall} \tag{4}$$

Where T_P is 'true positive', T_N is 'true negative', F_P is 'false positive', and F_N is 'false negative'.

4.1   Confusion Matrix Calculation

When evaluating the performance of a classification model, a confusion matrix is used. Actual target values are compared with the predictions made by the machine learning model in this matrix. The information provided by this gives us a complete view of how well our classification model is doing, as well as the kinds of errors it is making. Using a confusion matrix, it is straightforward to calculate precision, accuracy, recall, and the f-1 score.

4.2   Performance Comparison

In this subsection, we assess the performance of six ML models based on their accuracy, precision, recall, and f-1 scores. All of these characteristics are derived from the confusion matrix and are shown graphically in Fig. 3 (Fig. 4).

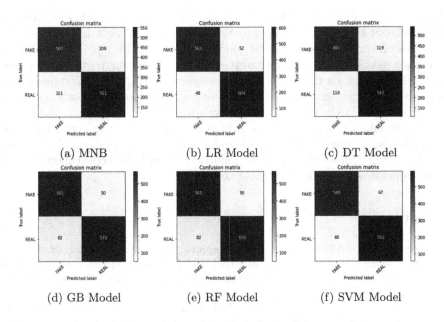

Fig. 3. Confusion matrix for different ML models

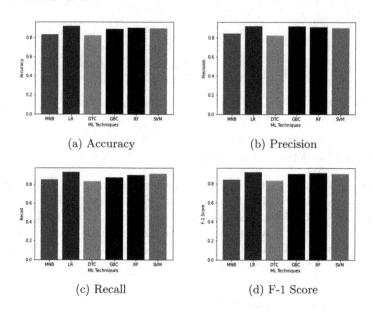

(a) Accuracy   (b) Precision

(c) Recall   (d) F-1 Score

Fig. 4. Statistical parameters calculation

5 Conclusion

Fake news is a serious issue that is spreading like wildfire as information becomes more accessible to the general population in a variety of ways. In every country, fake news has the ability to have a major impact on people's political and social life. As a consequence, we evaluated the accuracy, precision, recall, and F1-score of five ML models on the kaggle fake news dataset in this paper. Certain models, such as LR, performed substantially better on the datasets than others, such as DT, SVM, LR, MNB, RF, and GB. We will do more testing in the future using a variety of data sets and languages. Additionally, we will try to detect fake news using a broader variety of machine learning and deep learning models. To assist in identifying fake news in different countries, we will also collect more data on fake and real news in numerous languages.

References

1. Alharbi, J.R., Alhalabi, W.S.: Hybrid approach for sentiment analysis of twitter posts using a dictionary-based approach and fuzzy logic methods: study case on cloud service providers. Int. J. Semant. Web Inf. Syst. (IJSWIS) 16(1), 116–145 (2020)
2. Bahad, P., Saxena, P., Kamal, R.: Fake news detection using bi-directional LSTM-recurrent neural network. Procedia Comput. Sci. 165, 74–82 (2019). https://doi.org/10.1016/j.procs.2020.01.072. https://www.sciencedirect.com/science/article/pii/S1877050920300806

3. Bouarara, H.A.: Recurrent neural network (RNN) to analyse mental behaviour in social media. Int. J. Softw. Sci. Comput. Intell. (IJSSCI) **13**(3), 1–11 (2021)
4. Kaggle Dataset: Fake and real news dataset. https://kaggle.com/clmentbisaillon/fake-and-real-news-dataset
5. Gaurav, A., Gupta, B.B., Castiglione, A., Psannis, K., Choi, C.: A novel approach for fake news detection in vehicular ad-hoc network (VANET). In: Chellappan, S., Choo, K.-K.R., Phan, N.H. (eds.) CSoNet 2020. LNCS, vol. 12575, pp. 386–397. Springer, Cham (2020). https://doi.org/10.1007/978-3-030-66046-8_32
6. Gaurav, A., Singh, A.K.: Light weight approach for secure backbone construction for MANETs. J. King Saud Univ. Comput. Inf. Sci. **33**, 908–919 (2021)
7. Gupta, S., Gupta, B.B.: Robust injection point-based framework for modern applications against XSS vulnerabilities in online social networks. Int. J. Inf. Comput. Secur. **10**(2–3), 170–200 (2018)
8. Hamdi, T., Slimi, H., Bounhas, I., Slimani, Y.: A hybrid approach for fake news detection in twitter based on user features and graph embedding. In: Hung, D.V., D'Souza, M. (eds.) ICDCIT 2020. LNCS, vol. 11969, pp. 266–280. Springer, Cham (2020). https://doi.org/10.1007/978-3-030-36987-3_17
9. Hardalov, M., Koychev, I., Nakov, P.: In search of credible news. In: Dichev, C., Agre, G. (eds.) AIMSA 2016. LNCS (LNAI), vol. 9883, pp. 172–180. Springer, Cham (2016). https://doi.org/10.1007/978-3-319-44748-3_17
10. Jiang, T., Li, J.P., Haq, A.U., Saboor, A., Ali, A.: A novel stacking approach for accurate detection of fake news. IEEE Access **9**, 22626–22639 (2021). https://doi.org/10.1109/ACCESS.2021.3056079
11. Nasir, J.A., Khan, O.S., Varlamis, I.: Fake news detection: a hybrid CNN-RNN based deep learning approach. Int. J. Inf. Manag. Data Insights **1**(1), 100007 (2021). https://doi.org/10.1016/j.jjimei.2020.100007. https://www.sciencedirect.com/science/article/pii/S2667096820300070
12. Noor, S., Guo, Y., Shah, S.H.H., Nawaz, M.S., Butt, A.S.: Research synthesis and thematic analysis of twitter through bibliometric analysis. Int. J. Semant. Web Inf. Syst. (IJSWIS) **16**(3), 88–109 (2020)
13. de Oliveira, N.R., Medeiros, D.S.V., Mattos, D.M.F.: A sensitive stylistic approach to identify fake news on social networking. IEEE Signal Process. Lett. **27**, 1250–1254 (2020). https://doi.org/10.1109/LSP.2020.3008087
14. Ozbay, F.A., Alatas, B.: Fake news detection within online social media using supervised artificial intelligence algorithms. Phys. A Stat. Mech. Appl. **540**, 123174 (2020)
15. Deepak, S., Chitturi, B.: Deep neural approach to fake-news identification. Procedia Comput. Sci. **167**, 2236–2243 (2020). https://doi.org/10.1016/j.procs.2020.03.276. https://www.sciencedirect.com/science/article/pii/S1877050920307420
16. Sahoo, S.R., Gupta, B.B.: Multiple features based approach for automatic fake news detection on social networks using deep learning. Appl. Soft Comput. **100**, 106983 (2021). https://doi.org/10.1016/j.asoc.2020.106983. https://www.sciencedirect.com/science/article/pii/S1568494620309224
17. Shrivastava, G., Kumar, P., Ojha, R.P., Srivastava, P.K., Mohan, S., Srivastava, G.: Defensive modeling of fake news through online social networks. IEEE Trans. Comput. Soc. Syst. **7**(5), 1159–1167 (2020). https://doi.org/10.1109/TCSS.2020.3014135
18. Shu, K., Sliva, A., Wang, S., Tang, J., Liu, H.: Fake news detection on social media: a data mining perspective. ACM SIGKDD Explor. Newsl. **19**(1), 22–36 (2017)

19. Srinivasan, S., Dhinesh Babu, L.D.: A parallel neural network approach for faster rumor identification in online social networks. Int. J. Semant. Web Inf. Syst. (IJSWIS) **15**(4), 69–89 (2019)

20. Vosoughi, S., Roy, D., Aral, S.: The spread of true and false news online. Science **359**(6380), 1146–1151 (2018)

21. Wang, W.Y.: "Liar, liar pants on fire": a new benchmark dataset for fake news detection. arXiv preprint arXiv:1705.00648 (2017)

22. Wu, L., Liu, H.: Tracing fake-news footprints: characterizing social media messages by how they propagate. In: Proceedings of the Eleventh ACM International Conference on Web Search and Data Mining, pp. 637–645 (2018)

23. Zhang, L., Zhang, Z., Zhao, T.: A novel spatio-temporal access control model for online social networks and visual verification. Int. J. Cloud Appl. Comput. (IJCAC) **11**(2), 17–31 (2021)

24. Zhang, X., Ghorbani, A.A.: An overview of online fake news: characterization, detection, and discussion. Inf. Process. Manag. **57**(2), 102025 (2020)

25. Zhang, Z., Sun, R., Zhao, C., Wang, J., Chang, C.K., Gupta, B.B.: CyVOD: a novel trinity multimedia social network scheme. Multimedia Tools Appl. **76**(18), 18513–18529 (2017)

26. Zhao, J., Wang, H.: Detecting fake reviews via dynamic multimode network. Int. J. High Perform. Comput. Netw. **13**(4), 408–416 (2019)

27. Zhou, X., Zafarani, R.: A survey of fake news: fundamental theories, detection methods, and opportunities. ACM Comput. Surv. (CSUR) **53**(5), 1–40 (2020)

28. Zhou, Z., Gaurav, A., Gupta, B., Hamdi, H., Nedjah, N.: A statistical approach to secure health care services from DDoS attacks during COVID-19 pandemic. Neural Comput. Appl. 1–14 (2021). https://doi.org/10.1007/s00521-021-06389-6

29. Zhou, Z., Gaurav, A., Gupta, B.B., Lytras, M.D., Razzak, I.: A fine-grained access control and security approach for intelligent vehicular transport in 6g communication system. IEEE Trans. Intell. Transp. Syst. (2021)

Special Track: Information Spread in Social and Data Networks

Summarization Algorithms for News: A Study of the Coronavirus Theme and Its Impact on the News Extracting Algorithm

Lyudmila Gadasina[1](✉) ⓘ, Vladislav Veklenko[2], and Pasi Luukka[3] ⓘ

[1] St Petersburg University, 7/9 Universitetskaya nab., Saint Petersburg 199034, Russian Federation
l.gadasina@spbu.ru
[2] The Moscow Institute of Physics and Technology, 9 Institutskiy per., Moscow Region 141701 Dolgoprudny, Russian Federation
veklenko.vs@phystech.redu
[3] LUT University, Skinnarilankatu 34, 53850 Lappeenranta, Finland
pasi.luukka@lut.fi

Abstract. Extract summarization algorithms help identify significant information from the news by extracting meaningful sentences from the original text. The information background existing at the time of the news release often significantly affects its content. Such background can distort the text summarization algorithm working results. The study was conducted with the example of the theme "coronavirus" (COVID-19), which at the time of the study was one of the main topics in news feeds. Experiments were carried out on sports news articles, concerned football. This news area was selected because it is not related to medical topics. The TextRank algorithm for sport news extraction was applied in two ways. First, the key information from the source text of news was extracted. Then, a list of the COVID related words was created and the key information from news without considering words from this list was extracted. Our approach showed that mentioning a popular theme such as COVID that is not related to sports can have a negative impact on the text summarization algorithm. We suggest that to obtain accurate results of the algorithm operation, it is necessary to first compile a dictionary of terms related to the coronavirus theme and then exclude them when identifying the main content of news texts.

Keywords: Summarization algorithm · News · Coronavirus · Text · Extracting

1 Introduction

The problem of texts summarizing is important because of the ever-growing flow of information. Text summarization systems extract brief information from it. Using the resulting summary, the users can select the news related to their needs. Note that, influencing factors for content of news feeds are e.g. specifics of the area and information background relevant at the time of news are published. In usage of automatic text summarizing one needs to realize that temporary information background is influencing the

© Springer Nature Switzerland AG 2021
D. Mohaisen and R. Jin (Eds.): CSoNet 2021, LNCS 13116, pp. 351–360, 2021.
https://doi.org/10.1007/978-3-030-91434-9_30

results and also that results will be affected if it is applied to texts that do not take into account words related to such background.

At the present time coronavirus (COVID-19) is a major theme in news. Effect of coronavirus to media space has been interest of several researchers (see e.g. [1–4]. This has influenced e.g. the area of fakes news [1] and it has had an impact on financial markets [2, 3].

In this paper we are interested in examining: 1) Has coronavirus theme a significant semantic impact on the content of news; 2) Does coronavirus theme affect the results of summarizing news text algorithm?

2   Research Method

Research on Automatic Text Classification (ATC) can be traced back to seminal work of Luhn [5]. First studies were mainly centered on text categorization [6, 7]. Theoretical approaches for automatic indexing were developed in the 60's by several authors. For example in [6] proposed probabilistic approach using clue-words derived using frequencies with subject categories. Factor analysis is used in [7] and discriminant analysis for automatic indexing is used in [8]. Due to increased computational power in the late 90's research in the area started to shift from rules based on expert knowledge to learning based methods and machine learning based approaches started to emerge for solving text classification tasks. These include e.g. k-nearest neighbors [9, 10], decision trees [11], Bayesian classifiers [9, 11] and boosting methods such as AdaBoost [12].

There are two major categories of text summarization – Extractive and Abstractive Summarization [13]. Extractive summarization methods are based on extracting several parts, such as phrases and sentences from a text and stack them together to create a summary. These methods select the most significant words or sentences from the text. The other type, Abstract summation, uses advanced NLP techniques based on recurrent neural networks to create a new resume.

The aim of this article is to identify the influence of the current coronavirus news topic on the text summarization algorithm. Therefore, abstract summarization algorithms are irrelevant in this case, since they are based on memorizing certain sequences of words. This creates a serious obstacle for creating a model that does not take into account coronavirus theme. In our study we use the extractive summarization algorithm based on the graph approach. In [14, 15], extractive summarization is used to extract keywords. In our research, we need to extract the entire sentences to reveal the key idea embedded in the text. In [16], a comparison of PageRank algorithms and Shortest-path algorithms is made. Authors note that the PageRank algorithms do not depend on the language specifics, and the Shortest-path Algorithm generate summaries, which is not so "smooth" to read as a manually written summary.

In this study, a TextRank algorithm based on an extractive approach was constructed. This algorithm was originally proposed by Balcerzak et al. [17, 18]. It is based on the principles of the more famous PageRank algorithm [19], which is mainly used for ranking web pages in online search results. PageRank is a link analysis algorithm that assigns a numerical weighting to each element of a hyperlink document set, to measure its relative importance within the set. The algorithm can be applied to any collection of objects with mutual quotes and links. In [20], experiments were carried out on extraction to determine the reliability of web content (for identifying web content credibility). Although the method did not show top quality, the authors note that it can help with manual evaluation.

TextRank algorithm works with the same principle, but uses sentences instead of web pages. The similarity level between any two sentences is equivalent to the probability of clicking on a web page. In our study cosine distance was used for similarity calculations. Each word was matched with a numeric vector of a fixed dimension so that the words close in meaning correspond to the vectors close in meaning. Then to get a general sentence vector, the vectors of all the words in the sentence were summed up and divided by the number of words. The coefficients of similarity for each pair of sentences is stored in a square matrix, which is used to build the graph for analyzed text.

Thus, the stages of the TaxtRank algorithm are the following:

1. Split the text into separate sentences.
2. Find a vector representation for each sentence.
3. Calculate the similarities between sentence vectors and put them to the matrix.
4. Transform the similarity matrix into a graph with sentences as vertices and similarity estimates as edges.
5. Calculate the rank of sentences using a graph.
6. Form the final summary with a certain number of top-level ranking sentences.

3   Data Sources and Modelling

In our study, we experimented with a collection of texts from sports news articles, concerned football. We collected the data from a Russian-language sport information source [21]. We wanted to explore how the coronavirus theme affects the news feed in a particular area that is least related to medicine sphere. The corpus of 30,000 news published in the period from 07.09.2019 to 25.04.2020 was collected to create a dictionary containing vector representations of words to form a matrix of similarity of sentences. For this we used the standard libraries of Python programming language (such as nltk, nltk.tokenize, genism.models, rusenttokenize and so on) and developed our own functions for data collecting, processing, and algorithm constructing.

The next important step before building the model was to clean the texts. We created a function for clearing texts, which included the use of both the built-in stop-word library and regular expressions written for processing Russian-language texts, comprises:

- lowercase translation of texts,
- deleting all characters except Russian words,
- deleting stop words,
- deleting short words (less than 3 letters in length).

As a result, we compiled the dictionary with 91825 words mapped each word to a vector of dimension 100, using the Word2Vec library. In particular, the word coronavirus corresponds to the following top 10 words in the collected corpus, the degree of compliance (probability of being in one sentence) is indicated in brackets[1]: test (0.890), positive (0.858), passed (0.848), suspicion (0.809), passed (plural, 0.797), discovered (0.784), condition (0.781), relative (0.777), informal (0.774), confirmed (0.771).

Figure 1 created using the T-SNE dimension reduction method [22], shows groups of tokens (words) similar in meaning.

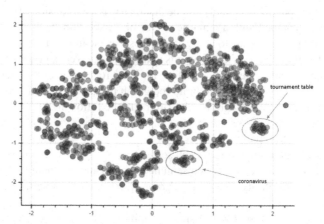

Fig. 1. Displaying word vectors with T-SNE algorithm. Examples of two groups of words are shown: words related to the "coronavirus" theme and words related to the "tournament table" theme.

Described procedure allows us to create a common vector for each sentence from analyzed texts. For these texts we built the graphs, using the TextRank algorithm discussed above to get a ranking of sentences.

Thus, our algorithm extracts the key information from the text where sentence vectors are based on all words from the created dictionary. To more thoroughly test the hypothesis about the impact of the coronavirus theme on the algorithm, we compiled a list of words that contained the major words associated with this virus: "coronavirus", "virus", "pandemic", "infection", "lethal", "disease", "patient", "death", "died", "discovered", "symptom", "suspicion", "quarantine"[2]. Then we implemented the same algorithm, but do not consider words from the created list in sentence vectors. We did not remove all sentences related to the virus theme from the news, but rank sentences according to the algorithm without taking into account the most popular words associated with coronavirus theme in them.

4   Results and Discussion

We tested each algorithm on original sports articles included a coronavirus theme. We considered the two most popular sentences of texts that should determine its content. To

---

[1] In the original corpus, these are Russian words.

[2] In the original corpus, these are Russian words in different forms (noun cases, verb conjugations etc.).

identify the most significant sentences from the text, we developed a web service that gives as an output results of both algorithms.

First, we illustrate our approach with the four following texts (analyzed texts in original language are presented in Appendix):

Text 1: {1} To complete the English championship, teams will have to play on neutral fields, the BBC reports. {2} The Season was interrupted because of the coronavirus pandemic. {3} It is reported that up to 10 stadiums will be needed to continue the tournament. {4} The Premier League Will also need up to 40,000 coronavirus tests for players and club employees. {5} Earlier, the clubs expressed their readiness to continue the championship when it is possible. {6} According to the source, the resuming football process tournaments in England will be lengthy. {7} Its timing has not yet been determined. {8} It is Planned that the matches will be held without spectators. {9} According to AR, the tournament can start on June 8. {10} According to Worldometer, the total number of people infected with coronavirus in the world has reached 3402886. {11} 239653 deaths were Recorded, and 1084606 people recovered.

Text 2: {1} Roma will suffer serious financial losses at the end of the 2019/20 season, Calciomercato reports. {2} According to the source, the Rome club will suffer losses of $110 million. {3} It is reported that this is due to the financial crisis caused by the coronavirus pandemic. {4} "Roma did not manage to reduce losses even though the salaries of players and staff were reduced. {5} According to the latest Worldometers data, there are 203,591 cases of coronavirus infection have registered in Italy. {6} 27682 deaths were recorded, and 71252 people recovered.

Text 3: {1} Spain's health Minister, Salvador Ilya, believes that football matches in the country will not be resumed until the summer. {2} Competitions in Spain were suspended due to the coronavirus pandemic on March 12. {3} – It is reckless to say that football will return before the summer – quotes AP El Salvador Ilya. {4} – We continue to monitor the evolution of the virus. {5} Recommendations will show how life can return in original areas of activity. {6} According to Worldometers, the world revealed 2977188 cases of infection with coronavirus. {7} 206 139 people died, 874587 recovered. {8} Spain is one of the most affected countries. {9} It recorded 226629 cases of infection, 23190 cases were fatal.

Text 4: {1} Borussia Monchengladbach released a special version of the club's t-shirt. {2} All the money will be used to support medical personnel fighting the coronavirus. {3} Purchase is only possible through an online store. {4} Almost all sporting events were postponed or canceled because of the coronavirus epidemic. {5} According to Worldometer on the night of May 1 to 2, the coronavirus was confirmed in 3,389,933 people worldwide. {6} Dead – 239029, recovered – 1076487. {7} According to Rospotrebnadzor, in Russia on the day of May 1, a total 114431 cases of coronavirus were registered in 85 regions. {8} Over the entire period, 1169 deaths were recorded, and 13220 people recovered.

Results from the created algorithm to four texts above are presented in Table 1. In the Table 1 two most important sentences are reported from both cases, news items for text including and excluding the coronavirus theme.

As we can see applying algorithm to news content without words related to coronavirus allows indicate in texts 1–3 a significant information. For example, the sentence

Table 1. Results of algorithm application to four texts.

News	Content with all words		Content without words related to coronavirus	
	The most important sentence in the text	The second in importance	The most important sentence in the text	The second in importance
Text 1	{2}	{6}	{9}	{6}
Text 2	{3}	{5}	{4}	{3}
Text 3	{4}	{8}	{8}	{1}
Text 4	{5}	{7}	{5}	{3}

{9} in news 1 gives information which can be used for decision making, but sentence {2} gives only part of important information. Applying algorithm to original texts in gives noisy information: sentences {5} and {4} respectively. We can note that in news 4 (see Table 1), the sentence related to the coronavirus theme that is not important within the meaning of the news remained in the top. In many cases sentences related to coronavirus remain after deleting the created word list. Sometimes the algorithm pays attention to other words in such sentences that may also be significant for it and it focuses on words that go in combination with words associated with the virus. Examples of such words: "world", "confirmed", "crisis" and others. These words can't be discarded from texts, because they can also appear in another context.

Result for Full Dataset. We analyzed all news for the period from 25.04.2020 to 02.05.2020: the total number of news is 921. Of these, 116 news contain the coronavirus theme. In 81 of the original news, the dominant topic is related to coronavirus theme. Of these, in 17 news are no longer the dominant topic is related to corona-virus theme after removing the most popular words associated with this theme.

Thus, 12.6% of news published during the analyzed period contains the virus theme. In almost 70% of cases, this theme is the key topic. The algorithm modification described above allows us to correct this result – the coronavirus theme is the key one in 55.1% of news containing it. Our result shows that in almost half of the cases, the virus theme is only the background context of current events feed of the topic football in Russia. This theme clutter up the news and does not add information.

To test coronavirus theme impact on news we used the following approach. We considered the most popular sentence in the text that can be used as a news headline or can help with creating a headline. We compared the actual news headline that was written by experts and the headline predicted by algorithms using the BLEU (Bilingual Evaluation Understudy) [23] and ROUGE (Recall-Oriented Understudy for Gisting Evaluation) [24] metrics. Table 2 shows the metrics values for original news and news without the most popular words associated with coronavirus theme. ROUGE metrics refers to the overlap of unigram (ROUGE-1), Bigrams (ROUGE-2) and longest common subsequence of texts were applied.

Table 2. Comparing metrics values for original news and news without words related to coronavirus.

Used metric	Metric value for content with all words	Metric value for content without words related to coronavirus
BLEU	0.216	0.220
ROUGE-1	0.128	0.132
ROUGE-2	0.047	0.049
ROUGE-L	0.120	0.124

Thus, we note an improvement in the quality of the algorithm after deleting words associated with the coronavirus. Our results show that the appearance of a new topic in the media that is not related to sports can negatively affect the quality of the extraction algorithm. In some cases, the words associated with the virus are information noise and do not define the main idea of the article.

The topic of coronavirus affects the operation of the TextRank algorithm. However, we note that the characteristics of the data used in the study affected the result. In particular, coronavirus theme distorted the corpus on which we trained the algorithms. Words related to this theme are present in the texts for training many times. For example, the word "коронавируса" (one of the forms of the word "coronavirus") was found in the texts 3540 times and ranked 51th in frequency of use (see Fig. 2). The word "коронавирусом" (another form of the word "coronavirus") was found 1703 times and was placed in 145 places. Our total dictionary comprise 91 825 words. It means that a lot of attention is paid to the coronavirus pandemic despite the sports theme of the news.

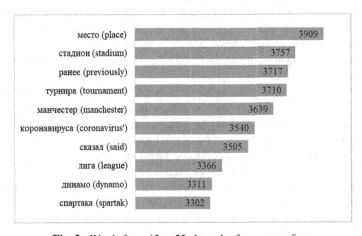

Fig. 2. Words from 45 to 55 places by frequency of use.

5 Conclusion

In this paper, we explored whether the coronavirus theme is noisy information or has a significant semantic impact on the content of news. We applied a graph-based ranking algorithm TextRank for sport news processing. We identified the key information from news containing the coronavirus theme. Our results show a negative impact on news headline extraction algorithms when there also exists a popular unrelated general topic. To identify significant information during periods when there is a dominant theme more properly, it is necessary to cleanse the source texts from the most popular words associated with this theme. Otherwise, the summarization algorithms may produce distorted results that are not relevant to users' needs.

Acknowledgments. The authors are grateful to participants at the Centre for Econometrics and Business Analysis (CEBA, St Petersburg University) seminar series for helpful comments and suggestions.

Appendix

Analyzed news examples in original (Russian) Language

Text 1: Для завершения чемпионата Англии командам придется играть на нейтральных полях, сообщает BBC. Сезон был прерван из-за пандемии коронавируса. Сообщается, что для продолжения турнира понадобится до 10 стадионов. Также премьер-лиге понадобится до 40 тысяч тестов на коронавирус для игроков и работников клубов. Ранее клубы выразили готовность продолжить чемпионат, когда это будет возможно. По данным источника, процесс возобновления футбольных турниров в Англии будет длительным. Его сроки пока не определены. Планируется, что матчи будут проходить без зрителей. По данным AP, турнир может начаться 8 июня. По данным Worldometer, общее число зараженных коронавирусом в мире достигло 3402 886 человек. Зафиксировано 239653 летальных исхода, 1084606 человек выздоровели.

Text 2: «Барселона» объявила, что клуб передает права на название стадиона «Камп Ноу» собственному фонду Barca Foundation. Фонд займется поиском спонсора, который получит право на имя арены на один сезон – 2020/21. Вырученные средства пойдут на борьбу с коронавирусом. Таким образом, стадион получит спонсорское название впервые в своей истории. Деньги от контракта пойдут на исследовательские проекты, связанные с борьбой с коронавирусом в Испании и во всем мире. Стадион «Камп Ноу» был открыт в 1957 году. Его вместимость составляет 99 354 зрителя. Он принимал матчи чемпионата Европы и мира, финал Лиги чемпионов, футбольный турнир Олимпиады-1992, а также концерты звезд мировой музыки.

Text 3: «Рома» понесет серьезные финансовые потери по итогам сезона-2019/20, сообщает Calciomercato. По информации источника, римский клуб понесет убытки в размере 110 миллионов долларов. Сообщается, что это связано с финансовым кризисом из-за пандемии коронавируса. «Роме» не удалось сократить потери даже несмотря на сокращение зарплаты игроками и персоналу. По последним данным Worldometers, в Италии зарегистрирован 203 591 случай заражения коронавирусом. Зафиксировано 27 682 летальных исхода, выздоровели 71 252 человека.

Text 4: Министр здравоохранения Испании Сальвадор Илья считает, что футбольные матчи в стране не будут возобновлены до лета. Соревнования в Испании были приостановлены из-за пандемии коронавируса 12 марта.

— Безрассудно говорить, что футбол вернется до лета, — цитирует AP Сальвадора Илью.

— Мы продолжаем следить за эволюцией вируса. Рекомендации покажут, как жизнь сможет вернуться в разных сферах деятельности. По данным Worldometers, в мире выявлено 2 977 188 случаев заражения коронавирусом. 206 139 человек умерли, 874 587 выздоровели. Испания — одна из наиболее пострадавших стран. В ней зафиксировано 226 629 случаев заражения, 23 190 случаев стали летальными.

References

1. Groza, A.: Detecting fake news for the new coronavirus by reasoning on the COVID-19 on-tology. arXiv preprint arXiv:2004.12330 (2020)
2. Lopatta, K., Alexander, E.-K., Gastone, L., Tammen, T.: To Report or not to report about coronavirus? The Role of Periodic Reporting in Explaining Capital Market Reactions dur-ing the Global COVID-19 Pandemic (2020). https://ssrn.com/abstract=3567778
3. Mamaysky, H.: Financial markets and news about the coronavirus 27 March 2020. https://ssrn.com/abstract=3565597. https://doi.org/10.2139/ssrn.3565597
4. Mejova, Y., Kalimeri, K.: Advertisers jump on coronavirus bandwagon: politics, news, and business. arXiv preprint arXiv:2003.00923 (2020)
5. Luhn, H.P.: The automatic creation of literature abstracts. IBM J. 159–165 (1958). https://doi.org/10.1147/rd.22.0159
6. Maron, M.E.: Automatic indexing: an experimental inquiry. J. ACM **8**(3), 404–417 (1961)
7. Borko, H., Bernick, M.: Automatic document classification. J. ACM **10**(2), 151–162 (1963)
8. Williams, J.: Discriminant analysis for content classification, IBM, Technical report no. RADC-TR-66-6 (1966)
9. Larkey, L.: Automatic essay grading using text categorization techniques. In: Proceedings of the 21st Annual International ACM SIGIR Conference on Research and Development in Information Retrieval, pp. 90–95 (1998)
10. Larkey, L.: A patent search and classification system, DL 99 (1999)
11. Lewis, D.D., Ringuette, M.: A comparison of two learning algorithms for text categorization. In: Third Annual Symposium on Document Analysis and Information Retrieval, vol. 33, pp. 81–93 (1994)
12. Wilbur, W.J., Kim, W.: The dimensions of indexing. In: AMIA Annual Symposium Proceedings, pp. 714–718 (2003)
13. Mani, I., Maybury, M.T.: Advances in Automatic Text Summarization, vol. 293. MIT Press (1999)
14. Litvak, M., Last, M.: Graph-based keyword extraction for single-document summarization. In: Proceedings of the workshop on Multi-source Multilingual Information Extraction and Summarization, pp. 17–24. Association for Computational Linguistics (2008)

15. Naidu, R., Bharti, S.K., Babu, K.S., Mohapatra, R.K.: Text summarization with automatic keyword extraction in Telugu e-Newspapers. In: Satapathy, S.C., Bhateja, V., Das, S. (eds.) Smart Computing and Informatics. SIST, vol. 77, pp. 555–564. Springer, Singapore (2018). https://doi.org/10.1007/978-981-10-5544-7_54

16. Thakkar, K.S., Dharaskar, R.V., Chandak, M.B.: Graph-based algorithms for text summarization. In: 2010 3rd International Conference on Emerging Trends in Engineering and Technology, pp. 516–519. IEEE (2010). https://doi.org/10.1109/ICETET.2010.104

17. Balcerzak, B., Jaworski, W., Wierzbicki, A.: Application of TextRank algorithm for credibility assessment. In: 2014 IEEE/WIC/ACM International Joint Conferences on Web Intelligence (WI) and Intelligent Agent Technologies (IAT), Warsaw, pp. 451–454 (2014). https://doi.org/10.1109/WI-IAT.2014.70

18. Mihalcea, R., Tarau, P.: Textrank: bringing order into text. In: Proceedings of the 2004 Conference on Empirical Methods in Natural Language Processing, Geneva, Switzerland, pp. 404–411 (2004)

19. Brin, S., Page, L.: The anatomy of a large-scale hypertextual Web search engine. Comput. Netw. ISDN Syst. **30**, 1–7 (1998). https://doi.org/10.1016/S0169-7552(98)00110-X

20. Li, W., Zhao, J.: TextRank algorithm by exploiting Wikipedia for short text keywords extraction. In: 2016 3rd International Conference on Information Science and Control Engineering (ICISCE), pp. 683–686. IEEE (2016). https://doi.org/10.1109/ICISCE.2016.151

21. "Sport-ekspress" – sportivnyj portal. ("Sport-Express" — sports portal.). https://www.sport-express.ru/. Accessed 25 Apr 2020

22. Maaten, L.V.D., Hinton, G.: Visualizing data using t-SNE. J. Mach. Learn. Res. **9**(Nov), 2579–2605 (2008)

23. Papineni, K., Roukos, S., Ward, T., Zhu, W.J.: Bleu: a method for automatic evaluation of machine translation. In: Proceedings of the 40th Annual Meeting of the Association for Computational Linguistics, pp. 311–318 (2002)

24. Lin, C. Y.: Rouge: a package for automatic evaluation of summaries. In: Text Summarization Branches Out, pp. 74–81 (2004)

Social Cohesion During the Stay-at-Home Phase of the First Wave of the COVID-19 Pandemic on Polish-Speaking Twitter

Andrzej Jarynowski[1,2(✉)], Alexander Semenov[3,4], Monika Wójta-Kempa[5], and Vitaly Belik[2]

[1] Interdisciplinary Research Institute, Wroclaw, Poland
ajarynowski@interdisciplinary-research.eu
[2] System Modeling Group, Institute for Veterinary Epidemiology and Biostatistics, Freie Universität Berlin, Berlin, Germany
{a.jarynowski,vitaly.belik}@fu-berlin.de
[3] Herbert Wertheim College of Engineering, University of Florida, Gainesville, FL, USA
asemenov@ufl.edu
[4] Center of Econometrics and Business Analytics, Saint Petersburg State University, Saint Petersburg, Russia
[5] Department of Health Humanities and Social Science, Wroclaw Medical University, Wrocław, Poland
monika.wojta-kempa@umw.edu.pl

Abstract. Catastrophic and urgent events, such as the COVID-19 pandemic, are known not only to polarize societies and induce selfish, individualistic behavior, but might also motivate altruistic behavior. We have analyzed COVID-19 perception using data collected from the Polish-language Internet from 15.01-30.06.2020, equaling 930,319 tweets. Deploying methods of computational social science and digital epidemiology, we aim to understand mechanisms of social consolidation and depolarization (measured by network modularity and sentiment) during the so-called "stay-at-home phase" of the COVID-19 pandemic. Mauss' theory of interaction or exchange of gifts, the theory of social capital, as well as Kaniasty's theory of mobilization and deterioration serve as a background for reflection on the Polish example during the first epidemic wave. Our study highlights the potential of social support and caretaking to reduce affective and behavioral polarization in social media

Keywords: COVID-19 · Risk perception · Polarization · Twitter · Social media · Social network analysis · Sentiment · Internet research

1 Introduction

Theoretical Background. Social interactions have a fundamental impact on the dynamics of infectious diseases (such as COVID-19), challenging public health

© Springer Nature Switzerland AG 2021
D. Mohaisen and R. Jin (Eds.): CSoNet 2021, LNCS 13116, pp. 361–370, 2021.
https://doi.org/10.1007/978-3-030-91434-9_31

mitigation strategies and possibly the political consensus. COVID-19 posed a paradox of social interaction. On the one hand, it polarized and disrupted social networks [18,34], which resulted in the formation of filter bubbles. On the other hand, it provides a stage for mutual help and support [25,26].

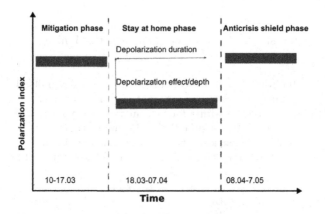

Fig. 1. Theoretical model of mobilization and deterioration during the first wave of COVID-19 in 2020.

Social media gives a voice to user concerns regarding COVID-19, society, and political systems in an interactive way. Perceived social support and a sense of society (Gemeinschaft vs. Gesellschaft [41]) are known to be mediated by social media [31]. The COVID-19 pandemic found Poland and other post-communistic countries totally unprepared in March 2020, while richer Western European countries limited the sharing of resources they possessed and controlled supply chains, disconnecting weaker EU members [10]. The new reality of the crisis has not been seen in Poland since 1981's martial state [23]. The COVID-19-related state of emergency was declared due to the necessity of supporting hospitals with food delivery, homemade personal protective equipment, etc. due to a shortage of resources and limitations in dealing with patients and the elderly. Mauss's theory of help-giving defines the concept of gifts to benefit other people. Pro-social interactions depend on "social embedding" [28], which has its own projection in virtual reality (e.g., social media). Interactions between goal-oriented actors could create a focus on the social environment [37] and build positive and negative attitudes towards help-giving.

Most studies on self-organized and collective movements on social media during the pandemic revealed its negative impact on the society [34], but this need not always be the case. These mechanisms provide emotional or material support, which could not appear in non-crisis situations due to the social distance/anti-homophily rule. In non-crisis times, people are unlikely to befriend people who are too different - above some similarity threshold-from each other, because building and maintaining social interactions is costly [30]. Social capital from these interactions [8] can lead to action (such as social support and cohesion [33]). Thus, help-giving principles and social consolidation [6] could be independent from political orientation and be projected, even induced, by social media, thereby moving society to a new equilibrium. We investigate the emergence of altruistic behavior characterized by higher-than-usual levels of social interactions (e.g., in social media) between conflicting (polarized) communities. Thus, in such a scenario, we should expect more inter-community links, which could lead to a suboptimal state from an evolutionary perspective [30]. Social embeddedness at the country level was significantly negatively associated with controlled COVID-19 R_t [22] during the first wave of infections. This implies that better-connected/-networked societies were more efficient in coping with the spread of the disease.

The aim of this study was to use social media (Twitter) data to explore polarization dynamics revealed by intercommunity links and sentiment analysis. Thus, using network modularity and sentiment to assess polarization dynamics could provide an overview of this ongoing phenomenon in society.

2 Data, Methods and Research Questions

Twitter has relatively low popularity in Eastern Europe (∼1.5 million active users out of ∼6 million registered accounts or less than 5% of the literate population in Poland [14]) and is mainly used by expats, journalists, and politicians. Consequently, it reflects the opinion of key influencers of national politics. During the lockdown, as in many other social media, the number of users significantly increased [40]. Moreover, professional news (traditional media) was ahead of the commentating reaction on Twitter in Poland [21]. This implies that Twitter is not a primary knowledge hub about COVID-19. For example, only 8% of respondents mentioned Twitter as a source of information, many times less than TV or nationwide Internet portals [49]. Despite many initial observations, mainly in the USA [38] the pandemic did not drive the polarization significantly deeper or weaker in the behavioral dimension [3,9]. Other studies on Twitter did not confirm that polarization decreased over time during the first waves of the pandemic [35,42]. In our study, based on the analysis of Polish Twitter data, we claim that we could observe and quantify the polarization. Our main hypothesis is that communities divided according to political beliefs (partisan), became less segregated during the mobilization phase, and returned to the previous level of polarization during the deterioration phase (Fig. 1). Thus, the main goal of our study is to indicate the depth and duration of depolarization on Twitter using the affective (emotional load) and behavioral (structure of the social connections) metrics.

The social network of Twitter users may be represented as a graph $G = (N, E)$ with a set of n nodes $N = \{1, ..., n\}$ and a set of m links (edges) $E \subset N \times N$, $|N| = n$ and $|E| = m$ with $|\cdot|$ denoting cardinality. We represent a Twitter user as a node, and we connect the nodes with a link if there is an interaction between the nodes. We considered three interactions: retweeting, replying, and quoting. Each interaction results in a different network; thus, we consider three networks corresponding to specific interactions.

We used network modularity [29] as an index of polarization, which has been widely used in measuring polarization [11,13]. Modularity is computed for the network and partitions of its nodes into communities. Modularity has a value in the range $[-1/2; 1]$ characterizing the partition. The maximization of modularity leads to the best separation of the network into communities; there are multiple algorithms for modularity maximization. In the current paper, we used the unsupervised weighted Louvain algorithm [7] and considered the networks as undirected.

The data collection includes tweets with the #Koronawirus hashtag posted from 15.01.2020 to 30.06.2020. We constructed networks based on the interactions that occurred within each week. We used word count-based sentiment analysis, with the Nencki affective word list [36,45] for the Polish language. We computed the proportion of the words (content of Tweets for weekly aggregated data) having connotation for selected emotions (with focus on optimism/happiness in our study). The baseline to understand the process is the pre-pandemic polarization and partisan level in Poland, which is reflected in social media [4,20,27].

3   Results

A starting point of our analysis was to reveal the partisan structure of our pandemic-related data sample. Previous studies on the Polish-language Twitter showed similar clusters emerging among partisans with sharp community divisions in general [27] almost the same as in the COVID-19 discourse (Fig. 2). After

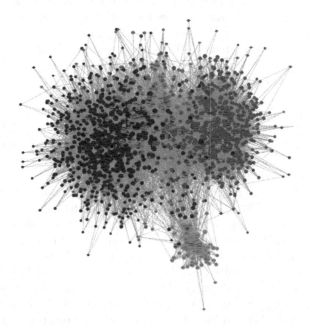

Fig. 2. A network of retweets with the #Koronawirus hashtag in the Polish language during 15.01-30.06.2020 (1,500 most central nodes by weighted degree centrality, color codes correspond to different communities). Blue—the ruling coalition, Orange—the mainstream opposition, Light blue—far right, Yellow—a Protestant minority. (Color figure online)

community detection, the vertices were colored the same color if they belonged to the same community. Out of 76,822 Twitter users engaged in coronavirus discourse, there is a clear polarization between major communities: ruling party, mainstream opposition, and minor but clear communities of the far-right party and Protestant movement. See Fig. 2. Notably, before SARS-CoV-2's introduction to Poland, there was no far-right community [20] that came onto the scene when the topic became popular and gained political power. The role of central nodes, such as celebrities, politicians, and scientists [24] could be important in building the discourse on Twitter. In this study, we focus on the "stay-at-home" phase; however, it is important to draw an epidemiological and infodemiological background before and after the investigated period. Based on the qualitative and quantitative analysis of multiple media sources in Poland from 10.03-07.05.2020 (which we define as the first wave of the epidemic), we can distinguish the following phases [15,21].

1. Mitigation phase (10-17.03.2020). Declaration of the main restrictions.
2. "Stay at home" phase (18.03-07.04.2020). Social consolidation.
3. Anti-crisis shield phase (08.04-7.05.2020). Anti-crisis shield declaration.

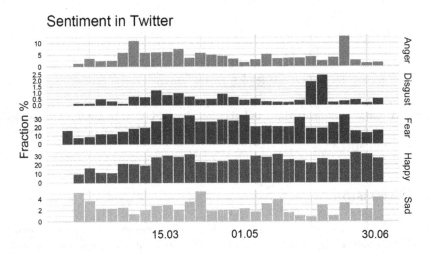

Fig. 3. Sentiment dynamics on Twitter week by week. The depolarization phase corresponds to 18.03−07.04.2020.

In the middle of the "stay-at-home phase," we observe a temporary minimum of network modularity, which implies temporary depolarization and mobilization (Fig. 4) in supporting each other (which is also seen in optimism in the sentiment analysis) over political divisions (Fig. 3). However, it lasted very shortly (just a few weeks) and the system relaxed back to reach the peak of polarization around the summer presidential campaign.

The main hashtags of the consolidating phenomenon were #widzialnareka (visible hand), #pomoc (help), #solidarność (solidarity), #WspierajSeniora (support senior), #PomocZakupowa (shopping help), #covidowesos (COVID SOS), #szyje maseczki (sewn masks), and #brawadlaWas (applause for you). The end of social consolidation happened when the presidential election campaign started around 10.04.2020, which was the 10-year anniversary of the Smoleńsk crash, the main polarizing event in the history of an independent Poland [17].

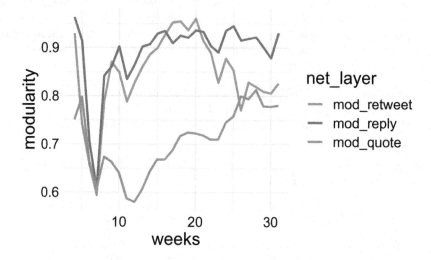

Fig. 4. Attempt to indicate the depolarization phase 18.03-07.04.2020. The modularity of the full networks and their weekly dynamics for types of communication (retweeting, reply, quotation).

4   Discussion

In critical moments of the pandemic, the social media were already incorporated into emergency planning and crisis management for monitoring as well as for crowd-sourcing and collaborative development [2] mainly for local communities. It provided opportunities for commenting and engagement by active users (76,822 unique users in the Polish COVID-19 context).

The unexpected scale of the pandemic and the loss of information control by the government have destroyed the existing "institutionalized society" in social media (with high polarization) and created a "communal society". However, in the COVID-19 era, social media has gained a bad reputation due to the spread of mis-/dis-information [44], creation of filter bubbles/echo chambers [5], etc. Most current research on communication through social networks during crises concentrates on the speed of spread [12] and the quality of information [43];

however, the most influential element—inter-community links—was not investigated to such an extent. A special role of social media as the moderator could be deployed to maintain social capital during the COVID-19 pandemic.

We showed that in the mobilization phase ("stay at home"), people demonstrated an altruistic attitude (experiencing positive emotions such as optimism (Fig. 3)) on Twitter. They could potentially take actions benefiting others in the physical world and the virtual reality of social media. We also discovered that sharing activities (for instance, retweeting) crosscut partisan lines (Fig. 4). Thus, we confirmed the short-term depolarization during the "stay-at-home phase" (Fig. 4, 3) in late March/early April. However, this phase of relative optimism and high interlinkage between all users was unstable. The system soon returned to its polarized equilibrium. Studies on inter-connected partisan communities formed in crisis situations in the pre-social media era have reported that interconnectivity (depolarization) was also unstable. For instance, in 1997 during the flood of the century in the Polish city of Wroclaw, a support society was also formed for a short time [39].

Limitations. We are aware that operationalization of polarization to network modularity and sentiment is a simplification, and we miss many aspects of this phenomenon. Historically in Poland, solidarity movements that gathered people of different statuses, religions, and political orientations in the 1980s (form of social capital [1,16]) played a role in the democratisation of this postcommunistic country, so the level of mobilisation observed in Polish social media during COVID-19 does not need to be observed in other countries to such an extent. Some civil initiatives via social media such as participatory epidemiology of COVID-19 vaccines have been only carried out in post-communistic societies only [16,19]. Moreover, the retrospective character of this study and the use of the biased medium of Twitter does not allow causative reasoning without further qualitative reflection [32].

5   Conclusion

In conclusion, social media provides new opportunities for fostering the dynamics of supportive actions [37] allowing the mobilisation of people in crises such as COVID-19. Massive, apolitical actions supporting the healthcare system or people with special needs are creating social cohesion. Social media platforms (particularly Twitter) are known for polarizing (by default) the communication ecosystem, because users see content filtered by this technological giant according to their preferences mostly from the other users they like, follow, and retweeted previously [47,48]. However, top-down policies and procedures on social networks are likely to fail [46], so the advocacy of bottom-up (civil initiatives of average users) could be critical in future resilience policy planning, aiming for depolarization. Thus, the catastrophic weather anomalies in Europe in 2021 could also lead to forming supportive Internet communities in societies.

Acknowledgement. This study was partially funded by the Deutsche Forschungsgemeinschaft (DFG, German Research Foundation, project number 458528774). Support from The Endowment Fund of St. Petersburg State University is gratefully acknowledged.

References

1. Åberg, M., Sandberg, M.: Social Capital and Democratisation: Roots of Trust in Post-communist Poland and Ukraine. Routledge, Abingdon (2017)
2. Alexander, D.: Social media in disaster risk reduction and crisis management. Sci. Eng. Ethics **20**(3), 717–733 (2014). https://doi.org/10.1007/s11948-013-9502-z
3. Baron, H., et al.: Can Americans depolarize? Assessing the effects of reciprocal group reflection on partisan polarization (2021). https://doi.org/10.31219/osf.io/3x7z8
4. Batorski, D., Grzywińska, I.: Three dimensions of the public sphere on Facebook. Inf. Commun. Soc. **21**(3), 356–374 (2018). https://doi.org/10.1080/1369118X.2017.1281329
5. Baumann, F., Lorenz-Spreen, P., Sokolov, I.M., Starnini, M.: Modeling echo chambers and polarization dynamics in social networks. Phys. Rev. Lett. **124**(4), 048301 (2020)
6. Bispo Júnior, J.P., Morais, M.B.: Community participation in the fight against COVID-19: between utilitarianism and social justice. Cad. Saude Publica **36**, e00151620 (2020)
7. Blondel, V.D., Guillaume, J.L., Lambiotte, R., Lefebvre, E.: Fast unfolding of communities in large networks. J. Stat. Mech: Theory Exp. **2008**(10), P10008 (2008)
8. Bourdieu, P.: The forms of capital. Willey (1986)
9. Boxell, L., Conway, J., Druckman, J.N., Gentzkow, M.: Affective polarization did not increase during the coronavirus pandemic. Technical report, National Bureau of Economic Research (2020). https://doi.org/10.3386/w28036
10. Danielewski, M.: W poszukiwaniu nadziei po mrocznym 2020. Zawiodło państwo PiS, obywatele zdali egzamin) (2020). https://oko.press/w-poszukiwaniu-nadziei-po-mrocznym-2020-zawiodlo-panstwo-pis-obywatele-zdali-egzamin/. Accessed 28 Dec 2020
11. Demszky, D., et al.: Analyzing polarization in social media: method and application to tweets on 21 mass shootings. In: Proceedings of the 2019 Conference of the North American Chapter of the Association for Computational Linguistics: Human Language Technologies, Volume 1 (Long and Short Papers), pp. 2970–3005 (2019)
12. Dragović, N., Vasiljević, Stankov, U., Vujičić, M.: Go social for your own safety! Review of social networks use on natural disasters-case studies from worldwide. Open Geosci. **11**(1), 352–366 (2019)
13. Guerra, P., Meira Jr., W., Cardie, C., Kleinberg, R.: A measure of polarization on social media networks based on community boundaries. In: Proceedings of the International AAAI Conference on Web and Social Media, vol. 7 (2013)
14. IAB: Przewodnik po Social Media w Polsce (2020). https://iab.org.pl/wp-content/uploads/2020/01/IAB-Przewodnik-po-Social-Media-w-Polsce-2019-2020.pdf. Accessed 04 Sept 2020
15. Jarynowski, A.: Monitorowanie percepcji ryzyka COVID-19 na Dolnym Ślasku za pomoca analizy śladu cyfrowego w internecie 15.01-05.08.2020. Instytut Badań Interdyscyplinarnych (2020)

16. Jarynowski, A.: Phenomenon of participatory "guerilla" epidemiology in post-communist European countries (2021). https://sites.utu.fi/bre/phenomenon-of-participatory-guerilla-epidemiology-in-post-communist-european-countries/. Accessed 10 July 2021
17. Jarynowski, A., Buda, A., Piasecki, M.: Multilayer network analysis of Polish parliament 4 years before and after Smoleńk crash. In: 2016 Third European Network Intelligence Conference (ENIC), pp. 69–76. IEEE (2016)
18. Jarynowski, A., Semenov, A., Belik, V.: Protest perspective against COVID-19 risk mitigation strategies on the German internet. In: Chellappan, S., Choo, K.-K.R., Phan, N.H. (eds.) CSoNet 2020. LNCS, vol. 12575, pp. 524–535. Springer, Cham (2020). https://doi.org/10.1007/978-3-030-66046-8_43
19. Jarynowski, A., Semenov, A., Kaminski, M., Belik, V.: Mild adverse events of sputnik V vaccine in Russia: social media content analysis of telegram via deep learning. J. Med. Internet Res. (2021). https://doi.org/10.2196/30529
20. Jarynowski, A., Wójta-Kempa, M., Belik, V.: Perception of "coronavirus" on the Polish Internet until arrival of SARS-CoV-2 in Poland. Nurs. Public Health 10(2), 89–106 (2020). https://doi.org/10.17219/pzp/120054
21. Jarynowski, A., Wójta-Kempa, M., Belik, V.: Trends in interest of COVID-19 on Polish Internet. Przeglad epidemiologiczny 74(2), 258–275 (2020). https://doi.org/10.32394/pe.74.20
22. Jarynowski, A., Wójta-Kempa, M., Płatek, D., Belik, V.: Social values are significant factors in control of COVID-19 pandemic-preliminary results (2020). https://www.preprints.org/manuscript/202005.0036/v1
23. Jarynowski, A., Wójta-Kempa, M., Płatek, D., Czopek, K.: Attempt to understand public health relevant social dimensions of COVID-19 outbreak in Poland. Soc. Reg. 4(3), 7–44 (2020)
24. Kamiński, M., Szymańska, C., Nowak, J.K.: Whose tweets on COVID-19 gain the most attention: celebrities, political, or scientific authorities? Cyberpsychol. Behav. Soc. Netw. 24(2), 123–128 (2021)
25. Kaniasty, K.: Kleska żywiołowa czy katastrofa społeczna? Gdańskie Wydaw. Psychologiczne (2003)
26. Kaniasty, K., Norris, F.H.: Mobilization and deterioration of social support following natural disasters. Curr. Dir. Psychol. Sci. 4(3), 94–98 (1995)
27. Matuszewski, P., Szabó, G.: Are echo chambers based on partisanship? Twitter and political polarity in Poland and Hungary. Soc. Media+ Soc. 5(2), 2056305119837671 (2019)
28. Mauss, M.: Essai sur le don forme et raison de l'échange dans les sociétés archaïques. L'Année sociologique (1896/1897-1924/1925) 1, 30–186 (1923)
29. Newman, M.E.: Modularity and community structure in networks. Proc. Natl. Acad. Sci. 103(23), 8577–8582 (2006)
30. Nowak, M.A.: Five rules for the evolution of cooperation. Science 314(5805), 1560–1563 (2006)
31. Oh, H.J., Ozkaya, E., LaRose, R.: How does online social networking enhance life satisfaction? The relationships among online supportive interaction, affect, perceived social support, sense of community, and life satisfaction. Comput. Hum. Behav. 30, 69–78 (2014)
32. Pousti, H., Urquhart, C., Linger, H.: Researching the virtual: a framework for reflexivity in qualitative social media research. Inf. Syst. J. 31(3), 356–383 (2020)
33. Putnam, R.D., et al.: Bowling Alone: The Collapse and Revival of American Community. Simon and Schuster, New York (2000)

34. Qureshi, I., Bhatt, B., Gupta, S., Tiwari, A.A.: Causes, Symptoms and consequences of social media induced polarization (SMIP). https://onlinelibrary.wiley.
com/pb-assets/assets/13652575/ISJ_SMIP_CFP-1586861685850.pdf
35. Rao, A., et al.: Political partisanship and antiscience attitudes in online discussions
about COVID-19: Twitter content analysis. J. Med. Internet Res. **23**(6), e26692
(2021)
36. Riegel, M., et al.: Nencki affective word list (NAWL): the cultural adaptation of
the Berlin affective word list-reloaded (BAWL-R) for Polish. Behav. Res. Methods
47(4), 1222–1236 (2015). https://doi.org/10.3758/s13428-014-0552-1
37. Sæbø, Ø., Federici, T., Braccini, A.M.: Combining social media affordances for
organising collective action. Inf. Syst. J. **30**(4), 699–732 (2020)
38. Sides, J., Tausanovitch, C., Vavreck, L.: The politics of COVID-19: partisan
polarization about the pandemic has increased, but support for health care
reform hasn't moved at all (2020). https://doi.org/10.1162/99608f92.611350fd,
publisher=PubPub
39. Sitek, W.: Wspólnota i zagrożenie: Wrocławianie wobec wielkiej powodzi: socjo-
logiczny przyczynek do analizy krótkotrwałej wspólnoty. Wydawawnictwo Uniwer-
sytetu Wrocławskiego (1997)
40. Statista: Most popular social media services during the coronavirus (COVID-19)
epidemic in Poland (2020). https://www.statista.com/statistics/1114857/poland-
leading-social-media-during-the-covid-19-pandemic/. Accessed 04 Dec 2020
41. Tönnies, F.: Gemeinschaft und gesellschaft. In: Studien zu Gemeinschaft und
Gesellschaft, pp. 27–58 (1887)
42. Valdez, S.: Online Polarization on Swedish Twitter during COVID) (2021).
https://www.sociology.su.se/english/about-us/events/stockholm-university-
computational-sociology/computational-sociology-group-meeting-online-
polarization-on-swedish-twitter-during-covid-1.529409. Accessed 10 July 2021
43. WHO: The Health System Response Monitor - Poland (2020). https://www.
covid19healthsystem.org/countries/poland/countrypage.aspx. Accessed 04 Dec
2020
44. WHO: WHO public health research agenda for managing infodemics. Technical
report, World Health Organization (2021). https://www.who.int/publications/i/
item/9789240019508
45. Wierzba, M., et al.: Basic emotions in the Nencki Affective Word List (NAWL BE):
new method of classifying emotional stimuli. PLoS One **10**(7), e0132305 (2015)
46. Wilkin, J., Biggs, E., Tatem, A.J.: Measurement of social networks for innovation
within community disaster resilience. Sustainability **11**(7), 1943 (2019)
47. Wojcieszak, M., Garrett, R.K.: Social identity, selective exposure, and affective
polarization: how priming national identity shapes attitudes toward immigrants
via news selection. Hum. Commun. Res. **44**(3), 247–273 (2018)
48. Wojcieszak, M., Winter, S., Yu, X.: Social norms and selectivity: effects of norms
of open-mindedness on content selection and affective polarization. Mass Commun.
Soc. **23**(4), 455–483 (2020)
49. Wójta-Kempa, M.: Ocena poziomu poinformowania na temat przebiegu i skutków
pandemii COVID-19. Technical report, Uniwersytet Wroclawski (2020)

Target Set Selection in Social Networks with Influence and Activation Thresholds

Zhecheng Qiang[1](✉) , Eduardo L. Pasiliao[2] , and Qipeng P. Zheng[1]

[1] Department of Industrial Engineering and Management Systems, University of Central Florida, Orlando, FL, USA

[2] Munitions Directorate, Air Force Research Laboratory, Shalimar, FL, USA

Abstract. Social media networks have gradually become the major venue for broadcasting and relaying information, thereafter making great influences in many aspects of our daily lives. With the mass adoption of the internet and mobile devices, social media users tend to follow and adopt their friends' or followers' thoughts and behaviors. Thus finding influential users in social media is crucial for many viral marketing, cybersecurity, politics, and safety-related applications. In this study, we address the problem through solving the influence and activation thresholds target set selection problem, which is to find the minimum number of seed nodes that influence all the users at time T. These time-indexed integer program models suffer from computational difficulties with binary variables at each time step. To this respect, this paper leverages computational algorithms, i.e., Graph Partition, Nodes Selection, and Greedy Algorithm to solve the models for large-scale networks. Computational results show that it is beneficial to apply the BFS Greedy algorithm for large scale networks. In addition, the results also indicate nodes selection methods perform better in the long-tailed networks.

Keywords: Networks · Integer programming · Target set selection · Greedy algorithm · Influence maximization

1 Introduction

Nowadays, the use of social media networks has become a necessary daily activity for people to interact with family and friends, access news, information and make decisions. Besides being a handy means for keeping in touch with friends and family, social media is more of a platform spreading the tremendous influence. Social media has great impact on businesses, politics, disease control and others. To this end, researchers have studied various practical problems in social media to better understand how the social media behaves and propagates the information. The problems include buzz prediction [7], volume prediction [24], infection prediction [3,17], source prediction [21], link detection [19], target set selection [6,8,9,14,26] and firefighter problem [2].

© Springer Nature Switzerland AG 2021
D. Mohaisen and R. Jin (Eds.): CSoNet 2021, LNCS 13116, pp. 371–380, 2021.
https://doi.org/10.1007/978-3-030-91434-9_32

1.1 Motivation and Problem Description

Here we focus on investigating the problem of target set selection. Formally, we define a social media network as a directed graph $G = (V, E)$, where users are defined as all nodes V and their friendships are defined as all edges E. Users are active when they repost the messages. Target set selection problem refers to find a minimum subset $S \subset V$, then all the nodes will be activated by S. The target set selection problem can be applied in the areas of viral marketing [11] and cyber security [4].

In the target set selection model, the seed nodes spread the influence until the diffusion process stops. The goal is to activate all the nodes. The influenced users refer to activated users who repost the messages. However, in reality, influenced users sometimes will repost the messages. But in most cases, even they are convinced or influenced by the message, users will not repost the message for certain reasons. In this case, influence not only refers to activation (repost) but also refers to the belief in the messages. Thus we build the influence and activation thresholds target set selection models to describe the situation. Here we introduce two thresholds, one is activation threshold (ϕ) and another one is influence threshold (θ). Users will be influenced first before be activated, thus we define $\theta <= \phi$. The goal of the model is to influence all the nodes.

Our models are time-indexed integer program models, which can be divided into two parts, the first part is the information propagation. There are two widely used propagation models, namely Independent Cascade Model (IC) [14] and Linear Threshold Model (LT) [12]. Independent Cascade Model assumes every node has a single chance to activate its neighbors. In Linear Threshold Model, each node will be influenced by each neighbor according to a weight. When the total weights from its neighbors is larger than a threshold ϕ, then the node will be activated. In this paper, we propose all the mathematical models based on the Linear Threshold Propagation Model. Here we set the weights as 1 for all the nodes. Thus an inactive node will become active if at least ϕ of its neighbors are active in the previous step. The second part is the influence dominating part, which means the users should be either active or influenced through having at least θ of activated neighbors at time T.

1.2 Literature Review

To investigate the influence and activation thresholds target set selection problem, the research of target set selection problem offers us some good insights. Kempe et al. [14] study the maximum active set (under the name target set selection) and show that it is NP-hard. They also provide a greedy algorithm within provable approximation guarantees based on the submodularity property of the objective function. Chen et al. [9] study the minimum target set selection problem and show the problem is hard to approximate within a polylogarithmic factor. Besides, he comes up a polynomial-time algorithm to find an optimal solution when the underlying graph is a tree. Ackerman et al. [1] propose a combinatorial model for the minimum target set selection and prove the combinatorial bounds for the perfect target set selection problem. Shakarian et al. [22] present a time-indexed formulation to find the minimum seeds for the target set selection and come up with a scalable heuristic based on the idea of shell decomposition. Spencer et al. [23] consider the problem of how to target individuals with subsidy in the network in order to promote pro-environmental behavior. It is also a target set selection problem and they use a time-indexed integer program formulation with as many time periods as the number of nodes in the network to tackle the problem. Günneç et al. [13] study the variation of the target set selection problem called least cost target set selection on social networks, and they propose greedy algorithm and dynamic programming algorithm to solve the problem for the tree structure network. Raghavan et al. [18] develop and implement a branch-and-cut approach to solve the weighted target set selection problem on arbitrary graphs.

1.3   Contribution and Organization

To our knowledge, the previous research involving target set selection focuses on the single threshold (activation threshold) target set selection. In this paper, we propose practical influence and activation thresholds target set selection mathematical model and its computational algorithms correspondingly.

The rest of the paper is organized as follows. Section 2 presents the novel Minimum Influence and Activation Thresholds Target Set Selection Model. In Sect. 3, we propose several computational algorithms for solving large scale networks. Section 4 shows the experimental results of the proposed model and its corresponding computational algorithms. Section 5 concludes the article.

2   Minimum Target Set Selection Model with Influence and Activation Thresholds

In this section, we introduce a time-indexed integer program to find the minimum number of influential seeds for the influence and activation thresholds target set selection problem. An artificial time index t taking values from 0 to T is introduced to model the order in which nodes become active. The messages could propagate at varying distances through different forms of social media. Cha et al. [5] observe that even for popular photos, only 19% of fans are more than 2 hops away from uploaders on Flickr.com. Ye et al. [25] find that, on Twitter, 37.1 percent message flows spread more than 3 hops away from the originators. Thus here we set T as 0,1,2,3, which means we only consider the cascades less than or equal to three time steps. The formulation uses a binary variable $x_{i,t}$ to represent the status of node i at time t, which is 1 if node i is active at time t and 0 otherwise. Here θ represents the influence threshold and ϕ represents the activation threshold. Nodes should always be influenced before activated, so we set $\theta <= \phi$. $N(i)$ represents the neighborhood of node i. The Minimum Influential Seeds Model is as follows:

$$\min_{\mathbf{x}} \quad \sum_{i \in V} x_{i,0} \tag{1}$$

$$s.t. \quad \theta x_{i,T} + \frac{1}{|N(i)|} \left\{ \sum_{j \in N(i)} x_{j,T} \right\} - \theta \geq 0 \qquad \forall i \in V \tag{2}$$

$$\frac{1}{|N(i)|} \left\{ \sum_{j \in N(i)} x_{j,t} \right\} + \phi x_{i,t} \geq \phi x_{i,t+1} \qquad \forall i \in V, t < T \tag{3}$$

$$\frac{1}{|N(i)|} \left\{ \sum_{j \in N(i)} x_{j,t} \right\} + \phi x_{i,t} \leq \phi - \epsilon + (1+\epsilon)x_{i,t+1} \quad \forall i \in V, t < T \tag{4}$$

$$\theta \in (0,1] \qquad \leftarrow \text{influence threshold}$$
$$\phi \in (0,1] \qquad \leftarrow \text{activation threshold}$$
$$\theta \leq \phi \qquad \leftarrow \text{node is influenced before activated}$$
$$N(i) = \{ \, j : (i,j) \in E \, \} \leftarrow \text{Neighborhood, adjacent nodes}$$

The objective function (1) aims to minimize the seed nodes activated at time 0. Constraints (2) are influential constraints, making sure that all the nodes should be either active or be influenced by at least θ of active neighbors at time T. Constraints (3) refer that a node i will stay inactive at time $t+1$ when it is not activated at time t. Constraints (4) restrict that a node will stay active if it is originally active, which means when $x_{i,t} = 1$, $x_{i,t+1}$ should be 1 as well. In addition, the constraints make sure that a node will become active at time $t+1$ when it is activated at time t, which means when $x_{i,t} = 0$, the influence from its neighbors is larger or equal than ϕ, then $x_{i,t+1} = 1$. Here we introduce two ϵ, the first ϵ restricts that node i should be active at time $t+1$ even if the influence from its' neighbors is ϕ. The second ϵ confirms that when $x_{i,t} = 1$ and all the neighbors of i are active, the node i being active at time $t+1$ still holds.

3 Computational Algorithms

The time-indexed integer program model proposed in Sect. 2 is computationally intractable unless in very small instances because of the large number of binary variables. However, social media networks are usually in an extremely large scale. Thus we apply multiple computational algorithms to tackle the influence and activation thresholds target set selection model for larger scale networks in this manuscript. More details will be discussed in the rest of this section.

3.1 Graph Partition

When the social media network is large-scale, solving the models exactly through Gurobi is very difficult. The most intuitive way is to solve multiple smaller subgraphs instead of one large graph. Here we use techniques from Modularity and

Community Structure [10] in networks to divide the large graph into several smaller subgraphs. Then we solve the models exactly separately for each subgraph.

3.2   Nodes Selection

When we're dealing with influence and activation thresholds target set selection problem for large-scale network, the large number of binary decision variables, which are $|V||T|$ in total, makes the problem difficult to solve. In order to accelerate the computational speed, we could reduce the decision variables through adding some constraints to restrict that some of the nodes are not selected or some of the nodes should be selected. Here we come up with two methods, one is to delete the leaf nodes, and another is to choose the nodes with high degree.

Leaf Nodes Deletion. Leaf nodes in a connected graph may not be seeded because they'll influence or activate at most one neighbor directly. Thus we add the constraints (5) to remove the option of activating leaf nodes. In other words, all the leaf nodes will not be seeded using the method.

$$x_{i,0} + 1 \leq |N(i)| \tag{5}$$

Degree Centrality Selection. Nodes with high degree have more potential to influence and activate other nodes. Therefore, we assume the high degree nodes must be seeded. Here $\frac{1}{|V|} \ll \rho < 1$ is defined as the criteria for choosing the seed nodes. When the total neighbors $|N(i)|$ of node i is larger than $\rho|V|$, the node i will be seeded. Thus we add the constraints (6) to the original model in order to choose the nodes with more than $\rho|V|$ neighbors as seed nodes.

$$x_{i,0} + \rho \geq \frac{|N(i)|}{|V|} \tag{6}$$

$$\rho = \epsilon \text{ implies that } N(i) = 1 \text{ for a connected graph}$$
$$\rho = 1 - \epsilon \text{ implies that } N(i) = V - 1$$

The larger the ρ, the nodes with higher degree will be selected as seed nodes. When ρ is ϵ, the nodes having neighbors will all be selected. When ρ is $1 - \epsilon$, only the node connecting to all the other nodes will be selected.

3.3   Greedy Algorithm

We propose the greedy algorithm for the Minimum Influence and Activation Thresholds Target Set Selection problem. The greedy algorithm selects the seed node with the largest number of inactive neighbors in each iteration and adds it to the seed node set S until the stop conditions have been met. Then we update the nodes threshold and active set in each iteration considering the propagation process. Here we update the threshold and active set based on the Breadth First Search (BFS).

For the BFS Search Greedy Algorithm shown in Algorithm 1, firstly we choose the seed node with the largest number of inactive neighbors and add it to the seed set S. Then we update the threshold and activation time step (T) of an inactive neighbor node by adding the influence sent from the activated seed node. BFS search starts at the tree root and explores all of the neighbor nodes at the present depth prior to moving on to the nodes at the next depth level. Here we use a queue Q to store the parent nodes which will spread the influence within propagation step P.

Algorithm 1. BFS Search Greedy Algorithm

Input: Graph $G = (V, E)$, Propagation step P
Output: Seed node set S

1: $A \leftarrow \emptyset$, $S \leftarrow \emptyset$, Q be queue
2: $threshold(i) = 0$, $T(i) = 0$, $ite(i) = 0$
3: **while** $\exists\, v \in V \setminus A$, $n_A(v) < \theta deg(v)$ **do**
4:      Pick $u \in V \setminus A$ with the most inactive neighbors
5:      $Q.enqueue(u)$
6:      $S = S \cup u$
7:      $A = A \cup u$
8:      $iteration = 0$
9:      **while** Q is not empty **do**
10:         $v = Q.dequeue()$
11:         $iteration = ite(v) + 1$
12:         **for** all ω in $N(v)$ **do**
13:             **if** ω not in A **then**
14:                 $threshold(\omega) = threshold(\omega) + \frac{1}{deg(\omega)}$
15:                 $T(\omega) = max(T(\omega), iteration)$
16:                 **if** threshold(ω)$>= \phi$ **then**
17:                     $A = A \cup \omega$
18:                     **if** $T(\omega) < P$ **then**
19:                         $Q.enqueue(\omega)$
20:                         $ite(\omega) = iteration$
21:                     **end if**
22:                 **end if**
23:             **end if**
24:         **end for**
25:     **end while**
26: **end while**
27: Return (S)

---

4   Computational Algorithms Comparison

In this subsection, we assess and draw comparisons between different computational algorithms introduced in the Sect. 3 for the minimum influence and activation thresholds target set selection model.

We consider a subset of real-life social networks as datasets in our experiment: Karate Club [27], Hamster Friendships Network [15], Facebook Network Dataset [16] and LastFM Social Network Dataset [20]. We set the time limit of 3600 s for bold methods in the following experiments. For the method of degree centrality, we set the ρ as 0.2.

The Karate network is a network of very small size and the result is shown in Table 1. For the Karate Club network, we could solve the model directly. However, the Leaf Node method could accelerate the computation slightly without sacrificing the performance. For larger size network of Hamster Dataset, the Graph Partition method couldn't generate a solution in one hour, so we don't include here in Table 2. For Leaf Node method, the model is not feasible which means we couldn't exclude all the leaf nodes as seed nodes for Minimum Influential Seeds model. We could see from the table the Original Model even performs better than the Degree Centrality method within the time limit of 3600 s. In addition, BFS Greedy method offers good solutions within much less time compared to the Original Model. The results of Facebook1 dataset are shown in Table 3. Facebook 1 is a dataset of low density. The Facebook 1 network has a large number of nodes with few friends. Thus it is easy for Gurobi to solve it directly. However, the Leaf node method has the shortest computation time for this dataset. The results of LastFM Asia Dataset are shown in Table 4, here Leaf Node performs the best compared with Original Model and Degree Centrality methods within the time limit of 3600 s.

Table 1. Karate network

Method	θ	ϕ	T	Seeded	Activated	Influenced	Obj	Time
Original Model	0.4	0.6	3	6	18	34	6	0.34
Graph Partition	0.4	0.6	3	7	32	34	7	0.09
Leaf Node	0.4	0.6	3	6	18	34	6	0.32
Degree Centrality	0.4	0.6	3	7	32	34	7	0.02
BFS Greedy	0.4	0.6	3	7	33	34	7	0.001

Table 2. Hamster dataset

Method	θ	ϕ	T	Seeded	Activated	Influenced	Obj	Time
Original Model	0.4	0.6	3	282	1543	1858	282	3600.47
Degree Centrality	0.4	0.6	3	291	1529	1858	291	3600.56
BFS Greedy	0.4	0.6	3	327	1766	1858	327	15.05

Table 3. Facebook 1 Dataset

Method	θ	ϕ	T	Seeded	Activated	Influenced	Obj	Time
Original Model	0.4	0.6	3	10	2888	2888	10	16.66
Graph Partition	0.4	0.6	3	10	2888	2888	10	20.17
Leaf Node	0.4	0.6	3	10	2888	2888	10	0.55
Degree Centrality	0.4	0.6	3	10	2888	2888	10	4.40
BFS Greedy	0.4	0.6	3	10	2888	2888	10	1.11

Table 4. LastFM Asia Dataset

Method	θ	ϕ	T	Seeded	Activated	Influenced	Obj	Time
Original Model	0.4	0.6	3	2850	6682	7624	2850	3601.25
Leaf Node	0.4	0.6	3	1498	6453	7624	1498	3601.52
Degree Centrality	0.4	0.6	3	3427	7624	7624	3427	3601.42
BFS Greedy	0.4	0.6	3	1675	7255	7624	1675	868.62

In summary, for the small size datasets, we could solve the problem directly using Gurobi. For the network of low density, especially when large portion of the nodes have few neighbors(long-tailed network), we could consider the Leaf Node method and Degree Centrality method. For the larger size datasets, normally the BFS Greedy will have better performance. The Graph Partition has poor performance and long computational time for the selected social media networks. For the Graph Partition method, it could result from the structure of network which is hard to divide into subgraphs. Furthermore, even it is divided properly, sometimes the size of the subgraph is still hard to solve directly.

5   Conclusion

The increasing popularity of social media networks has created the need for businesses, politicians and organizations to find influential users in social media to spread the influence. In this work, we have addressed the problem through developing the minimum influence and activation thresholds target set selection model. Our model allows us to find the minimum seed nodes that influence all the nodes at time T. In addition, we provide different computational algorithms to tackle the various datasets as well. They are Graph Partition, Leaf Node, Degree Centrality and BFS Greedy computational algorithms. Experiments in various datasets show that BFS Greedy is much more efficient than the other methods for large size datasets. Besides, leaf node deletion and degree centrality selection perform better in terms of long-tailed network.

References

1. Ackerman, E., Ben-Zwi, O., Wolfovitz, G.: Combinatorial model and bounds for target set selection. Theoret. Comput. Sci. **411**(44–46), 4017–4022 (2010)
2. Anshelevich, E., Chakrabarty, D., Hate, A., Swamy, C.: Approximation algorithms for the firefighter problem: cuts over time and submodularity. In: Dong, Y., Du, D.-Z., Ibarra, O. (eds.) ISAAC 2009. LNCS, vol. 5878, pp. 974–983. Springer, Heidelberg (2009). https://doi.org/10.1007/978-3-642-10631-6_98
3. Bourigault, S., Lamprier, S., Gallinari, P.: Representation learning for information diffusion through social networks: an embedded cascade model. In: Proceedings of the Ninth ACM International Conference on Web Search and Data Mining. WSDM 2016, pp. 573–582. ACM, New York (2016). https://doi.org/10.1145/2835776.2835817
4. Budak, C., Agrawal, D., El Abbadi, A.: Limiting the spread of misinformation in social networks. In: Proceedings of the 20th International Conference on World Wide Web, pp. 665–674. ACM (2011)
5. Cha, M., Mislove, A., Gummadi, K.P.: A measurement-driven analysis of information propagation in the Flickr social network. In: Proceedings of the 18th International Conference on World Wide Web, pp. 721–730 (2009)
6. Chen, C.-L., Pasiliao, E.L., Boginski, V.: A cutting plane method for least cost influence maximization. In: Chellappan, S., Choo, K.-K.R., Phan, N.H. (eds.) CSoNet 2020. LNCS, vol. 12575, pp. 499–511. Springer, Cham (2020). https://doi.org/10.1007/978-3-030-66046-8_41
7. Chen, G.H., Nikolov, S., Shah, D.: A latent source model for nonparametric time series classification. In: Advances in Neural Information Processing Systems, pp. 1088–1096 (2013)
8. Chen, M., Zheng, Q.P., Boginski, V., Pasiliao, E.L.: Reinforcement learning in information cascades based on dynamic user behavior. In: Tagarelli, A., Tong, H. (eds.) CSoNet 2019. LNCS, vol. 11917, pp. 148–154. Springer, Cham (2019). https://doi.org/10.1007/978-3-030-34980-6_17
9. Chen, N.: On the approximability of influence in social networks. SIAM J. Discret. Math. **23**(3), 1400–1415 (2009)
10. Clauset, A., Newman, M.E., Moore, C.: Finding community structure in very large networks. Phys. Rev. E **70**(6), 066111 (2004)
11. Domingos, P.: Mining social networks for viral marketing. IEEE Intell. Syst. **20**(1), 80–82 (2005)
12. Granovetter, M.: Threshold models of collective behavior. Am. J. Sociol. **83**(6), 1420–1443 (1978)
13. Günneç, D., Raghavan, S., Zhang, R.: Least-cost influence maximization on social networks. INFORMS J. Comput. **32**(2), 289–302 (2020)
14. Kempe, D., Kleinberg, J., Tardos, E.: Maximizing the spread of influence through a social network. In: Proceedings of the Ninth ACM SIGKDD International Conference on Knowledge Discovery and Data Mining. KDD 2003, pp. 137–146. ACM, New York (2003). https://doi.org/10.1145/956750.956769
15. Kunegis, J.: KONECT - the Koblenz network collection. In: Proceedings of the International Conference on World Wide Web Companion, pp. 1343–1350 (2013). http://dl.acm.org/citation.cfm?id=2488173
16. Leskovec, J., Mcauley, J.J.: Learning to discover social circles in ego networks. In: Advances in Neural Information Processing Systems, pp. 539–547 (2012)

17. Qiang, Z., Pasiliao, E.L., Zheng, Q.P.: Model-based learning of information diffusion in social media networks. Appl. Netw. Sci. **4**(1), 111 (2019)
18. Raghavan, S., Zhang, R.: A branch-and-cut approach for the weighted target set selection problem on social networks. INFORMS J. Optim. **1**(4), 304–322 (2019)
19. Rodriguez, M.G., Balduzzi, D., Schölkopf, B.: Uncovering the temporal dynamics of diffusion networks. arXiv preprint arXiv:1105.0697 (2011)
20. Rozemberczki, B., Sarkar, R.: Characteristic functions on graphs: birds of a feather, from statistical descriptors to parametric models (2020)
21. Shah, D., Zaman, T.: Detecting sources of computer viruses in networks: theory and experiment. SIGMETRICS Perform. Eval. Rev. **38**(1), 203–214 (2010). https://doi.org/10.1145/1811099.1811063
22. Shakarian, P., Eyre, S., Paulo, D.: A scalable heuristic for viral marketing under the tipping model. Soc. Netw. Anal. Min. **3**(4), 1225–1248 (2013). https://doi.org/10.1007/s13278-013-0135-7
23. Spencer, G., Howarth, R.: Maximizing the spread of stable influence: leveraging norm-driven moral-motivation for green behavior change in networks. arXiv preprint arXiv:1309.6455 (2013)
24. Tsur, O., Rappoport, A.: What's in a hashtag?: content based prediction of the spread of ideas in microblogging communities. In: Proceedings of the Fifth ACM International Conference on Web Search and Data Mining, pp. 643–652. ACM (2012)
25. Ye, S., Wu, S.F.: Measuring message propagation and social influence on Twitter.com. In: Bolc, L., Makowski, M., Wierzbicki, A. (eds.) SocInfo 2010. LNCS, vol. 6430, pp. 216–231. Springer, Heidelberg (2010). https://doi.org/10.1007/978-3-642-16567-2_16
26. Yun, G., Zheng, Q.P., Boginski, V., Pasiliao, E.L.: Information network cascading and network re-construction with bounded rational user behaviors. In: Tagarelli, A., Tong, H. (eds.) CSoNet 2019. LNCS, vol. 11917, pp. 351–362. Springer, Cham (2019). https://doi.org/10.1007/978-3-030-34980-6_37
27. Zachary, W.W.: An information flow model for conflict and fission in small groups. J. Anthropol. Res. **33**(4), 452–473 (1977)

Extended Abstracts

Social Activity and Decentralized Applications in Blockchain-Based Social Networks

Cheick Tidiane Ba[✉], Galdeman Alessia, Matteo Zignani, and Sabrina Gaito

Department of Computer Science, University of Milan, Via Celoria, 18, Milan, Italy
cheick.ba@unimi.it

Blockchain-Based Online Social Networks. Blockchain-based online social networks (BOSNs) are social platforms built on a blockchain that enables a reward system for high-quality content. The key idea is to reward users with cryptocurrencies, according to the popularity of the content they produce and/or promote. An attractive point for users is the possibility of exchanging cryptocurrencies for traditional currencies, services and goods. From a researchers' viewpoint, BOSNs are an appealing socio-technological complex system where users' social activities are influenced by the technological infrastructure; and internal socio-economic dynamics are linked with external financial trends. Another interesting aspect is that these platforms usually share the blockchain with other apps, called "decentralized apps" (DApps in short). While BOSNs are interesting, there is still limited work: some aspects have been studied, such as innovative characteristics [1], the static network structure [2], economic and rewards aspects [3]. Even though BOSNs provide high and detailed volumes of temporal data, there is a lack of work focused on network dynamics and temporal aspects.

Objective and Data. Our goal is to study dynamic aspects of complex systems such as BOSNs. Here, we focus on Steemit [4], a successful platform launched in 2016 that emerged in the BOSN ecosystem. The basic idea of Steemit is that user activity should be rewarded. Every week, content producers and content promoters of the most successful posts are rewarded with cryptocurrency tokens. According to BOSNs principles, Steemit relies on its blockchain - Steem - for data storage and validation. The blockchain is also home to many DApps such as Dtube, a video sharing platform, and Splinterlands, a trading card game. In this work we focus on social activity, to understand how user activity evolves and changes in these innovative social media platforms. To do so, we analyze data covering a four-year period of users' activities, from Dec. 6, 2016, up to Jan. 6, 2021. The obtained dataset covers user actions, stored as a collection of *operations*. Specifically, we focus on social operations, i.e. user activity typical of traditional social media platforms. We look at: *i)* `comment` operations, that record published content (posts) and comments *ii)* `vote` operations that track user votes, and influence content popularity and rewards; and *iii)* `custom_json`, a type of operation designed to store different actions, that we can differentiate by their "id" field, e.g. "follow".

D. Mohaisen and R. Jin (Eds.): CSoNet 2021, LNCS 13116, pp. 383–384, 2021.
https://doi.org/10.1007/978-3-030-91434-9

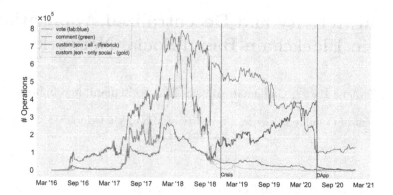

Fig. 1. Time plots for operations. X-axis: time in months. Y-axis: operations per day. Vertical lines for important events.

Results. We first build time series of the daily user activities carried out by Steemit users (Fig. 1). The first observation is that, for all the social operations considered (`comment`, `vote`, `custom_json`), we observe a first period of growth; then, all social actions drop in volume, as time passes. The drop was followed by an announcement, by Steemit's Founder Ned Scott, explaining that the BOSN was in a crisis.[1] A second observation is that `custom_json` operations grow and then drop again around the beginning of June 2020. As we can see in Fig. 1, social operations stored as `custom_json` operations (characterized by "id"s "follow", "unfollow", "reblog", "mute", "ignore") decline like other social operations. The new `custom_json` operations are mostly performed by other DApps operating on the blockchain. Indeed, an analysis of operations' id confirms that the majority of actions performed were made by one of the most popular DApps, Splinterlands. Furthermore, the successive drop coincides with the DApp transferring to a new Blockchain, Hive.[2] We can observe a shift in the way the blockchain is being used, combining social activity and popular decentralized apps. From a focus on supporting social interactions, Steemit is moving to an app ecosystem where social interactions are becoming more marginal.

References

1. Guidi, B.: When blockchain meets online social networks. Pervasive Mob. Comput. **62**, 101131 (2020)
2. Guidi, B., Michienzi, A., Ricci, L.: A graph-based socioeconomic analysis of steemit. IEEE Trans. Comput. Soc. Syst. 1–12 (2020)
3. Li, C., Palanisamy, B.: Incentivized blockchain-based social media platforms: a case study of steemit. In: Proceedings of the 10th ACM Conference on Web Science, pp. 145–154 (2019)
4. Steemit: Steemit Whitepaper (2020). https://steem.com/steem-whitepaper.pdf

---

[1] steemit.com/steem/@ned/2fajh9-steemit-update.
[2] peakd.com/splinterlands/@splinterlands/important-splinterlands-update.

Vulnerabilities Assessment of Deep Learning-Based Fake News Checker Under Poisoning Attacks

Lelio Campanile[1], Pasquale Cantiello[2], Mauro Iacono[1],
Fiammetta Marulli[1(✉)], and Michele Mastroianni[1]

[1] Department of Maths and Physics, Universitá degli Studi della Campania
"L. Vanvitelli", Caserta, Italy
{lelio.campanile,mauro.iacono,fiammetta.marulli,
michele.mastroianni}@unicampania.it
[2] Istituto Nazionale di Geofisica e Vulcanologia, Naples, Italy
pasquale.cantiello@ingv.it

Abstract. In this work, we envise an effective case study concerning a data and a model poisoning attack, consisting in evaluating how much a poisoned word embeddings model could affect the reliability of a deep neural network-based Fake News Checker; furthermore, we plan to train three different word embeddings models among the most performing in the Natural Language Processing field, in order to investigate which of these models can be considered more resilient and robust when such kind of attacks are applied.

Keywords: Natural Language Processing · Fake news · Adversarial attacks · Data poisoning attacks · Deep neural networks resilience

1 Introduction

Adversarial machine learning (AML) can act by prompting a Deep Neural Network with intentionally manipulated inputs, with the aim of fooling it and reducing its accuracy in accomplishing its task.

Adversarial examples, that represent purposely designed inputs able to exploit vulnerabilities of machine and deep learning models, got great popularity in performing adversarial attacks against image classification systems[1].

As for the Natural Language Processing (NLP) field, adversarial attacks are more difficult to be acted: while an image perturbation can lead to small changes of pixel values, that are hardly perceived by human eyes, small perturbations applied on texts are easier to sense. A text perturbation could consist, for example, in replacing characters or words within a sentence, thus generating invalid words, meaningless sequences or syntactically incorrect sentences. Therefore, perturbations on natural language texts are easily perceived.

---

[1] Goodfellow, I. J., Shlens, J., & Szegedy, C. (2014). Explaining and harnessing adversarial examples.

© Springer Nature Switzerland AG 2021
D. Mohaisen and R. Jin (Eds.): CSoNet 2021, LNCS 13116, pp. 385–386, 2021.
https://doi.org/10.1007/978-3-030-91434-9

2   The Envised Case Study

Recent studies have evidenced that also NLP systems can be fooled by AML attacks[2]. More precisely, they are vulnerable to poisoning attacks acted by the means of knowledge transferred, in the shape of pre-processed models, as an input during the training step of a deep learning model While small perturbations expressly introduced in a training data set could be easily recognized, a model poisoning, performed by the means of altered data models and typically provided in the transfer learning step to a deep neural network, are harder to be detected. A notable instance of a scenario is represented by the *poisoned word embeddings models* attack[3]. So, in this work we envise an effective case study to analyze this kind of attack to NLP systems; more precisely, we considered a NLP application of great interest in the current times, provided by the Fake News Detectors (FNDs). These systems typically apply several NLP techniques to analyze the features of a set of news and information published and disseminated, mainly through the web. FNDs represent an ideal target to adversarial attacks, since malicious users are interested to bypass news checker, to go on spreading unnoticed misleading information. We considered a model poisoning attack consisting in poisoning a word embeddings model, used to train a deep neural network-based Fake News Checker; we trained two different word embeddings models, based respectively on fastText[4] and BERT[5] algorithms; we performed experiments both on the trusted and the poisoned versions of these WEs models, in order to investigate the vulnerability level of these NLP systems to these poisoning attacks and the level of resilience of the WEs models to these kinds of traps.

3   Conclusion and Future Works

This work explores the effects of a poisoned word embeddings model attack to fool a deep neural network implementing a Fake News Checker, that generate alerts when potential Fake News are detected. The rise of the False Positives rate, while reducing the reliability of a Fake News Checker, is among the goals of such attacks. We considered, as case study, a Fake News Checker, that performs a stylometric analysis of texts written in natural language, to analyze the severity of a WEs poisoning attack and how much it can affect the accuracy in recognizing real and fake news. We evaluated two different WEs models, to build a vulnerability assessment of NLP systems to poisoning attacks.

The research described in this work is funded and performed within the activities of the Research Program "Vanvitelli V:ALERE 2020 - WAIILD TROLS", financed by the University of Campania "L. Vanvitelli", Italy.

---

[2] Zhang, W. E., Sheng, Q. Z., Alhazmi, A., & Li, C. (2020). Adversarial attacks on deep-learning models in natural language processing: A survey. ACM Transactions on Intelligent Systems and Technology (TIST), 11(3), 1–41.

[3] Zhao, Z., Dua, D., & Singh, S. (2017). Generating natural adversarial examples.

[4] https://fasttext.cc/docs/en/crawl-vectors.html.

[5] https://blog.google/products/search/search-language-understanding-bert/.

Transfer Learning and Loan Default Prediction

Tzvi Feinberg[1]([✉]), Alexander Semenov[1], Yongpei Guan[1], Dmitry Grigoriev[2], and Artem Prokhorov[3]

[1] University of Florida, Gainesville, FL 32611, USA
{tzvifeinberg,asemenov}@ufl.edu, guan@ise.ufl.edu
[2] Center of Econometrics and Business Analytics, Saint Petersburg State University, St. Petersburg, Russia
[3] University of Sydney, Sydney, Australia
artem.prokhorov@sydney.edu.au

Abstract. Predicting probability of default for potential loan customers is of the utmost importance to banks and other financial institutions. Nowadays, most financial institutions assess borrowers using machine learning algorithms. However, they may require large amounts of training data to make accurate predictions. Small financial institutions may not be able to collect large training datasets, and would benefit from the large datasets or pretrained models provided by larger financial institutions. This paper employs the use of transfer learning with neural networks to predict probability of default for new borrowers. We explore multiple architectures of deep neural networks trained on a large dataset and transfer the learned knowledge to a smaller dataset.

Keywords: Probability of default · Machine learning · Neural network · Transfer learning

1 Introduction

Accurately predicting probability of default can help firms minimize the risk they take with new borrowers by reliably determining how likely one is to default and even how much loss they'll incur due to said default. Machine learning classifiers have been employed in the past, including in multitudes of combinations. Neural networks are fairly new to this particular application. The banking industry/financial institutions rely heavily on credit scoring to control risk and boost profits [4]. Competition from online institutions is leading banks to innovate using machine learning based automatic credit evaluation systems. Models require large amounts of financial data to train properly, but new businesses lack sufficient data [2]. Often when analyzing credit risk, sufficient lending history data is unavailable or unattainable for satisfactory analysis and can land both lenders and borrowers in precarious credit outcomes [1]. Transfer learning is shown to be commercially valuable for financial risk assessment [2].

Analyzing a dataset that is too small to predict accurate results on its own can make it impossible for newer firms to compete with existing conglomerates.

© Springer Nature Switzerland AG 2021
D. Mohaisen and R. Jin (Eds.): CSoNet 2021, LNCS 13116, pp. 387–388, 2021.
https://doi.org/10.1007/978-3-030-91434-9

Our goal is to use transfer learning to improve accuracy of results by training on much larger datasets than the ones for which we predict results. In this paper, we experiment with transfer learning using neural networks, seeking both novel approaches and sufficiently accurate results.

2   Datasets and Methods

Our original dataset came from a financial institution in Australia. It consists of 12,982 data points and 216 features, including target feature "Isdefault". Our goal is to use this dataset to predict probability of default at a higher accuracy than previous attempts.

The second dataset that we analyzed is much larger and publicly available on Kaggle [3]. It consists of 2,260,701 data points and 151 features, including target feature "loan_status". Our hope in analyzing this second dataset is to utilize model learning on a much larger range of data in the hopes that it would have better predictive capabilities than those same methods used on a smaller dataset, simply due to the model having more information to learn from. The motivation behind transfer learning is to be able to take learning from one place and apply what was learned in another, to transfer the learning, so to speak.

This is accomplished in several different ways, three of which were experimented with in our study. We used a large dataset to train a binary classifier neural network, and a small dataset to tune it. In the first method we trained the neural network on the large dataset and tested it on the small one. The second method consisted of initial training on the large dataset, and then supplementary training at a much lower learning rate (between .0001 and .01 of the original training) on the small dataset before testing on the small dataset. The third method utilized the freezing of initial layers of the neural network (in our case all but the final) after training it on the large dataset, to then retrain only the final layer by refitting it with data from the small set before testing on the small dataset. The neural network itself was built using keras from the tensorflow library and it contained 4 dense layers, with descending weights 200, 100, 50, and 1. The first three layers use a RelU activation, while the last one is a sigmoid.

References

1. Beydoun, G.: Transfer learning in credit risk
2. Li, W., Ding, S., Chen, Y., Wang, H., Yang, S.: Transfer learning-based default prediction model for consumer credit in China. J. Supercomputing **75**, 862–884 (2019)
3. Papouskova, M., Hajek, P.: Two-stage consumer credit risk modelling using heterogeneous ensemble learning. Decis. Support Syst. **118**, 33–45 (2019)
4. Plawiak, P., Abdar, M., Acharya, U.R.: Application of new deep genetic cascade ensemble of SVM classifiers to predict the australian credit scoring. Appl. Soft Comput. J. **84**, 105740 (2019)

Author Index

Printed in the United States
by Baker & Taylor Publisher Services